Polymer Characterization

# POLYMER CHARACTERIZATION

by
ELISABETH SCHRÖDER

and
GERT MÜLLER
KARL-FRIEDRICH ARNDT

With 116 Figures and 90 Tables

Hanser Publishers, Munich Vienna New York

Distributed in the United States of America by
Oxford University Press, New York
and in Canada by
Oxford University Press, Canada

Authors:
Prof. em. Dr. Sc. nat. Elisabeth Schröder, Leipzig
Dozent Dr. rer. nat. Gert Müller, Merseburg
Dozent Dr. rer. nat. Karl-Friedrich Arndt, Merseburg

The translation was performed by H. Liebscher and revised by John Haim

First published by Akademie-Verlag Berlin 1982, 2. revised edition by Akademie-Verlag Berlin 1989

Distributed in USA by
Oxford University Press
200 Madison Avenue, New York, N.Y. 10016

Distributed in Canada by
Oxford University Press, Canada
70 Wynford Drive, Don Mills, Ontario M3C IJ9

Distributed in all other countries by
Carl Hanser Verlag
Kolbergerstr. 22
D-8000 München 80

The use of general descriptive names, trademarks, etc. in this publication, even if the former are not especially identified, is not to be taken as a sign that such names, as understood by the Trade Marks and Merchandise Marks Act, may accordingly be used freely by anyone.
While the advice and information in this book are believed to be true and accurate at the date of going to press, neither the authors nor the editors nor the publisher can accept any legal responsibility for any errors or omissions that may be made. The publisher makes no warranty, express or implied, with respect to the material contained herein.

CIP-Kurztitelaufnahme der Deutschen Bibliothek

**Schröder, Elisabeth:**
Polymer characterization / by Elisabeth Schröder and Gert
Müller ; Karl-Friedrich Arndt. [The transl. was performed by
H. Liebscher and rev. by John Haim]. – Munich ; Vienna ; New
York : Hanser, 1988
Einheitssacht.: Leitfaden der Polymercharakterisierung 〈engl.〉
ISBN 3-446-14986-4
NE: Müller, Gert:; Arndt, Karl-Friedrich:
ISBN 3-446-14986-4 Carl Hanser Verlag, Munich, Vienna, New York
ISBN 0-19-520800-5 Oxford University Press
Library of Congress Catalog Card Number 89-19955
© Akademie-Verlag Berlin 1989
Licence: Carl Hanser Verlag München · Wien
Printed in GDR by VEB Druckerei „Thomas Müntzer", 5820 Bad Langensalza

# Preface

The synthesis and characterization of polymers are closely related and, to some extent, interdependent fields of knowledge. Thus, the width of molecular weight distribution in radical polymerization is determined by the coupling constant which depends on the manner in which the growth reaction is terminated and, the distribution of the chemical composition of copolymers is determined by reactivity parameters. Deviations of experimental data from theoretical values of these properties enable conclusions to be drawn with regard to the reactions that have taken place. Polymer characterization also includes monitoring of the polymer formation reaction and examination of the end product after processing since both steps determine the properties of the polymer to a considerable degree.

Relevant literature does not always reflect the economic importance of this. Although books and monographs on the theory and practice of polymer synthesis are available in several languages, systematic educational and exercise material introducing polymer characterization is completely absent. Usually the subject is mentioned only in passing in the literature on synthesis.

From our experience in teaching at the „Carl Schorlemmer" Technical University, Leuna—Merseburg, we have reached the conclusion that the fundamentals of polymer characterization present some difficulties, even to advanced students. For this reason, we have availed ourselves of the proven method of combining theory and practice in our attempts to introduce tyros to the problems of characterization via experiment.

This book is intended for students and graduates at universities, technical colleges and polytechnics. It concentrates on methods used to determine microstructural parameters of the individual molecule. There are four chapters, the first of which deals with the determination of molecular weight. The second deals with the determination of molecular weight distribution and chemical heterogeneity and is followed by a chapter on methods of investigating microstructure, e.g. tacticity, sequence length distribution, branching and network structure. The final chapter describes various specialized methods of investigating polymers. All the chapters could of course be extended considerably but we have restricted ourselves in order to retain the introductory nature of the book.

Every method included in the book is preceded by a brief introduction to the theory and an account of the experimental background. Specimen examples and their evaluation are described in detail so that they may be repeated. The examples were selected on experimental, instrumental and didactic considerations and have generally proved successful in polymer characterization courses for polymer chemists and physicists. In general, they can be applied to other polymers without significant changes.

We hope that this book will prove a useful source of information to specialists in industry and institutes and that it will be instrumental in encouraging more people to use polymer characterization.

# Contents

| | | |
|---|---|---|
| **1.** | **Molecular Weight Determination** | 15 |
| 1.1. | Introduction | 15 |
| 1.2. | Molecular Weight Determination by End Group Analysis | 18 |
| 1.2.1. | Introduction | 18 |
| 1.2.2. | Methods of Measurement and Determination | 19 |
| 1.2.2.1. | Chemical Methods of Determining End Groups | 19 |
| 1.2.2.1.1. | COOH End Group Determination | 19 |
| 1.2.2.1.1.1. | Volumetric Determination | 19 |
| 1.2.2.1.1.2. | Determination of COOH via Cation-Anion Complexes | 20 |
| 1.2.2.1.2. | $NH_2$ End Group Determination | 21 |
| 1.2.2.1.3. | SH End Group Determination | 22 |
| 1.2.2.1.4. | OH End Groups | 22 |
| 1.2.2.2. | Radiochemical Methods for End Group Determination | 24 |
| 1.2.2.3. | Physical Methods | 25 |
| 1.2.3. | Examples | 29 |
| 1.2.3.1. | Molecular Weight Determination of a Linear Saturated Polyester by Chemical Determination of the COOH and OH End Groups | 29 |
| 1.2.3.1.1. | Carboxyl Group Determination using Sodium Alcoxide in an Anhydrous Medium | 29 |
| 1.2.3.1.2. | OH End Group Content — Hydroxyl Value (OH.V.) | 30 |
| 1.2.3.2. | Infrared End Group Determination in Polystyrene | 31 |
| *Bibliography* | | |
| 1.3. | Determination of the Number-average Molecular Weight of Polymers by Osmotic Measurement — Membrane Osmosis | 33 |
| 1.3.1. | Introduction | 33 |
| 1.3.2. | Principles | 34 |
| 1.3.3. | Experimental Background | 37 |
| 1.3.4. | Example | 40 |
| *Bibliography* | | |
| 1.4. | Determination of Number-average Molecular Weight of Polymers by Osmotic Measurement — Vapour-pressure Osmosis | 45 |

| | | |
|---|---|---|
| 1.4.1. | Introduction | 45 |
| 1.4.2. | Principles | 45 |
| 1.4.3. | Experimental Background | 47 |
| 1.4.4. | Example | 50 |
| *Bibliography* | | |

| | | |
|---|---|---|
| 1.5. | Determination of Weight-average Molecular Weight of Polymers by Light Scattering Measurements | 53 |
| 1.5.1. | Introduction | 53 |
| 1.5.2. | Principles | 54 |
| 1.5.3. | Experimental Background | 60 |
| 1.5.4. | Examples | 62 |
| *Bibliography* | | |

| | | |
|---|---|---|
| 1.6. | Sedimentation and Diffusion | 72 |
| 1.6.1. | Introduction | 72 |
| 1.6.2. | Principles | 73 |
| 1.6.3. | Experimental Background | 80 |
| 1.6.4. | Example | 86 |
| 1.6.5. | Determination of Diffusion Coefficient | 91 |
| *Bibliography* | | |

| | | |
|---|---|---|
| 1.7. | Determination of the Viscosity-average Molecular Weight $M_\eta$ of Polymers | 93 |
| 1.7.1. | Introduction | 93 |
| 1.7.2. | Principles | 93 |
| 1.7.3. | Experimental Background | 96 |
| 1.7.4. | Example | 99 |
| 1.7.4.1. | Molecular Weight Determination | 99 |
| 1.7.4.2. | Derivation of a $[\eta]$-$\bar{M}$-relation | 102 |
| *Bibliography* | | |

| | | |
|---|---|---|
| **2.** | **Determination of Molecular Weight Distribution and Chemical Heterogeneity** | 106 |
| 2.1. | Molecular Weight Distribution | 106 |
| 2.1.1. | Introduction | 106 |
| 2.1.2. | Principles of Separation by Fractionation | 107 |
| 2.1.3. | Principles of Solubility Fractionation | 108 |
| 2.1.4. | Experimental Conditions for Solubility Fractionation | 112 |
| 2.1.4.1. | Experimental Set-up for Solubility Fractionation | 112 |
| 2.1.4.2. | Optimum Process Parameters for Solubility Fractionation | 113 |
| 2.1.4.2.1. | Criteria for Selection of Optimum Solvent/Precipitant-Systems ($S/P_r$-Systems) | 113 |
| 2.1.4.2.2. | Optimum $\gamma^*$-Gradient of the Elution Mixture | 114 |
| 2.1.4.2.3. | Fractionation Step-width | 115 |

## Contents

| | | |
|---|---|---|
| 2.1.4.2.4. | Charging the Supporting Material | 115 |
| 2.1.4.2.5. | Contact Time and Mode of Operation | 116 |
| 2.1.4.3. | Processing and Characterization of Fractions | 117 |
| 2.1.4.4. | Separation Efficiency Investigations | 118 |
| 2.1.4.5. | Constructing the Distribution Curve | 118 |
| 2.1.4.6. | Determination of Molecular Inhomogeneity and Polydispersity | 120 |
| 2.1.5. | Distribution Functions of Molecular Weight and Their Mathematical Treatment | 122 |
| | | |
| 2.2. | Chemical Heterogeneity | 131 |
| 2.2.1. | Introduction | 131 |
| 2.2.2. | Principles of Solubility Fractionation | 133 |
| 2.2.3. | Selection of $S/P_r$-Systems | 133 |
| 2.2.4. | Experimental Foundations | 135 |
| 2.2.5. | Determination of Chemical Heterogeneity by Thin-Layer Chromatography ($TLC$) | 136 |
| | | |
| 2.3. | Examples | 138 |
| 2.3.1. | Determination of the Integral Distribution Function $I(M)$ of PVC by Solubility Fractionation (FUCHS Method) | 138 |
| 2.3.2. | Determination of Chemical Heterogeneity of VC-Methyl Acrylate (VCMA) Copolymers by Column Elution Fractionation with $S/P_r$-gradients | 141 |
| 2.3.3. | Determination of the Distribution Functions of a HDPE from Elution Fractionation Data | 145 |

*Bibliography*

| | | |
|---|---|---|
| 2.4. | Determination of Molecular Weight Distribution by Turbidimetric Titration | 151 |
| 2.4.1. | Introduction | 151 |
| 2.4.2. | Principles | 152 |
| 2.4.2.1. | Fundamentals of Light Scattering in Disperse Systems | 152 |
| 2.4.2.2. | Coagulation of Precipitated Polymer Particles During Turbidimetric Titration | 153 |
| 2.4.3. | Empirical Evaluation | 154 |
| 2.4.3.1. | Relation between Solubility and Molecular Weight | 154 |
| 2.4.3.2. | Fundamentals of Calculating Molecular Weight Distribution from Turbidity Curves | 156 |
| 2.4.4. | Absolute Method for the Evaluation of Turbidity Curves | 159 |
| 2.4.5. | Experimental Background | 160 |
| 2.4.6. | Examples | 161 |
| 2.4.6.1. | Determination of the Relationship between Solubility and Molecular Weight of Emulsion PVC | 161 |
| 2.4.6.2. | Determination of the Molecular Weight Distribution of a Commercial Emulsion PVC by Turbidimetric Titration | 165 |

*Bibliography*

| | | |
|---|---|---|
| 2.5. | Gel Permeation Chromatography | 166 |
| 2.5.1. | Introduction | 166 |
| 2.5.2. | Principles | 166 |
| 2.5.3. | Experimental Background | 172 |
| 2.5.4. | Optimization | 175 |
| 2.5.5. | Examples | 176 |
| 2.5.5.1. | Determination of Molecular Weight Distribution of a Polystyrene Sample | 176 |
| 2.5.5.2. | Determination of Molecular Weight Distribution of an Ethylene-Vinyl acetate Copolymer | 179 |
| Bibliography | | |
| | | |
| 3. | **Microstructural Investigations** | 183 |
| 3.1. | Introduction | 183 |
| 3.2. | Identification of High-polymer Materials | 184 |
| 3.2.1. | Introduction | 184 |
| 3.2.2. | Comminution of Polymer Samples | 184 |
| 3.2.3. | Separation of Plasticizers and their Quantitative Determination | 185 |
| 3.2.4. | Identification of Plasticizers | 185 |
| 3.2.5. | Separation of Inorganic Fillers and Quantitative Determination | 187 |
| 3.2.6. | Identification of Stabilizers | 189 |
| 3.2.7. | Identification of Polymers | 190 |
| 3.2.7.1. | Preliminary Tests for Quantitative Identification | 191 |
| 3.2.7.1.1. | Exposure to Open Flames and Heating in a Glow Tube | 191 |
| 3.2.7.1.2. | Depolymerization | 191 |
| 3.2.7.2. | Determination of Simple Physical Characteristic Values | 196 |
| 3.2.7.2.1. | Solubility | 196 |
| 3.2.7.2.2. | Determination of Melting Point and/or Melting Range and of Density | 196 |
| 3.2.7.3. | Determination of Chemical Characteristics | 197 |
| 3.2.7.3.1. | Saponification Value (S.V.) | 197 |
| 3.2.7.3.2. | Hydroxyl Value (OH.V.) and Acid Value (A.V.) | 198 |
| 3.2.7.4. | Chemical Determination | 198 |
| 3.2.7.4.1. | Chemical Determination of Heteroelements and Functional Groups | 198 |
| 3.2.7.4.1.1. | Detection of Nitrogen, Chlorine, Fluorine, Sulphur and Oxygen | 198 |
| 3.2.7.4.1.1.1. | Detection of Nitrogen | 198 |
| 3.2.7.4.1.1.2. | Detection of Chlorine | 199 |
| 3.2.7.4.1.1.3. | Detection of Fluorine | 199 |
| 3.2.7.4.1.1.4. | Detection of Sulphur | 199 |
| 3.2.7.4.1.1.5. | Detection of Oxygen | 200 |
| 3.2.7.4.2. | Group Test and Functional Groups | 200 |
| 3.2.7.4.2.1. | Detection of Ester Groups using Hydroxamic Acid | 200 |
| 3.2.7.4.2.2. | Detection of Phenols | 200 |
| 3.2.7.4.2.3. | Detection of Formaldehyde | 201 |
| 3.2.7.4.2.4. | Detection of the Epoxide Group | 201 |
| 3.2.7.4.2.5. | Detection of the Urethane Group | 201 |

## Contents

| | | |
|---|---|---|
| 3.2.7.4.3. | Specific Tests | 202 |
| 3.2.7.4.3.1. | Detection of Polystyrene | 202 |
| 3.2.7.4.3.2. | Detection of Polyvinyl Alcohol | 202 |
| 3.2.7.4.3.3. | Identification of Saturated Aliphatic Polymeric Chlorinated Hydrocarbons | 202 |
| 3.2.7.4.3.4. | Identification of Cellulose and Its Derivatives | 203 |
| 3.2.7.3.5. | Detection of Urea with Urease | 204 |
| 3.2.7.4.3.6. | Detection of Polyamides | 204 |
| 3.2.7.4.3.7. | Reaction of Iodine with polymers | 204 |
| 3.2.8. | Separation Procedures for Polymers | 206 |
| *Bibliography* | | |
| | | |
| 3.3. | Double Bonds in Polymers | 207 |
| 3.3.1. | Introduction | 207 |
| 3.3.2. | Methods for Determination of Double Bonds | 210 |
| 3.3.2.1. | Chemical Methods | 210 |
| 3.3.2.2. | Physical Methods | 211 |
| 3.3.3. | Examples of Double Bond Determination | 213 |
| 3.3.3.1. | Volumetric Determination of Double Bond Content and Composition of Buna-S with Iodine Monochloride | 213 |
| 3.3.3.2. | Gravimetric Determination of Double Bond in Aged PVC (ROSSMANN's method) | 214 |
| 3.3.3.3. | IR-Spectrometric Double Bond Determination: Isomer Analysis of Polybutadiene | 215 |
| 3.3.3.4. | UV-Spectrometric Determination of Polyene Sequences in Degraded PVC | 217 |
| *Bibliography* | | |
| | | |
| 3.4. | Sequence Length Distribution in Copolymers | 220 |
| 3.4.1. | Introduction | 220 |
| 3.4.2. | Characteristic Values of Sequence Distribution | 222 |
| 3.4.3. | Experimental Determination of Sequence Distribution | 225 |
| 3.4.4. | Examples | 228 |
| 3.4.4.1. | NMR Determination of Sequence Distribution using Ethylene-Vinyl acetate Copolymers with Low Vinyl acetate Content [9] as an example | 228 |
| 3.4.4.2. | IR Determination of Sequence Distribution using Styrene-Acrylonitrile Copolymers [12] as an Example | 232 |
| *Bibliography* | | |
| | | |
| 3.5. | Tacticity Determination in Polymers | 234 |
| 3.5.1. | Introduction | 234 |
| 3.5.2. | Formation of Tactic Polymers; Characteristic Bond Distribution Values in the Polymer Chain | 235 |
| 3.5.3. | Experimental Determination of Tacticity | 236 |
| 3.5.3.1. | NMR Spectroscopy | 236 |
| 3.5.3.2. | IR Spectroscopy | 240 |

| | | |
|---|---|---|
| 3.5.3.3. | Further Experimental Methods of Ascertaining Tactic Structure | 243 |
| 3.5.4. | Examples of Determination of Tactic Constituents | 244 |
| 3.5.4.1. | NMR Determination of Isotactic, Syndiotactic and Heterotactic Triades using Polymethyl methacrylate as an Example | 244 |
| 3.5.4.2. | IR Determination of the Proportion of Isotactic and Syndiotactic Monomer Linkages using Polyvinyl Chloride as an Example | 246 |

*Bibliography*

| | | |
|---|---|---|
| 3.6. | Determination of Branching in Polymers | 250 |
| 3.6.1. | Introduction | 250 |
| 3.6.2. | Formation of Branched Structures in Polymers | 252 |
| 3.6.3. | Determination of Branching | 254 |
| 3.6.3.1. | Determination of Long-Chain Branching | 254 |
| 3.6.3.2. | Determination of Ester Linked Side Chains | 259 |
| 3.6.3.3. | Determination of Total Branching | 259 |
| 3.6.3.3.1. | IR Method | 259 |
| 3.6.3.3.2. | NMR Method, $^1$H-NMR-Spectroscopy | 261 |
| 3.6.3.4. | Specific Methods for the Determination of Short-Chain Branching | 261 |
| 3.6.3.4.1. | $^{13}$C-Nuclear Magnetic Resonance Spectroscopy | 261 |
| 3.6.3.4.2. | Radiolysis Gas Chromatography | 263 |
| 3.6.4. | Examples | 265 |
| 3.6.4.1. | Determination of the Contraction Factor g from Viscosity Measurements | 265 |
| 3.6.4.2. | Determination of the Branching Number of a Low Density Polyethylene | 266 |
| 3.6.4.3. | NMR Determination of the Number of Methyl Groups in an Ethylene-butene-1 Copolymer | 269 |

*Bibliography*

| | | |
|---|---|---|
| 3.7. | Characterization of Network Polymers | 271 |
| 3.7.1. | Introduction | 271 |
| 3.7.2. | Characteristic Structural Parameters and Their Quantitative Determination | 273 |
| 3.7.2.1. | Determination of Network Density by Elastometric Methods | 273 |
| 3.7.2.2. | Swelling Measurements | 276 |
| 3.7.2.3. | Solubility Investigations | 277 |
| 3.7.2.4. | Determination of the Mean Distance and Distance Distribution between Two Junction Points | 278 |
| 3.7.3. | Experimental Basis | 280 |
| 3.7.3.1. | Elastometry | 280 |
| 3.7.3.2. | Swelling Measurements | 282 |
| 3.7.3.3. | Determination of the Chemical Potential ($\Delta\mu_A$) of Cross-linked Polymers | 283 |
| 3.7.3.4. | Sol-Gel-Analysis | 284 |
| 3.7.3.5. | Relative and Qualitative Methods of Network Characterization | 284 |
| 3.7.4. | Examples | 285 |
| 3.7.4.1. | Optical Measurement of the Degree of Swelling of Peroxide-Cross-linked PMMA | 285 |

| | | |
|---|---|---|
| 3.7.4.2. | Determination of the Parameters of the MOONEY-RIVLIN Equation of Isocyanate-Cross-linked Butadiene Prepolymers using Stress-strain Measurement | 286 |
| 3.7.4.3. | Determination of the Network Density ($v$) and of the $M_c$-value of Sulphur-Cross-linked Natural Rubber using Compression Measurement | 288 |
| 3.7.4.4. | Sol-Gel Analysis of Sulphur-Cross-linked Polybutadiene-cis-132 | 290 |
| 3.7.4.5. | Determination of the Constants of the FLORY-REHNER Equation for Sulphur-Cross-linked Natural Rubber | 291 |
| 3.7.4.6. | Determination of $d_c$ and $H(d_c)$ of Peroxide-Cross-linked Natural Rubber by Measurement of the Freezing-point Depression of the Swelling Agent | 292 |

*Bibliography*

| | | |
|---|---|---|
| **4.** | **Special Techniques** | **296** |
| 4.1. | Determination of Theta Conditions | 296 |
| 4.1.1. | Introduction | 296 |
| 4.1.2. | Principles | 296 |
| 4.1.3. | Experimental Background | 300 |
| 4.1.4. | Examples | 303 |
| 4.1.4.1. | Theta Temperature | 303 |
| 4.1.4.2. | Theta Composition | 305 |

*Bibliography*

| | | |
|---|---|---|
| 4.2. | Determination of Solubility Parameter | 306 |
| 4.2.1. | Introduction | 306 |
| 4.2.2. | Principles | 307 |
| 4.2.3. | Experimental Background | 309 |
| 4.2.3.1. | Estimation of Solubility Parameter | 310 |
| 4.2.3.2. | Determination of Solubility Parameter from Swelling Measurements | 311 |
| 4.2.3.3. | Determination of Solubility Parameter from the Limiting Viscosity | 312 |
| 4.2.3.4. | Solubility Parameter and Theta Temperature | 313 |
| 4.2.3.5. | Turbidimetric Solubility Parameter Determination | 313 |
| 4.2.3.6. | Solubility Parameter Determination by Inverse Gas Chromatography | 315 |
| 4.2.4. | Examples | 316 |

*Bibliography*

| | | |
|---|---|---|
| 4.3. | Compatibility of Polymers | 319 |
| 4.3.1. | Introduction | 319 |
| 4.3.2. | Principles | 320 |
| 4.3.3. | Experimental Background | 326 |
| 4.3.4. | Examples | 328 |
| 4.3.4.1. | Determination of Phase Equilibrium in the Carbon Tetrachloride/Polystyrene/Polyisobutene System at 18 °C | 328 |

4.3.4.2. Determination of the Compatibility
of the Toluene/Polystyrene/Amorphous Polypropylene System by
Viscosity Measurements .............................................. 330

*Bibliography*

5. **Authors Index** ..................................................... 333

6. **Subject Index** ..................................................... 338

# 1. Molecular Weight Determination*)

## 1.1. Introduction

The molecular weight characterizes the size of a molecule. While for low-molecular substances a certain uniform molecular weight is characteristic, all synthetic and many natural macromolecular substances are polydisperse in their molecular weight or polymolecular. This means that they have a range of molecular weight distribution. The molecular weights measured by various methods are mean values which differ according to the statistical weighting factor $g_i$ incorporated in them. The most important $g_i$, for the molecular weight distribution are

$n_i$ = number of moles

$w_i = n_i M_i$ = total weight of all molecules $i$ of molecular weight $M_i$

$c_i = w_i/V_{\text{solution}}$ = weight of dissolved substance per unit volume of solution

$z_i = w_i M_i = n_i M_i^2$

In general distribution curves can be well characterized by their moments and, for practical reasons, the mean values should be defined using of these moments. The moment $\mu$ of order $q$ of a property (here molecular weight $M$) is defined by equation (1)

$$\mu_g^{(q)}(M) = \frac{\sum_i g_i M_i^q}{\sum_i g_i} \tag{1}$$

All products and quotients of moments, which have the dimension of the property under consideration, are mean values of the property. The mean values $M_n$, $M_w$, $M_z$ of the molecular weights are defined as the first moments of the distribution

---

*) According to recommendations in the International System of Units (SI) the correct expression is "relative molecular mass". Since this expression has not been adopted generally in polymer literature "molecular weight" will be used.

functions with the respective weighting factor — $g_i = n_i$, $w_i$ or $z_i$.*) When considering the relationships between individual weighting factors averages of molecular weight can be represented as quotients of moments of different orders. Thus, it is possible to calculate the weight average molecular weight if the numerical molecular weight distribution is known.

The following relationships hold:

number-average molecular weight

$$M_n = \mu_n^{(1)}(M) = \frac{\sum_i n_i M_i}{\sum_i n_i} = \frac{\sum_i w_i}{\sum_i w_i M_i^{-1}} = \frac{\mu_w^{(0)}(M)}{\mu_w^{(-1)}(M)} \tag{2}$$

weight-average molecular weight

$$M_w = \mu_w^{(1)}(M) = \frac{\sum_i w_i M_i}{\sum_i w_i} = \frac{\sum_i n_i M_i^2}{\sum_i n_i M_i} = \frac{\mu_n^{(2)}(M)}{\mu_n^{(1)}(M)} = \frac{\sum_i z_i}{\sum_i z_i M_i^{-1}} = \frac{\mu_z^{(0)}(M)}{\mu_z^{(-1)}(M)} \tag{3}$$

z-average molecular weight

$$M_z = \mu_z^{(1)}(M) = \frac{\sum_i z_i M_i}{\sum_i z_i} = \frac{\sum_i w_i M_i^2}{\sum_i w_i M_i} = \frac{\mu_w^{(2)}(M)}{\mu_w^{(1)}(M)} = \frac{\sum_i n_i M_i^3}{\sum_i n_i M_i^2} = \frac{\mu_n^{(3)}(M)}{\mu_n^{(2)}(M)} \tag{4}$$

The average molecular weights considered so far are moments of integer order. An important molecular weight average, the viscosity average, $M_\eta$ is an exponent average. The power is given by the reciprocal order of the moments:

$$M_\eta = \left( \frac{\sum_i w_i M_i^a}{\sum_i w_i} \right)^{1/a} \tag{5}$$

$a$ = exponent of the KUHN-MARK-HOUWINK equation

Since the viscosity average molecular weight involves $w$ as weighting factor, it is a weight average. It is distinguished from the simple weight average molecular weight $M_w$, by its order $q = a \neq 1$ and by its power $1/q = 1/a$ (n.b.: if $a = 1$, then $M_\eta = M_w$!).

Individual methods of molecular weight determination can be distinguished qualitatively according to the statistical averages obtained and also according to the relationship between quantity to be measured and molecular weight.

*) If it is clear that average values of molecular mass are being discussed, then the bar usually placed over the $M$ to indicate this will be omitted.

A distinction is made between:
— absolute methods,
— equivalent methods,
— relative methods.

With absolute methods, the quantity to be measured is directly related to the molecular weight without having to make any assumptions about chemical and/or physical structures. For equivalent methods, the chemical structure must be known. Relative methods measure properties which depend both on chemical and physical structure, and which require for evaluation a calibration relationship between quantity to be measured and molecular weight.

Table 1 summarizes the most important methods of determining molecular weight and divides them into absolute ($A$), equivalent ($E$) and relative methods ($R$).

In Sections 1.2.—1.4., methods of determining number-average molecular weights are explained. This quantity is of particular interest in kinetic studies.

*Table 1*: Principal methods of determining molecular weight [1]

| Method | Type | Molecular weight range (g/mol) | Mean values of the molecular weight |
| --- | --- | --- | --- |
| Membrane osmometry | $A$ | $10^4 \ldots 10^6$ | $M_n$ |
| Ebullioscopy | $A$ | $<10^4$ | $M_n$ |
| Cryoscopy | $A$ | $<10^4$ | $M_n$ |
| Isothermal, distillation | $A$ | $<10^4$ | $M_n$ |
| Vapour pressure osmometry | $A*)$ | $<10^4$ | $M_n$ |
| End group determination | $E$ | $<10^5$ | $M_n$ |
| Light scattering | $A$ | $10^2 \ldots 10^8$ | $M_w$ |
| Sedimentation equilibrium | $A$ | $<10^6$ | $M_w, M_z, M_{z+1}$ |
| Sedimentation equilibrium with density gradient | $A$ | $>10^5$ | different averages |
| Combination of sedimentation/ diffusion coefficient | $A$ | $10^3 \ldots 10^8$ | different averages |
| Solution viscosity | $R$ | $10^2 \ldots 10^8$ | $M_\eta$ |
| Melting viscosity | $R$ | $10^2 \ldots 10^7$ | $M_w$ |
| Gel permeation chromatography | $R$ | $10^2 \ldots 10^7$ | $M_{GPC}$ (The index $GPC$ does not indicate a recognized argument) |

*) In practice the methods used involue calibration.

[1] ELIAS, H.-G., BARREIS, R. and J. G. WATTERSON, Fortschr. Hochpolymeren-Forsch. **11** (1973), 11

Light scattering methods (Section 1.5.) determine not only the weight-average molecular weight but can also provide information on molecular geometry, size, macromolecular interactions and thermodynamic parameters of polymer solutions. In an analogous manner, X-ray scattering can be used to determine the size and shape of dissolved polymer molecules. Molecular weight determination by electron microscopy is highly effective where a very dilute polymer solution is concentrated and the individual molecules are then counted. Sedimentation and diffusion of macromolecules in solvents are also dependent on molecular weight and can be measured in an ultracentrifuge and utilized for molecular weight determination (see Chapter 1.6.).

A simple low cost commonly used method of investigating molecular weights of polymers is that of viscosity measurements (see Chapter 1.7.).

## 1.2. Molecular Weight Determination by End Group Analysis

### 1.2.1. Introduction

In general, end group analysis is aimed at determining the end group content quantitatively. It takes advantage of the fact that the groups at the end of a polymer chain are frequently of different constitution to the "internal" chain links and are thus accessible to quantitative analytical determination. Examples include COOH and $NH_2$ groups in polyamides, COOH and OH groups in polyesters or terminated prepolymers SH groups and terminal double bonds and certain heteroatoms provided that the molecular weights of the polymers are less than $5 \cdot 10^4$ g/mol.

The technical importance of end group analysis in the manufacture of polyesters, polyurethanes, liquid rubber and other products is obvious. It is also important in research into structure-property relationships because end groups frequently exert considerable influence on properties.

Under favourable conditions, end group analysis can also be used for determining molecular weight since the latter can be obtained if the number of end groups in a given weight of polymer is known. End group analysis is therefore an equivalent method for obtaining the number average molecular weight $M_n$.

$$M_{n_{end}} = \frac{q \cdot m}{n_{end}} \qquad (1)$$

$q$ = number of end groups, $m$ = weight of the polymer in $g$, $n_{end}$ = number of moles of the end groups in $m$ gram of polymer

Determination of $M_n$ thus amounts to determine of the molecular concentration of end groups. This, however, calls for a knowledge of the type and number of all end groups and their quantitative determination. Since the number of moles of end groups decreases with increasing molecular size, the sensitivity of the $M_{n_{end}}$ determination diminishes. With the usual analytical methods, the upper limit is about $5 \cdot 10^4$ g/mol whereas radiochemical methods enable molecular weights up to $10^6$ g/mol to be determined reliably. When comparing $M_n$ values obtained by end group analysis and osmotic methods conclusions can sometimes be drawn regarding molecular branching and, thus, the mechanism of formation. Because of the great number of possible side reactions, such results require confirmation by other methods.

In spite of such reservations end group analysis has become routine practice — not least because of its simplicity.

## 1.2.2. Methods of Measurement and Determination

Functional end group content in polymers can be determined by chemical, radiochemical and physical methods. The choice depends largely on the sensitivity of the method and hence the range of molecular weights which can be determined.

### 1.2.2.1. Chemical Methods of Determining End Groups

Chemical determination of end groups is based on the complete reaction of reactive end groups with low-molecular weight reagents and measurement of reagent consumption by common analytical methods. In a polymer, the end groups behave as similar groups in low molecular weight compounds as long the solutions are very dilute and the solvents or neighbouring groups do not exert any influence or cause degradation. Chemical determination of functional end groups can be effected directly or after conversion to a substance that can be analyzed more conveniently. Examples of direct end group determination include COOH, $NH_2$ and SH groups in polyamides, polyesters or liquid thioplasts. Indirect methods are used primarily for determining OH-groups.

#### 1.2.2.1.1. COOH End Group Determination

##### 1.2.2.1.1.1. Volumetric Determination

COOH end group content can be determined universally by direct titrating with sodium alkoxide solutions in anhydrous media. When using alcoholic potassium hydroxide with hydrolyzable substances such as polyesters saponification or hydrolysis reactions may also occur since water is formed in the neutralization of the COOH groups. Water-containing solvent mixtures are only used when free dicarboxylic acid anhydrides are present because the latter react with water to form the respective dicarboxylic acids.

Separate determination of COOH end groups and dicarboxylic acid anhydride in polyesters, requires two titrations — one in a water-free medium for the detection of COOH end groups and any free acid, and the other in an aqueous medium after heating for determination of anhydride [1]. Consequently solvent selection and test conditions are substance-specific. Table 2 contains data for the most important types of compounds. Titration end points can be determined using an indicator or by potentiometric or conductometric techniques.

Potentiometry is especially useful in simultaneous determinations where COOH end groups protolysis constant are concerned.

*Table 2*: Experimental conditions for volumetric determination of COOH

| Substance | Solvent | Titrant | Indication |
|---|---|---|---|
| Homopolyamide | benzyl alcohol (175 °C) | 0.1 N KOH in ethylene glycol | visually with phenolphthalein potentiometrically conductometrically |
| Copolyamides | ethanol — $H_2O$ = 72:28 (volume parts) (boiling point) | 0.1 N KOH | same as above |
| Polyethylene glycol terephthalate | aniline 130 ± 3 °C | 0.1 N KOH in ethanol at 40 ± 3 °C | visually with thymolphthalein or electrometrically |
| Linear, saturated polyesters | acetone:alcohol = 3:1 (anhydrous) | 0.05 N sodium alcoholate | potentiometrically |
|  | acetone: $H_2O$ = 5:1 (65 °C) | 0.05 N sodium alcoholate | potentiometrically |

## 1.2.2.1.1.2. Determination of COOH via Cation-Anion Complexes

COOH end groups in polymers can be detected very precisely and determined by means of dyeing methods used in surfactant analysis. The macromolecules containing COOH-groups react directly or via an ion pair with oppositely charged dye ions to form cation-anion complexes with an associated colour change.

# Molecular Weight Determination by End Group Analysis

Rhodamine 6 G (colour index 45160) is used as a reagent for the identification of COOH. 3 mg of this substance are dissolved in 2.5 ml of phosphate buffer (pH 10 ... 12) and then extracted immediately with 100 ml of benzene at 25 °C in the absence of light and oxygen. The dye extract has a shelf life of several months when stored over solid NaOH in the dark. For COOH detection and determination 0.02 ... 0.2 g of polymer is dissolved in 15 ... 20 cm³ of benzene and the same amount of dye solution is added. A change in colour from yellow to pink indicates COOH-groups. A lower limit for this is set by the acidic strength of the end groups. For the relatively weak acid groups, usually present in polymers, the minimum $n_{end}$ is about $10^{-5}$ [2].

## 1.2.2.1.2. $NH_2$ End Group Determination

As an end group in linear polymers $NH_2$ occurs commercially only in polyamides. Where it is determined by titration with 0.1 N HCl solutions. A mixture of phenol:ethanol:$H_2O$ = 45:30:25 is used as solvent for homopolyamide while for copolyamides, the solvent shown in Table 2 is employed. In contrast to COOH end group determination in polyamides, acidimetric $NH_2$ determination is influenced by the degree of polymerization. Its sensitivity diminishes with increasing degree of polymerization because when titrating with strong acids the dissociation equilibrium of the amino group is influenced by the dissociation of CONH-groups in the chain and the type of solvent (weak acid). In a potentiometric titration direct relationship between degree of polymerization $P$ and concentrations at amino ($c_A$) as well as of amide ($c_D$) groups according to

$$c_D = (P - 1) c_A \qquad (2)$$

means that the potential decreases with increasing $P$. The upper limit of the potentiometric method is thus reached at $P = 100$. Spectrometric $NH_2$-group determination after reaction of the amino-group with fluoro-2,4-dinitrobenzene in the presence of sodium bicarbonate is considerably more sensitive:

Polyamide—$NH_2$ + F—⟨⟩—$NO_2$ + $NaHCO_3$ ⟶ Polyamide—NH—⟨⟩—$NO_2$
                    $NO_2$                                    $NO_2$
                         + NaF + $CO_2$ + $H_2O$                    (3)

A yellow polyamide is formed whose solution in o-cresol can be measured spectrometrically [1].

### 1.2.2.1.3. SH End Group Determination

SH-groups act as end groups in liquid thioplasts and in prepolymers of butadiene. For their analytical determination, advantage is taken of their tendency to form salts and of their tendency to oxidize [3]. Both argentometry and iodometry are used. $H_2S$ and elementary sulphur interfere with salt formation with silver ions; oxidation reactions are restricted to primary SH-group in end groups etc.

Argentometric determination is usually carried out according to the KOLTHOFF method, involving amperometric titration with a rotating platinum electrode [4]. A mercury reference electrode is used covered with a layer of saturated potassium chloride solution containing potassium iodide and mercury (II) iodide. The two electrodes are connected via a salt bridge filled with saturated potassium nitrate solution and 3 per cent agar-agar. Examples of solvents for the sample include nitrobenzene, tetrahydrofuran and ethanol. The diffusion current is measured with a microammeter and the amount of 0.05 N $AgNO_3$ solution used is ascertained graphically (see also 1.2.3.). Frequently SH end groups are determined iodometrically using starch as an indicator. 2-phase titration is frequently necessary and thus excess iodine solution is normally added and then back-titrated with sodium thiosulphate solution. Assuming equal end groups of the linear chain and neglecting possible chain branchings we have

$$M_n = \frac{66.14}{\% \text{ SH}} \tag{4}$$

### 1.2.2.1.4. OH End Groups

Despite progress in modern analytical chemistry, methods for determining OH end groups in polymers are still only marginally satisfactory. The causes lie predominantly in the behaviour of the OH-group itself; the susceptibility to hydrolysis of its reaction products, the low selectivity between isomers, insufficient sensitivity and reliability and, finally, in the tendency of the OH-group to associate which affects IR determinations.

In polymers the problems are greater because of their limited solubility and the effects produced by neighbouring groups. Chemical methods of determining OH-groups are still preferred to physical methods because detection sensitivity is usually higher and less equipment is necessary.

In chemical methods, the hydroxyl groups are converted, using suitable reagents, and the reaction products are either isolated and investigated separately (indirect method) or the extent of reaction, and thus the hydroxyl group content, is

Table 3: Methods and experimental conditions for OH end group determination of polymers [1]

| Polymer | Reaction type | Reagent | Catalyst | Solvent | Method of Detection |
|---|---|---|---|---|---|
| Polyester, saturated | esterification | acetic anhydride | toluene-p-sulphonic acid | ethyl acetate | volumetric of reagent consumption |
| Polyethylene glycol terephthalate | esterification | 3,5-dinitrobenzoyl chloride | pyridine | nitrobenzene | potentiometric titration of reagent consumption with glycolic NaOH |
| Polyethylene glycol terephthalate | esterification | o-sulphobenzoic anhydride | none | nitrobenzene | potentiometric determination of reagent consumption with 0.1 N NaOH |
| Polyethylene oxide | esterification | phthalic anhydride | pyridine | pyridine | potentiometric determination of reagent consumption |
| Polyethylene oxide | esterification | 3,5 dinitrobenzoyl chloride | dimethyl formamide | dimethyl formamide | isolating the ester by $CHCl_3$ extraction and colorimetric determination |
| Polyethylene oxide | esterification | acetyl chloride | Zn powder | toluene | determination of reagent consumption by checking IR absorption |
| polyethylene oxide | formation of urethane | phenyl isocyanate | tin(II) octoate | benzene, toluene, dioxan | determination of reagent consumption by gas chromatography |
| Prepolymers of butadiene | conversion with $LiAlH_4$ | $LiAlH_4$ | | tetrahydrofuran | volumetric determination of the reaction product |

determined from the amount of reagent consumed. In spite of the considerable number of possible reactions of OH end groups, the methods normally employed involve the following:
— catalyzed esterification with acid anhydrides or acid chlorides;
— conversion with alkyl and aryl isocyanates into the respective urethanes and
— reaction of active hydrogen with GRIGNARD reagent, lithium aluminium hydride or trimethyl silyl chloride.

The reactions are detailed by KAISER [3] and others. Table 3 summarizes the methods and experimental conditions used for polymers. Obviously the most practical are esterification reactions leading to compounds which can be determined colorimetrically. From the point of view of sensitivity, reactions with isocyanates are noteworthy since they not only proceed rapidly and completely but also because reagent consumption can be determined by gas chromatography. Methods based on reactions of active hydrogen are not recommended because of their poor selectivity.

OH end groups can be detected analogously to carboxyl groups (see 1.2.2.1.1.2.) and determined using rhodamine 6 G as a colouring agent following reaction with anhydrides of dibasic acids or chloro-sulfonic acid. ,,Rose-Bengal" reagent can be used for the direct dyeing of polymers containing hydroxyl groups (PALIT [5]). In the presence of traces of bases a rose colour is produced.

## 1.2.2.2. Radiochemical Methods for End Group Determination

Their high sensitivity makes radiochemical methods of analysis of particular importance for end group determination. Their application is limited, however, because of the necessary safety measures and equipment involved and special facilities are generally required.

The preferred radiochemical method of analysis is isotopic labelling, involving either the use of isotope-containing initiators during polymerization or reactions of polymers with labelled reagents or exchange methods. Because of the readily exchangeable H-atoms, deuterium and tritium exchange methods are popular for end group analysis. They measure all exchangeable H-atoms. Like the above mentioned methods they are thus rather ion-specific involving reaction of any active hydrogen. For example, in polyesters only the sum of COOH and OH end groups is measured so that their separate determination calls for a combination of the $D_2O$ exchange method and a volumetric COOH-group determination. The example of end group determination in polyethylene glycol terephthalate included in [1] gives a detailed description of the equipment procedure and evaluation.

## 1.2.2.3. Physical Methods

Of the physical methods of end group determination and, hence, of molecular weight determination, UV, IR and NMR procedures are described in publications listed in the bibliography. Their use depends on a high sensitivity and a proportional signal response to the concentration of end groups. The latter is frequently restricted by insufficient band separation, preparation problems or impurities especially traces of water. In spite of great progress in equipment and procedures the upper limit for $M_n$ determination by physical methods is about $7 \cdot 10^4$ g/mol in the most favourable case in IR spectroscopy. Otherwise it is far below this limit.

Problems and results of infrared-spectroscopic $M_n$ determinations are discussed in various reviews [6—8]. The absorbence of a well resolved strong band is measured by the area method and providing the molar absorption coefficient is known, the molecular weight can be evaluated directly. Alternatively a calibration curve is required. Intensity may be also compared with a weak band of the chain units especially with overtone and combination bands [6]. A summary of IR bands suitable for $M_n$ determination of polymers is given in Table 4 together with molar absorption coefficients and appropriate preparation techniques. The absorption coefficients are dependent on the equipment and the selected systems are of low accuracy. They can only provide a guide to the probable sensitivity.

For the number-average molecular weight $M_{n_{end}}^{IR}$ determined by IR the following equations holds if the two end groups are the same:

$$M_{n_{end}}^{IR} = \frac{q \cdot M_{end} \cdot c}{c_{end}} = 2M_{end} \cdot \varepsilon \cdot l \cdot c = k \cdot c \tag{5}$$

$M_{end}$ = molecular weight of the end groups, $c_{end}$ = end group concentration in $g \cdot l^{-1}$, $c$ = total polymer concentration in $g \cdot l^{-1}$, $\varepsilon$ = absorption coefficient of the end group band in $g^{-1} cm^{-1} l$, $l$ = layer thickness in cm

According to equation (5) the range of application of IR-determination of $M_n$ is governed by $\varepsilon$ (Table 4) and the polymer concentration. The latter is usually 3 ... 5 percentage by weight in solutions the preferred method. The sensitivity of end group determination can be increased by using KBr discs or cast and fused films but interference caused by intramolecular and intermolecular interactions and traces of water and scattering effects must be taken into consideration. When working with films, light scattering will occur even when scattering on the film surface is eliminated by immersion in a solvent. Chemical modification to increase $M_{end}$ in equation (5) in order to augment the sensitivity of physical methods is not generally advised because accurate quantitative monitoring of the reaction itself is possible and suffices for end group determination. Table 4 shows

Table 4: IR bands of polymer end groups and experimental conditions

| Polymer | Characteristic band of end groups | Wave number of band maxima (cm$^{-1}$) | Molar absorption coefficient $\varepsilon$ (m$^2 \cdot$ mol$^{-1}$) | Sample preparation | Maximum determinable $M_{n_{end}}$ (g/mol) | Bibliography |
|---|---|---|---|---|---|---|
| Polystyrene (initiated with bromobenzoyl peroxide) | $\nu_{C=O}$ | 1720 | | 3% in CCl$_4$ | $3 \cdot 10^4$ | [9] |
| Polystyrene (initiated with diacyl peroxides) | $\nu_{C=O}$ | 1724 | | 4% in CCl$_4$ | $7 \cdot 10^4$ | [10] |
| Polyacrylonitrile (redox polymer with NaHSO$_3$) | $\nu_{s\,S-O}$ | 1043 (strong) 1200 (weak) | | KBr disc | $5-7 \cdot 10^4$ | [11] |
| Polycarbonate | $\nu_{O-H}$(phenol) $\gamma w_{(phenyl\,end\,gr.)}$(CH) | 3597 638 | | 5% in methylene chloride | | [1] |
| Polyethylene glycol | $\nu_{O-H}$(trans) $\nu_{O-H}$ $\nu_{O-H}$(assoc.) $H_2O_{(a)}$ | 3640 3605 3505 3705 | 13 13 26 15 | solution in CCl$_4$ | $4 \cdot 10^4$ | [6] |

## Molecular Weight Determination by End Group Analysis

| | | | | |
|---|---|---|---|---|
| Polyester of adipic acid + hexane diol | $\nu_{O-H}$(free) | 3640 | 13 | solution in CCl$_4$ | $3.5 \cdot 10^3$ | [6] |
| | $\nu_{O-H}$(assoc.) | 3555 | 26 | | | |
| | $\nu_{O-H}$(COOH-free) | 3540 | 14 | | | |
| | $\nu_{C=O}$(1st harmonic) | 3455 | 0.8 | | | |
| | $\nu_{O-H}$(COOH assoc.) | 3300 | 31 | | | |
| | H$_2$O$_{(a)}$ | 3705 | 15 | | | |
| Polyethylene glycol terephthalate | $\nu_{O-H}$(COOH) (assoc.) | 3300 | | film | $1 \cdot 10^4$ | [6] |
| | $\nu_{O-H}$(assoc.) | 3550 | | | | |
| | H$_2$O$_{(a)}$ | 3630 | | | | |
| | phenylene (1st harmonic) | 4080 | | | | |

that the position of the band maximum (e.g. the band of the associated OH-groups) is influenced by neighbouring groups whereas more distant chain links do not exert any influence on the end group vibrations. The strong property of association possessed by OH and COOH groups is reduced by working at higher temperatures. Considerably more complex is the problem of band separation when the bands overlap to a high degree as can occur with OH-vibrations. Assuming the BEER-LAMBERT law apptres, i.e. for highly dilute solutions and monochromatic light the absorbance of the individual components behave additively for a given wave number:

$$\frac{A_{\text{sum.}}(\tilde{v}_i)}{l} = \varepsilon_1 \tilde{v}_1 c_1 + \varepsilon_2 \tilde{v}_2 c_2 \tag{6}$$

For determining the concentration of neighbouring groups, the concentrations and absorption coefficients of the groups causing the superposition must be known.
They may be determined by:
— graphical or mathematical band separation,
— absorption measurement at $n$ wave numbers and construction of a linear equation system ($n$ = number of superimposed bands),
— ascertaining band shape factors,
— band separation by compensation with radiation polymers.

For graphical or mathematical band analysis, band profile and symmetry must be known. Calculations are frequently carried out with curve analyzers or, more usually, computers.

For the analysis of poorly resolved spectra (both NMR and IR spectra), special computer methods have been developed. Visible resolution of the spectra using maxima, minima and inflection points can indicate the number of peaks, their positions, width at half height and intensities. Computer resolution has the advantage that all points in the spectrum are utilized. The data is used for constructing a table with $3n$ or $4n$ variables:
— position of the maximum,
— width at half height,
— intensity; and possibly the
— line shape factors.

After calibration with appropriate materials, the intensity at the band maximum can be evaluated and utilized in determining the concentration of the functional groups.

A similar procedure is vividly demonstrated in [6] using IR for determining the molecular weight of polyethylene glycol. Another example of computer aided band resolution, aimed at determining the concentration of end groups based on the number of chain links of the main chain, is given in Chapter 3.6.

The principle of band compensation by comparison with a polymer sample free from end groups but otherwise identical and of having equal layer thickness is explained in the same Chapter. Perturbations due to differences in layer thickness of the two films are eliminated by the use of wedge-shaped compensation films. When the end groups depend on the initiators, they can be compensated by suitable radiation polymers.

### 1.2.3. Examples

#### 1.2.3.1. Molecular Weight Determination of a Linear Saturated Polyester by Chemical Determination of the COOH and OH End Groups

##### 1.2.3.1.1. Carboxyl Group Determination using Sodium Alkoxide in an Anhydrous Medium

0.5 ... 1.0 g of the polyester are dissolved in 25 ml of acetone and 12.5 ml of absolute ethanol are added and immediately titrated potentiometrically with 0.05 N sodium methylate solution in the cell:

GE/measuring solution/$KNO_3$ (saturated), 2% of agar-agar/KCl (saturated), $Hg_2Cl_2$ (saturated)/Hg

Instead of the glass electrode (GE), a bismuth electrode can also be used. During titration, the solution is magnetically stirred. Near to the equivalence point, aliquots of 0.1 ml are added.

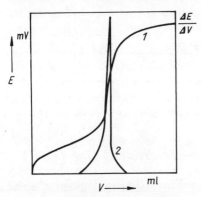

*Fig. 1*: Potentiometric titration curves
Curve 1: $E(V)$, Curve 2: $\Delta E/\Delta V$ against $V$

The approach of the equivalence point is indicated by an increase of the potential difference. Titration is continued with two further aliquots past the equivalence point. A pure polyester exhibits an S-shaped titration curve (see Fig. 1, curve 1). If the polyester still contains free dicarboxylic acid (e.g. maleic acid from the condensation) three potential steps will be observed because of the different protolysis constants of the carboxyl groups. The first step corresponds to neutralization of the first carboxyl group of the free dicarboxylic acid, the second step results from neutralization of the carboxyl end group of the polyester and the third corresponds to neutralization of the second carboxyl group of the dicarboxylic acid. Mention should be made of the simultaneous determination of two polyesters and the use of tetramethyl ammonium hydroxide or tetrabutyl ammonium hydroxide as titrant [13].

Evaluation can be effected graphically or by calculation. Usually, the measured potential values ($E$) are recorded as a function of the titrant volumes ($V$) or plotted graphically (Fig. 1, curve 1). In case of a symmetrical curve, the abscissa value at the point of inflection corresponds to the end point ($V_E$), which can also be determined from the position of the maximum when plotting the ratio $\frac{\Delta E}{\Delta V}$ against $V$ (Fig. 1, curve 2). A mathematical evaluation method has been described by HAHN [14]. The acid value $A.V.$ (defined as mg KOH per g of substance) and $M_{n_{end}}$ are obtained from $V_E$ using equations (7) and (8)

$$A.V. = \frac{V_E \cdot 2.8 \cdot F}{m} \tag{7}$$

$$M_{n_{end}} = \frac{q \cdot 5.6 \cdot 10^4}{A.V.} = \frac{q \cdot m \cdot 1000 \text{ (g/mol)}}{V_E \cdot N \cdot F} \tag{8}$$

$V_E$ = Titrant consumption in ml; $m$ = amount of polymer in g; $F$ = factor of the 0.05 Na-methylate solution; $N$ = normality of the sodium methylate solution.

The method can be used for molecular weights of up to $2.5 \cdot 10^4$ g/mol. Commercial polyesters may contain molecules with 2 OH and COOH end groups as well as free acid and acid anhydride. COOH end group determination must thus be complemented by an OH end group determination as a check. For the determination of free acid and acid anhydride reference is made to [1].

### 1.2.3.1.2. *OH End Group Content — Hydroxyl Value (OH.V.)*

OH end group content or OH.V. is ascertained from the consumption of acetic anhydride during acetylization in ethyl acetate in the presence of toluenesulphonic acid [15]. In this esterification method, free dicarboxylic acid, acid anhydride and COOH end groups must be considered. They are detected by the apparent acid value "$A.V._a$" which is determined by titrating the polyester in ethanol: benzene (1:1 solution) with 0.1 N ethanolic KOH using a suitable indicator [16]. During titration with ethanol as solvent, the dicarboxylic acid anhydride is converted into semi-ester so that only one carboxyl group has to be considered.

For the determination of $OH.V.$ and of the $M_n$-value, 1.000 g of the polyester is heated with 5.00 ml of acetylating reagent in a 100 ml conical flask covered with a watch glass at 50 °C ± 1 degree for 10 min in an oven. It is then shaken and left for 5 min at 50 °C. After cooling for 5 min, 2 ml of water and 10.00 ml of pyridine-water mixture (1:1) are added, the flask is then stoppered and, after a further 5 min, titration is carried out at room temperature with 0.5 N of methanolic KOH using phenolphthalein as an indicator ($b$, ml). The same procedure is repeated for a blank ($a$, ml). The acetylating reagent is obtained by dissolving 14.4 g of tolulene-p-sulphonic acid in 360 ml of ethyl acetate and adding 120 ml of acetic anhydride dropwise with continuous stirring.

The $OH.V.$ is calculated using equation (9) and, thus, $M_{n_{end}}$ by equation (10) (cf. equation (8))

$$OH.V. = \frac{(a - b) \cdot 28.05}{m} + A.V._a \tag{9}$$

$$M_{n_{end}} = \frac{q \cdot 5.6 \cdot 10^4}{OH.V.} \; (\text{g mol}^{-1}) \tag{10}$$

## 1.2.3.2. Infrared End Group Determination in Polystyrene

In the radical polymerization of styrene with the diacyl peroxide of p-azobenzene carboxylic acid, catalyst fragments of the type

are incorporated as end groups in the macromolecule. Thus, the latter contains ester end groups which absorb at 1724 cm$^{-1}$ ($v_{C=O}$), 1275 cm$^{-1}$ and 1115 cm$^{-1}$ (C—O arylester bands). The absorptions of the bands are proportional to concentration and inversely proportion to the viscosity average molecular weight. Thus molecular weights up to $\sim 7 \cdot 10^4$ g/mol can be determined.

For $M_{n_{end}}$ determination, the polystyrene is separated into molecularly uniform fractions (see Chapter 2.). A 4 per cent solution of the fractions in CCl$_4$ is made and measured in a twin beam IR-spectrometer with a cell path length of

0.35 mm over the range from 700 cm$^{-1}$ to 2000 cm$^{-1}$. For quantitative determination, the 1724 cm$^{-1}$ band is used. The end group content of the sample is read off a calibration curve (Fig. 2).

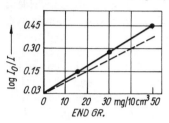

*Fig. 2*: Calibration curve for IR end group determination in PS 1724 cm$^{-1}$ band of mixtures of the reference with PS. Dashed line: without addition of PS.

The latter is constructed from absorbence measurements of the $v_{C=O}$-band in mixtures of thermal or radiation-polymerized PS with 4-benzene azobenzoic acid-$\beta$-phenyl ester using the same experimental conditions as for the sample. The synthesis of the low-molecular weight model substance is described in [10].

For evaluation it is assumed that each polystyrene molecule has the two end groups the same: in the present case this holds true approximately for fractions within the molecular weight range between $1.5 \cdot 10^4$ g/mol and $5.0 \cdot 10^4$ g/mol. For the conversion of the calibration curve of the model substance into that of the actual end groups of formula I, the concentrations are multiplied by the factor 0.6819 resulting from the ratio of the molecular weights of end group and model substance. For $M_{n_{end}}^{IR}$, equation (5) with $q = 2$, $c = 40$ g $\cdot$ l$^{-1}$ and $M_{end} = 225.23$ g $\cdot$ mol$^{-1}$, gives:

$$M_{n_{end}}^{IR} = \frac{1.8018 \cdot 10^4}{c_E} \text{ g} \cdot \text{mol}^{-1} \tag{11}$$

$c_E$ = end group concentration in g $\cdot$ l$^{-1}$

Some test results are shown in Table 5.

*Table 5*:

| $I$ (mm) | $I_0$ (mm) | $\log \dfrac{I_0}{I}$ | $c_{end}$ (g $\cdot$ l$^{-1}$) | $M_{n_{end}}^{IR}$ (g $\cdot$ mol$^{-1}$) |
|---|---|---|---|---|
| 121,5 | 142,2 | 0,06834 | 0,769 | 23 400 |
| 125,6 | 142,0 | 0,05331 | 0,600 | 30 000 |
| 132,0 | 142,3 | 0,03262 | 0,369 | 48 800 |

# Bibliography

[1] SCHRÖDER, E., FRANZ, J. and HAGEN, E., *"Ausgewählte Methoden zur Plastanalytik"*, Akademie-Verlag Berlin 1976, p. 374
[2] PALIT, S. R. and MUKHERJEE, A. R., J. Polymer Sci. **58** (1962), 1243
[3] KAISER, R., *"Quantitative Bestimmung organischer funktioneller Gruppen"*, Akademische Verlagsanstalt Frankfurt/Main 1966
[4] KOLTHOFF, J. M., STRICKS, W. and MORRAN, L., Analytic Chem. **88** (1954), 366
[5] PALIT, S. R., Analytic Chem. **33** (1961), 1441
[6] LANGBEIN, G., Kolloid Z. — Z. Polymere **200** (1964), 10
[7] HUMMEL, D. O. and SCHOLL, F., *"Atlas der Kunststoff-Analyse"*, Vol. I, Carl-Hanser-Verlag München 1968
[8] DECHANT, J., *"Ultrarotspektroskopische Untersuchungen an Polymeren"*, Akademie-Verlag Berlin 1972
[9] PFANN, H. F., WILLIAMS, V. Z. and MARK, H., J. Polymer Sci. **1** (1946), 14
[10] KÄMMERER, H., ROCABOY, F., STEINFORT, K. and KERN W., Makromolekulare Chem. **53** (1962), 80
[11] JAMADERA, R., J. Polymer Sci. **50** (1961), 4
[12] FIJOLKA, P., Plaste und Kautschuk **10** (1963), 582
[13] SCHULZ, G., Dissertation, Humboldt-Universität Berlin 1967
[14] HAHN, F. L., Z. anal. Chem. **163** (1958), 169
[15] TGL 26125 sheet 17 (Standard specification)
[16] TGL 14087 (Standard specification)

## 1.3. Determination of the Number-average Molecular Weight of Polymers by Osmotic Measurement — Membrane Osmosis

### 1.3.1. Introduction

Osmotic molecular weight determination is one of the most important methods of polymer characterization. It determines the number-average molecular weight $M_n$, which is of major importance in kinetic investigations. In the case of copolymers and polymer mixtures it is a simple means of measurement and evaluation compared with the weight-average because the $M_n$ value of a multicomponent system is the sum of the $M_n$ values of the components.

$$M_n^{AB} = M_n^A + M_n^B \tag{1}$$

with

$$M_n^A = \frac{\sum_i n_i M_i^A}{\sum_i n_i} \qquad M_n^B = \frac{\sum_i n_i M_i^B}{\sum_i n_i}.$$

Osmometry belongs to the group of absolute methods based on colligative properties, that is to say, properties which measure the activity of the solvent in a solution of the polymer. The chemical potential of the solvent in a polymeric solution is dependent not only on the number of dissolved particles but also on pressure and temperature and differs from that of the pure solvent. The potential difference can be balanced isothermally by increasing the pressure in the solution and isobarically by increasing the temperature of the solution. Depending on the measuring technique osmotic methods are divided into membrane osmometry and vapour-pressure osmometry.

## 1.3.2. Principles

In membrane osmometry, the hydrostatic overpressure in a polymer solution is measured. It occurs when two solutions of different concentration, e.g. polymer solution and pure solvent, are separated by a membrane, which should be ideally semipermeable, that is to say, impermeable to the dissolved molecules. The chemical potential of the solvent is lower in the polymer solution than in the pure solvent. The solvent molecules diffuse into the polymer solution through the membrane until the potential difference is zero. With conventional membrane osmometers this occurs when the hydrostatic pressure of a column of liquid of height $\Delta h$ is equal to the osmotic pressure $\pi$ caused by the potential difference $\Delta \mu_A$. The relationship between $\Delta \mu_A$ and $\pi$ can be derived via the pressure dependence of the chemical potential. At osmotic equilibrium, the following must hold at constant temperature:

$$(\mu'_{0A})_{P_0} = (\mu''_A)_{P_0} + \pi \tag{2}$$

$' =$ refers to solvent, $'' =$ refers to solution, $p_0 =$ external pressure

The change in solvent potential in the solution compartment when the pressure is increased by $\pi$ is given by the integral from $p_0$ to $p_0 + \pi$ of the partial derivation of the potential with respect to pressure.

$$(\mu'_{0A})_{P_0} = (\mu''_A)_{P_0} + \int_{P_0}^{P_0+\pi} \left(\frac{\partial \mu''_A}{\partial p}\right)_T dp$$

$$= (\mu''_{0A} + RT \ln a_A)_{P_0} + \int_{P_0}^{P_0+\pi} \left(\frac{\partial \mu''_A}{\partial p}\right)_T dp + RT \int_{P_0}^{P_0+\pi} \left(\frac{\partial \ln a_A}{\partial p}\right)_T dp \tag{3}$$

Partial differentiation of the standard potential $\mu_{0A}$ and the activity $a$ with respect to $p$ gives

$$\left(\frac{\partial \mu_{0A}}{\partial p}\right)_T = \bar{V}_{0A}, \qquad \left(\frac{\partial \ln a_A}{\partial p}\right)_T = \frac{\bar{V}_A - \bar{V}_{0A}}{RT} \tag{4}$$

where the $\bar{V}$ represent partial molar volumes. Combining these with equation (3) gives

$$(\mu'_{0A})_{P_0} = (\mu''_{0A} + RT \ln a_A)_{P_0} + \bar{V}''_{0A}\pi + RT\frac{(\bar{V}''_A - \bar{V}''_0)\pi}{RT} \tag{5}$$

Since the standard potentials in the two compartments are equal, the relation reduces:

$$\bar{V}''_A \pi = -RT \ln a_A = -\Delta\mu_A \tag{6}$$

According to the FLORY and HUGGINS theory the GIBBS' free energy of mixing of a polymer solution is given by

$$\Delta G_m = RT(n_A \ln \Phi_A + n_B \ln \Phi_B + \chi n_A \Phi_B) \tag{7}$$

$\chi$ = HUGGINS interaction parameter, $n$ = number of moles, $\Phi$ = volume fraction

The chemical potential of the solvent in the polymeric solution is obtained by partial differentiation of $\Delta G_m$ with respect to $n_A$. It should be remembered that $\Phi_A$ and $\Phi_B$ are functions of $n_A$. Thus:

$$\Delta\mu_A = \mu_A - \mu_{0A} = RT\left[\ln \Phi_A + \left(1 - \frac{V_A}{V_B}\right)\Phi_B + \chi\Phi_B^2\right] \tag{8}$$

$V_A, V_B$ = molar volumes

Expanding $\ln \Phi_A$ as far as the second term in a TAYLOR series:

$$\ln \Phi_A = \ln(1 - \Phi_B) = -\Phi_B - \Phi_B^2/2 \tag{9}$$

and substituting the modified equation (8) into equation (6), gives

$$\pi = -\frac{RT}{\bar{V}_A}\left[-\left(\frac{V_A}{V_B}\right)\Phi_B - \left(\frac{1}{2} - \chi\right)\Phi_B^2\right] \tag{10}$$

Since $\Phi_B = c_B/\varrho_B$ and $V_B = M_B/\varrho_B$

(where $\varrho$ = density), the ratio of osmotic pressure to concentration is:

$$\frac{\pi}{c_B} = \frac{RT}{M_B} + RT \frac{\left(\frac{1}{2} - \chi\right) c_B}{\varrho_B^2 \bar{V}_A} \tag{11}$$

In the case of polydisperse samples, the osmotic pressure is the sum of the osmotic pressure $\pi_i$ of all components $i$ with molecular weight $M_i$ and concentration $c_{B,i}$. Formula (11) is applicable to osmotic partial pressures. Summation will show that, in the case of polydisperse samples, $M_B$ can be substituted by the number-average molecular weight $M_n$. In the case of concentrated polymer solutions (greater than 1 per cent by weight) the TAYLOR series cannot be truncated after the 2nd term. The reduced osmotic pressure must then be represented by a power series of concentration:

$$\frac{\pi}{c_B} = \frac{RT}{M_n} + Bc_B + Cc_B^2 + \ldots = RT \left(\frac{1}{M_n} + A_2 c_B + A_3 c_B^2 + \ldots\right) \tag{12}$$

The virial coefficients, $B$. $C$ etc., of the osmotic pressure, are measures of the intermolecular forces of interaction in the polymeric solution. Usually $B$ is given in N cm$^4$ g$^{-2}$. Frequently, $B/RT$ is defined as the 2nd osmotic virial coefficient (e.g. in light scattering measurements) with the units cm$^3$ mol g$^{-2}$. $B$ can be easily determined by measurements in dilute solution ($C \rightarrow 0$) and from this HUGGINS' interaction parameter $\chi$ can be obtained. From the temperature dependence, the enthalpy of dilution $\Delta H_A$ and entropy of dilution $\Delta S_A$ can be calculated:

$$\chi = \frac{1}{2} - \frac{B \varrho_B^2 \bar{V}_A}{RT} \tag{13}$$

$$\Delta S_A = -(\partial \Delta \mu_A / \partial T)_{p, n_B} = -\left[\frac{R}{M_n} + c_B \left(\frac{\partial B}{\partial T}\right)_{p, n_B}\right] c_B \bar{V}_A \tag{14}$$

$$\Delta H_A = \left(\partial \left(\frac{\Delta \mu_A}{T}\right) \middle/ \partial T\right)_{p, n_B} T^2 = \left[\left(\frac{\partial B}{\partial T}\right)_{p, n_B} \cdot T - B\right] c_B^2 \bar{V}_A$$

$$\left(\frac{\partial \bar{V}_A}{\partial T} = 0, \frac{\partial \bar{V}_A}{\partial p} = 0, \frac{\partial c_B}{\partial T} = 0\right) \tag{15}$$

Using equation (6), to calculate the potential difference, we get approximately:

$$\Delta \mu_A = -\pi \bar{V}_A = -\left[\frac{RT}{M_n} + Bc_B\right] c_B \bar{V}_A \tag{16}$$

$$\Delta S_A = \frac{\bar{V}_A(T_1)\,\pi(T_1) - \bar{V}_A(T_2)\,\pi(T_2)}{T_1 - T_2} \tag{17}$$

$$\Delta H_A = \frac{\bar{V}_A(T_1)\,\pi(T_1)/T_1 - \bar{V}_A(T_2)\,\pi(T_2)/T_2}{1/T_1 - 1/T_2} \quad (T_1 > T_2) \tag{18}$$

## 1.3.3. Experimental Background

Membrane osmometers consist generally of a thermostatic measuring cell divided into two compartments by a semipermeable membrane. Regenerated cellulose (saponified cellulose acetate) is the preferred membrane material for organic solvents while for aqueous solvents cellulose acetate and cellulose nitrate are utilized. Synthetic polymer membranes, e.g. polyurethane, polyacrylonitrile, polyethylene, have so far been used only in development work. In automatic devices, the membranes are usually clamped horizontally or vertically. In the conventional apparatus, the solution compartment is provided with a capillary for measuring

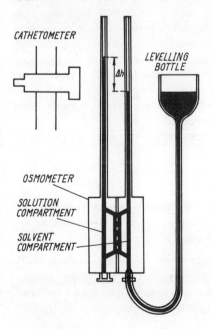

*Fig. 3*: Schematic representation of a WULF-KAPELLE two-chamber osmometer

the capillary rise; frequently a reference capillary is connected to the solvent compartment to compensate for the effects of temperature variations and wetting (Fig. 3).

Osmotic pressure can be measured by static, dynamic and compensation methods. In the static method, the initial elevation difference is "zero". Equilibrium occurs due to diffusion of solvent into the solution compartment. The change in capillary rise with respect to time is recorded. The inverse procedure may be adopted, that is to say, an elevation above the final expected value of the hydrostatic pressure is preset. In both cases, the same value should ultimately be obtained. A reliable value is usually obtained by combining the two methods and interpolating between the two curves, both of which vary exponentially with time $t$. This method is only suitable for polymer fractions with a molecular weight greater than 20,000 g mol$^{-1}$. About 4 days are allowed for the system to approach equilibrium.

In dynamic measuring methods, the rate of influx of the solvent is measured well before equilibrium is attained. The values obtained in this way are subject to inaccuracies due to extrapolation to $t = \infty$.

With the compensation method, the change in the rise in the solution capillary is balanced by continuously changing the difference in height until osmotic equilibrium is reached. A combination of dynamic and compensation methods was adopted by WULF and KAPELLE [1]. The height difference is changed continuously until the meniscus in the solution capillary no longer appears to move. This process corresponds to the compensation method. The variation of the permeation speed of the solvent with time in the solution compartment is then observed and a correction term calculated which, depending on the direction of the meniscus change (increase or decrease), is added to or subtracted from the compensation value.

The change in the osmotic pressure varies exponentially with time approximating to a linear relationship as equilibrium is approached.

The correction term is determined by equation (19):

$$\Delta h_{\mu_A} = \Delta h_0 + \lim_{\Delta t \to \infty} \left( \frac{1}{\zeta} \frac{\Delta h_t}{\Delta t} \right) \tag{19}$$

$\Delta h_{\mu_A}$ = height difference at equilibrium, $\Delta h_0$ = height difference at the beginning of the measurement, $\Delta h_t$ = change in the height difference during time $\Delta t$, $\Delta t$ = time interval between two measurements, $\zeta$ = an equilibration constant (KUHN [3])

HELLFRITZ osmometers are normally utilized for static measurements [2]. Automatic electronic membrane osmometers frequently operate by the compensation method. With rapid measuring osmometers volume changes of $10^{-6}$ cm$^3$ suffice to produce automatically within tenths of a second a hydrostatic counterpressure which prevents flow of the solvent. The hydrostatic pressure can be controlled, for example, by an air bubble in a capillary connected to the solvent compartment. Release of solvent from the levelling bottle is controlled optically. When the air bubble remains stationary (i.e., no solvent penetrates through the membrane), then the hydrostatic pressure determined by the position of the levelling bottle is equal to the osmotic pressure. Another type of commercial osmometer operates without automatic compensation. A pressure measuring system is connected to the solvent compartment which is sealed off from the atmosphere. Permeation of solvent into the solution chamber, induces a partial vacuum preventing further dilution of the solution when the difference in pressure is equal to the osmotic pressure.

The membrane pore size determines the lowest molecular weight which can be measured ($2 \cdot 10^4$ g/mol or less in favourable cases). The membranes, which are conditioned to the solvent and the measuring temperature in stages, are serviceable for a few months but they lose their permeability by ageing and shrinking so that long equilibration times are required. The thinner the membrane, the more it is swollen by the solvent and the higher is the diffusion coefficient for the self-diffusion of the solvent molecules, leading to shorter the equilibration times. New membranes can be checked conveniently by measuring polymer standards.

Crood sealing of the osmometer body and constant temperature are important pre-conditions for satisfactory osmotic measurements. It is difficult to find and exclude the following sources of faults:
— Balloon effect (pressure-dependent elastic bulging of the membrane leading to a change in the chamber volume and, hence in the hydrostatic height difference).
— Membrane asymmetry (caused by different capillary radii or surface-active substances in the solution).
— Adsorption on the membrane. This fault can be avoided if the membrane is conditioned to the sample before measurement.
— Partial permeation of the dissolved substance. The osmotic pressure falls with time. The value of $M_n$ is error, values being too large even if measurements are extrapolated to zero time.

The upper limit of measurement of membrane osmometry is of the order of $10^6$ g/mol and is determined by the accuracy of measurement of the small pressure differences (allowing an error of 30%). The concentration of the polymeric solution cannot be raised arbitrarily to increase the quantity to be measured because higher concentrations render evaluation of the measurement more difficult since higher virial coefficients cannot be neglected. At higher concentrations, the polymer molecules form an entanglement network with a density which can be characterized by measurements of the osmotic pressure at concentrations above a critical value. The possibility of association requires that the number-average molecular weights be measured in different solvents and at different temperatures. If these agree, association can be excluded. The number-average molecular weight is not as sensitive to the existence of associates as the weight-average molecular weight. The presence of association can also be inferred if the $\dfrac{\pi}{c_B} - c_B$ — curve exhibits a minimum at low concentrations.

## 1.3.4. Example

Membrane-osmometric molecular weight determination is exemplified using a PVC-sample. The second virial coefficient, which allows appraisal of the solution state in the solvent, cyclohexanone, at 25 °C is also calculated. Measurement in a WULF-KAPELLE osmometer is described first. Since the reduced osmotic pressure $\pi/c_B$ in polymers is dependent on concentration, heights must be measured at different concentrations.

For this purpose 4 solutions are prepared with concentrations in the range 2 ... 7 g/l. The purified distilled solvent is put into the compartment connected to the levelling bottle while the solution is poured into the second compartment. When fitting the capillaries care must be taken to ensure that they are tightly seated and no air bubbles are trapped. The apparatus is then placed in a thermostat. The menisci in the two capillaries should be at the same level. After short temperature equilibration, they are balanced once more by removing solution or solvent with a hypodermic syringe having a fine hollow needle. The solution capillary is cleaned inside to remove adhering liquid to eliminate errors due to solution drips because of the short equilibration time. The cathetometer is focussed on the solution meniscus. Solvent diffusing into the capillary causes the meniscus to rise slowly. By lowering the levelling bottle, the diffusion of the solvent into the solution compartment is counteracted. The cathetometer is again focussed on the solution meniscus. In this way, the levelling bottle and, thus, the height of the solvent are lowered in stages until the height difference reaches a value $\Delta h_0$ approximately equal to the equilibrium value. Then movement of the meniscus is noticeable only after five to ten minutes. At intervals of ten to fifteen minutes, the height difference $\Delta h_t$ and the relevant time difference $\Delta t$ are tabulated. About four to six measurements are required for evaluating the correction term graphically. Before measuring a new concentration, the solution compartment is rinsed with two to three ml of the new solution.

Because of the very short measuring times, required to determine the differences in heights by the above method, the balloon effect must be taken into consideration. Every change in the height difference due to external interference

# Membrane Osmosis

disturbs the membrane's equilibrium position. The consequent volume change is not restored to the initial condition immediately. The membrane will "flow" gradually into the equilibrium position corresponding to the hydrostatic pressure.

The correction term is determined graphically. The quantities calculated are summarized in Table 6.

*Table 6*: Data for determining the correction term

| $t$ (min) | $h_A$ (cm) | $h_{AB}$ (cm) | $\Delta h_0$ (cm) | $\Delta h_t$ (cm) | $\dfrac{\Delta h_t}{\Delta t}$ ($10^{-3}$ cm min$^{-1}$) | $\dfrac{\Delta h_t}{\zeta \Delta t}$ (cm) | $\Sigma \Delta h_t$ (cm) |
|---|---|---|---|---|---|---|---|
| 0  | 15,3425 | 9.7060 | 5.6365 | 0      | 0    | 0    | 0      |
| 14 |         | 9.6660 |        | 0.0400 | 2.86 | 3.50 | 0.0400 |
| 24 |         | 9.6475 |        | 0.0185 | 1.85 | 2.27 | 0.0585 |
| 34 |         | 9.6350 |        | 0.0125 | 1.25 | 1.53 | 0.0710 |
| 64 |         | 9.6170 |        | 0.0180 | 0.60 | 0.735| 0.0890 |

$h_A$ = height of the solvent in the capillary, $h_{AB}$ = height of the solution in the capillary, $\zeta$ = equilibration constant; here $8.16 \cdot 10^4$ min$^{-1}$.

The equilibration constant $\zeta$ is a measure of the permeability of the membrane. When pure solvent is in the solution compartment and a height difference is set, solvent permeates through the membrane. $\zeta$ can then be calculated from the time dependence of the height difference.

$$\zeta = \frac{\log p(t=0) - \log p(t)}{t} = \frac{\log \Delta h(t)}{t} \tag{20}$$

(Evaluation using equation (19) does not require knowledge of the equilibration constant since it only determines the slope of the extrapolation line.)

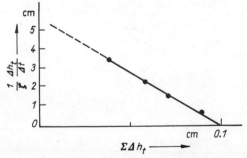

*Fig. 4*: Graphical determination of the correction term (PVC in cyclohexanone; $c_B = 6.0$ gl$^{-1}$) $\Delta h_{\mu_A} = \Delta h_0 + \lim \Sigma \Delta h_t = (5.64 + 0.10)$ cm $= 5.74$ cm

If the values of $\dfrac{\Delta h_t}{\zeta \Delta t}$ are plotted against $\Sigma \Delta h_t$, a straight line is obtained (Fig. 4). Extrapolation of the expression $\dfrac{\Delta h_t}{\zeta \Delta t}$ to zero gives a correction term which for an increase (decrease) in the height is added to (substracted from) the value $\Delta h_0$. The value $\Delta h_{\mu_A}$ resulting from this corresponds to the height difference at equilibrium.

The osmotic pressure is given by equation (21):

$$\pi = \Delta h_{\mu_A}\, \varrho_A g + h_m(\varrho_{AB} - \varrho_A)\, g \tag{21}$$

$\Delta h_{\mu_A}$ = total height, $\varrho$ = density (index $AB$ = solution), $g = 9.81$ m/s², $h_m$ = distance of the lower meniscus from the membrane

For dilute solutions in which $\varrho_A \approx \varrho_{AB}$, the second term can be omitted from equation (21).

For the determination of molecular weight, the ratio of osmotic pressure to concentration $\pi/c_B$ is plotted against the concentration $c_B$ and extrapolated to $c_B = 0$. According to equation (12), $M_n$ can be calculated from the intercept and the 2nd virial coefficient from the slope. Table 7 lists the height differences at equilibrium of PVC-solutions of different concentrations from which $\pi$ is calculated using equation (21).

From the plot of $\pi/c_B$ against $c_B$, the intercept is

$$\left(\dfrac{\pi}{c_B}\right)_{c_B \to 0} = 6.31 \text{ N cm g}^{-1};$$

the number-average molecular weight is calculated to be

$$M_n = \dfrac{RT}{\left(\dfrac{\pi}{c_B}\right)_{c_B \to 0}} = 3.9 \cdot 10^4 \text{ g mol}^{-1}$$

Table 7: Equilibrium heights and osmotic pressures

| $c_B$ (g$l^{-1}$) | $\Delta h_{\mu_A}$ (cm) | $\pi$ ($10^{-3}$ N cm$^{-2}$) | $\dfrac{\pi}{c_B}$ (N cm g$^{-1}$) |
|---|---|---|---|
| 2.6 | 2.095 | 19.45 | 7.48 |
| 3.5 | 2.89  | 26.84 | 7.67 |
| 4.7 | 4.28  | 39.74 | 8.46 |
| 6.0 | 5.74  | 53.30 | 8.88 |
| 6.7 | 6.655 | 61.80 | 9.22 |

# Membrane Osmosis

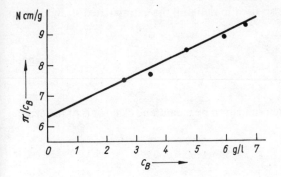

*Fig. 5*: Reduced osmotic pressure as a function of concentration

The slope of the straight line is the $B$-value or (when divided by $RT$) the $A_2$-value.

$$B = \frac{\Delta\left(\frac{\pi}{c_B}\right)}{\Delta c_B} = 0.42 \cdot 10^3 \text{ N cm}^4 \text{ g}^{-2}$$

$$A_2 = \frac{B}{RT} = 1.7 \cdot 10^{-3} \text{ cm}^3 \text{ mol g}^{-2}$$

In the majority of analytical laboratories, automatic or rapidly measuring osmometers are employed. Fig. 6 shows the recorder plot of measurements with a Knauer (Berlin-West) osmometer which measures the partial vacuum electronically in a seated solvent compartment. The equilibration times are between 2 and 20 minutes, depending on the solvent and the permeability of the membrane. The partial vacuum reading is calibrated manumetrically so that height differences from 2.5 to 40 cm can be read directly. Only a small amount of solution is required (about 2 cm³) so that the device is particularly suitable for the measurement of fractions. As already shown the osmotic pressure is calculated from the height differences taken from the recorder curves, and the osmotogram drawn.

Higher virial coefficients have so far been ignored. Frequently, it is useful to assume that $A_3 = 0.25 A_2^2$ and to plot $\left(\dfrac{\pi}{c_B}\right)^{0,5}$ versus $c_B$. Then, measured points of a higher concentration are also used in defining the extrapolated straight line and the intercept

$$\left(\frac{\pi}{c_B}\right)^{0,5}_{c_B \to 0} = \left(\frac{RT}{M_n}\right)^{0,5} \tag{22}$$

can be determined more exactly.

The relative error of the $M_n$ determination by membrane osmosis is of the order of 5 to 10%.

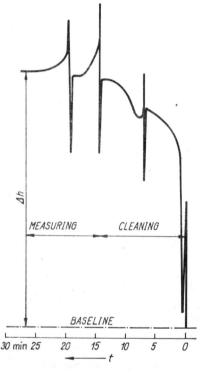

*Fig. 6*: Recorder plot for the measurement of the osmotic pressure using an electronic osmometer

## Bibliography

[1] WULF, K. and KAPELLE, R., Faserforsch. u. Textiltechn. **14** (1963), 157
[2] HELLFRITZ, H., Makromolekulare Chem. **7** (1951), 184
[3] SCHULZ, G. V. and KUHN, Wo. H., Makromolekulare Chem. **50** (1961), 37

## 1.4. Determination of Number-average Molecular Weight of Polymers by Osmotic Measurement — Vapour-pressure Osmosis

### 1.4.1. Introduction

Membrane osmosis cannot be used to determine the number-average molecular weight of polymer molecules less than $2 \cdot 10^4$ g/mol, because dissolved molecules permeate through the membrane into the solution compartment. Isothermal distillation, cryoscopy, ebullioscopy and vapour-pressure osmosis can be used to determine such molecular weights. Since its discovery in 1930, vapour-pressure osmometry has become a standard method, especially for prepolymer characterization, as the standard and availability of equipment have improved.

### 1.4.2. Principles

The same theoretical considerations which apply to membrane osmosis apply also to vapour-pressure osmosis. In both the difference in chemical potential between the pure solvent and the solvent in the solution is measured. In vapour-pressure osmometry, two thermistors are situated in a saturated solvent atmosphere (solvent vapour (vapour-pressure $p(T)$)-air-mixture). One drop of solvent is applied to one thermistor and one drop of a solution of known concentration of the sample is applied to the other. The vapour-pressure of the solvent in the solution is lower than that of the pure solvent. The solvent from the vapour compartment condenses on to the drop of solution. The heat of condensation thereby released heats the solution and increases its vapour pressure. At equilibrium, the vapour pressures of the two drops are equal and the temperature of the drop of solution has increased by $\Delta T$. $\Delta T$ is related to the activity of the solvent in the solution by (1):

$$\Delta T = -\frac{RT^2}{LM_A} \ln a_A \tag{1}$$

L = latent heat of evaporation of the solvent (J g$^{-1}$)

From equations (6) and (10) in Chapter 1.3., $\ln a_A$ is

$$\ln a_A = -\left(\frac{V_A}{V_B}\right)\Phi_B - (1/2 - \chi)\,\Phi_B^2 \tag{2}$$

$V$ = molar volume, $\Phi$ = volume fraction

Since $\Phi_B = c_B/\varrho_B$ and in general $V = \dfrac{M}{\varrho}$ substituting in equation (2) gives:

$$\ln a_A = -\frac{M_A c_B}{M_B \varrho_A} - A_{2,v} M_A \frac{c_B^2}{\varrho_A}$$

$$A_{2,v} = (1/2 - \chi)\frac{1}{\varrho_B^2 V_A} \tag{3}$$

Substituting equation (3) into equation (1) and converting mass per unit volume, $c_B$, into the form commonly used in vapour-pressure osmometry, i.e. mass fraction $c$ ($c \approx c_B/\varrho_A$), gives the equation for the molecular weight of the dissolved substance (the number-average in the case of polydispersed samples) and the second virial coefficient. Here it is assumed that the concentration of the solution is low enough for higher virial coefficients to be negligible.

$$\frac{\Delta T}{c} = K_e^{th}\left(\frac{1}{M_n} + A_{2,v}\varrho_A c\right) \tag{4}$$

$A_{2,v}$ = second virial coefficient of vapour-pressure osmosis in cm³ mol g⁻² (the virial coefficients of vapour-pressure osmosis are not identical with these of membrane osmosis because measurements are carried out isobarically), $c$ = is in g/kg of solution, $K_e^{th} = \dfrac{RT^2}{L}$

Since, in accordance with thermodynamic considerations and as indicated in equation (4), a relationship exists between the measured quantity (temperature difference) and molecular weight, vapour-pressure osmosis could be held to be an absolute method of molecular weight determination. Experiments with substances of known molecular weight, however, show that deviations of the measured values occur which can be corrected by using an experimental rather than the theoretical calibration constant. This experimental calibration constant $K_e^{exp}$ comprises the theoretical calibration constant $K_e^{th}$ and an apparatus constant $k_e$ the latter accounts for the complex processes of heat conduction and diffusion which occur in the vapour phase and in the two drops.

Some authors [1—3] have calculated suitable constants for their own equipment and derived the thermal balance of the drop of solution. The essential contributions to the heat losses are:

— heat conduction by the gas phase

$$\dot{Q}_g = -S k_g \Delta T_{10} \tag{5}$$

$S$ = LANGMUIR shape factor

$$S = 4\pi r_z \frac{r_z}{r_z - r_{dr}}$$

$r_z$ = radius of a sphere with a surface area equivalent to that of the measuring cell, $r_{dr}$ = radius of the drop of solution, $k_g$ = coefficient of thermal conductivity of the solvent-vapour-air mixture, $\Delta T_{10}$ = temperature difference between solution drop and cell

— thermal radiation

$$\dot{Q}_s = -4\sigma A\varepsilon T_0^3 \, \Delta T_{10}^4 \tag{6}$$

$\sigma$ = STEFAN-BOLTZMANN constant, $A$ = surface area of drop, $\varepsilon$ = emissivity of the drop surface

— thermal conduction by the connections to the thermistors

$$\dot{Q}_d = -2\pi \frac{r_{Pt}^2}{l_{Pt}} k_d \, \Delta T_{10} \tag{7}$$

$r_{Pt}$ = radius of the platinum connecting wires, $l_{Pt}$ = length of the platinum connecting wires, $k_d$ = coefficient of thermal conductivity of platinum

If the drop of solution cools, the solution is no longer in thermodynamic equilibrium with the environment. The consequent difference in chemical potential causes further condensation of solvent on the drop of solution in order to balance the heat losses by the heat of condensation. The consequence is a decrease in the concentration of the solution.

$$\dot{Q}_k = L\dot{m}_L \tag{8}$$

Where $\dot{m}_L$ is the amount of solvent condensing in unit time measured, for example, by determining the size of the solution drop.

The concentration of the solution and hence, the measured signal become time-dependent. The higher the concentration of the solution and the smaller the molecular weight of the dissolved substance, the higher is the decrease in temperature with time. Apart from the heat input due to condensation, the current passing through the two thermistors also heats the drops. Thus, the temperature difference between the two thermistors, need not correspond to the temperature difference between the solution and solvent. Instructions for commercial vapour-pressure osmometers suggest that the signal should be read after a certain time. It is more accurate, however (especially for measurements on substances of $M_n < 10^3$ g/mol) to allow for the time-dependence of the quantity by extrapolating to time $t = 0$.

Recent results from systematic investigations have shown that the experimentally determined calibration constant decreases with increasing molecular weight of the calibrating substance.

## 1.4.3. Experimental Background

The methodology of vapour-pressure osmometry is fundamentally different from that of membrane osmosis. In the former case, the quantity to be measured is the temperature difference between the pure solvent and the polymer solution. The temperature change is less than 0.1 degree. Such small temperature changes can be conveniently measured using thermistors. For temperature differences less than 3 degrees the thermistor resistance is given by

$$R(T) = R(T_0) \exp\left[B\left(\frac{1}{T} - \frac{1}{T_0}\right)\right] \tag{9}$$

$B$ = thermistor constant

The change in resistance resulting from the temperature change is

$$\Delta R = -R(T)\frac{B}{T^2}\Delta T \tag{10}$$

Thermistors suitable for osmotic measurements have a resistance of about $10^5\ \Omega$ at 25 °C. The resistance change is about $5\ \Omega$ for a temperature change of $1 \cdot 10^{-3}$ deg. The bridge circuits (Fig. 7) used for resistance measurements in commercial vapour-pressure osmometers are capable of measuring changes in resistance of up to $5 \cdot 10^{-4}\%$.

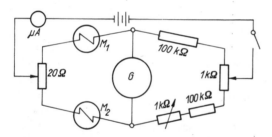

Fig. 7: WHEATSTONE bridge
M 1 — Thermistor with solvent drops, M 2 — Thermistor with solution drops, G — Central-zero galvanometer

The choice of the measuring temperature depends on the solvents used. In general, it is considered that the most favourable measuring range is about 60 deg. below the boiling point of the solvent.

Careful purification of the solvent can be dispensed with for low volatile impurities when solvents of the same quality are used in calibration and measurement. On the other hand, highly volatile impurities in the solvent are detrimental and lead to a slow stabilization of the system.

Reproducibility of results depends largely on the arrangement of the measuring cell which is maintained within $\pm 10^{-3}$ deg. per hour by an electrical control. Glass cells possess the advantage that they dampen temperature variations because of their poor thermal conductivity. Filter paper wicks, inserted into the measuring cell, dip into the solvent thus inducing rapid saturation of the vapour compartment with solvent. The time taken to reach equilibrium (2 ... 30 min) by systems, with temperature difference independent of time depends on various factors (solvent, substance, temperature). When applying the drops to the thermistor beads, a minimum size should be maintained otherwise the results will not be readily reproducible.

Commercial vapour-pressure osmometers consist of the vapour-pressure osmometer head, cell temperature control and the bridge for measuring resistance.

Fig. 8 shows the construction of the vapour-pressure osmometer head. The measuring cell consists of a temperature controlled aluminium block housing a

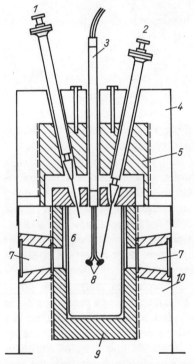

*Fig. 8*: Schematic representation of a vapour-pressure osmometer head (Manufacturer: Knauer, Berlin · West)
1 — Syringe in rest position, 2 — Syringe applying a drop to the thermistor, 3 — Measuring probe with two matched thermistors, 4 — Detachable head containing the heating for the syringes, 5 — Metal block for thermostating the syringes, 6 — Glass beaker with filter, 7 — Thermal isolation filters for observation windows, 8 — Marched measuring vapour wick probe thermistors, 9 — Measuring cell (thermostatted aluminium block), 10 — Casing

beaker with solvent. The measuring compartment saturated with vapour is sealed externally by means of a teflon gasket. The thermistors are situated in the centre of the beaker. In the upper casing block, openings are provided through which syringes can be inserted into the measuring cell for applying solvent and solution to the thermistor beads. The absolute error in molecular weight determination by vapour-pressure osmometry depends on the accuracy with which the calibration constant is determined.

Other errors result from:
— Extrapolation to $c = 0$,
— association or dissociation of the dissolved molecules,
— non-volatile low-molecular weight impurities in the sample which considerably reduce the average molecular weight.

In the case of greater heat losses, the time-dependence of the temperature difference must be recorded and extrapolated to $t = 0$.

### 1.4.4. Example

In the example, the determination of the number-average molecular weight and the second virial coefficient of a polystyrene will be described.

The beaker in the measuring cell is filled with solvent, toluene in this case, and wicks are inserted. The boiling point of toluene under normal conditions is 383.7 K so that the most suitable temperature setting is 323 K. Adequate temperature constancy in the measuring cell is reached after about 2 h for temperatures below 330 K and after about 2 ... 6 h for temperatures above 330 K.

In order to determine the number-average molecular weight by vapour-pressure osmometry, the calibration constant of the instrument must be known under the actual measuring conditions. The temperature differences of 4 solutions of the calibration substance are measured in the concentration range from 10 to 40 g/kg [e.g. benzil ($M = 210.24$ g mol$^{-1}$)] in toluene. Although measuring probe thermistors with the same nominal resistance and temperature coefficient are selected, there is still a resistance difference which must be balanced before the measurement. For the zero setting of the instrument, one drop of solvent is applied to both of the thermistors and, after equilibrium is reached, the zero point is set at on the measuring bridge meter. When solvent is applied again to one thermistor bead and, after carefully flushing, a drop of solution is applied to the second bead, a temperature difference measured by the bridge circuit occurs. For each concentration, the measurement must be repeated several times. If necessary, the dilution effect must be corrected by extrapolation to $t = 0$, from a plot of temperature difference against time.

In accordance with equation (4) et seq. and taking $\Delta R$ proportional to the temperature difference:

$$\frac{\Delta R}{c} = (K_e^{exp})' \left( \frac{1}{M_B} + A_{2,v} \varrho_A c \right) \qquad (K_e^{exp})' = K_e^{th} k_e K' \tag{11}$$

$K'$ = conversion factor between $\Delta R$ and $\Delta T$ [equation (10)].

$\Delta R$ of solutions of different concentrations is measured and the measured quantity divided by the concentration. $\Delta R/c$ is extrapolated graphically to $c = 0$. $(K_e^{exp})'$ can then be calculated.

$$(K_e^{exp})' = \left( \frac{\Delta R}{c} \right)_{c \to 0} M_B \tag{12}$$

The solutions of the polystyrene sample are measured and the $\Delta R/c$, extrapolated to $c = 0$, is determined graphically.

The molecular weight $M_n$ is then obtained from

$$M_n = \frac{(K_e^{exp})'}{\left( \dfrac{\Delta R}{c} \right)_{c \to 0}} \tag{13}$$

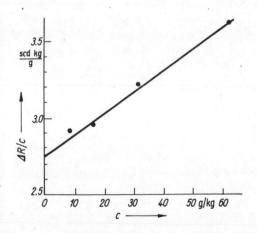

*Fig. 9*: Extrapolation of $\dfrac{\Delta R}{c}$ to infinite dilution ($c = 0$)

The slope of the straight line $\frac{\Delta R}{c}$ versus $c$ is proportional to an apparent second virial coefficient $A_{2,v}$ which can be converted into the second osmotic virial coefficient $A_{2,0}$ if the theoretical calibration constant is known [4]:

$$A_{2,0} = A_{2,v} - \left(\frac{V_A}{\overline{M}_n^2}\right)\left\{\left(\frac{RT}{L} - \frac{1}{2}\right)\left(\frac{K_e^{\exp}}{K_e^{th}}\right)^2 + \frac{K_e^{\exp}}{K_e^{th}} - \frac{1}{2}\right\} \tag{14}$$

For molecular weights greater than $3 \cdot 10^3$ g/mol, the additional term is so small that it can be neglected and $A_{2,v}$ is equal to the second osmotic virial coefficient. From measurements of the temperature difference (or the quantity $\Delta R$ which is proportional to it) of the calibration substance benzil, the calibration constant $(K_e^{\exp})'$ is calculated:

$(K_e^{\exp})' = 1.6 \cdot 10^4$ scd kg mol$^{-1}$

(scd = scale divisions)

The data for the polystyrene sample are summarized in Table 8 and plotted in Fig. 9.

Table 8: Data for the polystyrene sample in toluene

| $c$ (g kg$^{-1}$) | 7.8 | 15.6 | 31.1 | 62.2 |
|---|---|---|---|---|
| $\Delta R$/scd | | | | |
| 1st measurement | 21.9 | 47.2 | 97.5 | 224 |
| 2nd measurement | 23.6 | 42.8 | 103.0 | 221 |
| 3rd measurement | 22.9 | 45.0 | 100.5 | 230 |
| Mean value | | | | |
| $\overline{\Delta R}$/scd | 22.8 | 45.0 | 100.0 | 225 |
| $\frac{\overline{\Delta R}}{c}$ (scd kg g$^{-1}$) | 2.92 | 2.88 | 3.22 | 3.62 |

From the graph $\left(\frac{\overline{\Delta R}}{c}\right)_{c \to 0} = 2.76$ scd kg g$^{-1}$.

From equation (13):

$$M_n = \frac{1.6 \cdot 10^4 \text{ scd kg g}}{2.76 \text{ mol scd kg}} = 5.8 \cdot 10^3 \text{ g/mol}$$

Taking $A_{2,v} = A_{2,0}$ and using equation (11):

$$A_{2,v} = \frac{d\left(\frac{\overline{\Delta R}}{c}\right)/dc}{(K_e^{\exp})' \varrho_A}$$

Putting in the respective values will result in

$A_{2,v} = 10 \cdot 10^{-4}$ mol cm$^3$ g$^{-2}$

The number-average molecular weights can be determined within 3 ... 5% by means of vapour-pressure osmometry.

The error in the second virial coefficient is higher (10%).

## Bibliography

[1] WACHTER, A. H. and W. SIMON, Analytic. Chem. **41** (1969), 90
[2] CHYLEWSKI, Ch. and W. SIMON, Helv. chim. Acta **47** (1964), 515
[3] BERSTED, B. H., J. appl. Polymer Sci. **17** (1973), 1415; **18** (1974), 2399
[4] KAMIDE, K., TERAKAWA, T. and U. UCHIKI, Makromolekulare Chem. **177** (1976), 1447

## 1.5. Determination of Weight-average Molecular Weight of Polymers by Light Scattering Measurements

### 1.5.1. Introduction

Light scattering is a method not only for measuring the weight-average molecular weight and also provides a variety of information about the macromolecule in solution and its interaction with the solvent [1, 2]. The amount of experimental work is greater than that involved in osmotic methods (e.g. more complicated purification procedures) but is justified by the wealth of information obtained. The spatial expansion, i.e. radius of gyration and/or end-to-end distance of the dissolved macromolecules and their geometric shape can be determined from the angular dependence of the intensity of the light scattered by a polymer solution. The concentration dependence of the scattering intensity is used to calculate the second virial coefficient and other thermodynamic data. All effects which exert an influence on the angular and concentration dependence of the scattering intensity, e.g. particle size distribution, composition of copolymers [3], structure of the concentrated polymeric solution, branchings in the macromolecules, size and structure of molecular aggregations, can be investigated by light scattering. It should be noted that light scattering allows an overall description of the measured polymers. Descriptions of the internal structure of the polymer molecule based on measured mean values can only be made in connection with structural parameters (bond length, bond angle, restricted rotation) derived from models. For the evaluation of light scattering measurements, numerous assumptions and simplifications have to be made and their applicability to the system has to be checked.

## 1.5.2. Principles

When a light beam of intensity $I_0$ (incident beam) passes through a medium it is weakened by absorption and scattering. It is assumed that the primary light is monochromatic. When the frequency and phase of the scattered light are the same as those of the incident light, the scattering is said to be coherent or elastic. Brownian motion and diffusion of the scattering macromolecules cause the frequency spectrum of the scattered light to change (DOPPLER shift). This is known as "quasi-elastic light scattering" or "light scattering spectroscopy". The DOPPLER shift is associated with the speed and direction of the movement of the molecules so that diffusion coefficients, particle radii and other quantities can be calculated. For this purpose, the so-called autocorrelation function must be determined which represents the averaging over the product of the scattered intensities at different times in the microsecond to millisecond range ($C(t) = \langle I(t=0) \, I(t) \rangle$). For example from the time constant of the function $C(t)$, the translational diffusion coefficient [4] is obtained. Below, we confine ourselves to the derivation of the theoretical fundamentals of elastic scattering.

Primary (incident) and secondary rays (scattered light) are distinguished by intensity, direction of propagation and polarization. When the electromagnetic wave impinges on a single molecule, a dipole moment is induced by the electric field $E$ of the light wave. The induced dipole, which vibrated with the frequency of the exciting light wave, is the source of the scattered radiation.
Relevant equations are

$$P = \alpha E \tag{1}$$

$$E = E_0 \cdot \cos 2\pi v t \tag{2}$$

$$P = \alpha \cdot E_0 \cos 2\pi v t \tag{3}$$

$P$ = dipole moment, $\alpha$ = polarizability, $v$ = frequency of the exciting light

# Determination of Weight-average Molecular Weight

The radiated dipole energy $I_s$ is determined by the change of the dipole moment with time:

$$I_s = \frac{2}{3c^3} \left| \frac{\overline{d^2 P}}{dt^2} \right|^2 \tag{4}$$

$c$ = light velocity ($3 \cdot 10^8$ ms$^{-1}$)

Substituting equation (3) in (4) results in RAYLEIGH's formula:

$$I_s = \frac{8}{3} \pi I_0 \left( \frac{2\pi}{\lambda_0} \right)^4 \alpha^2 \tag{5}$$

$$\lambda_0 = c/v, \ I_0 = \frac{c}{8\pi} E_0^2$$

The scattered rays emitted from the molecules interfere with each other. Providing the distances between the centres of scattering are sufficiently large (diluted solution) and the particle diameter $D$ is less than $\lambda/20$, interference can be neglected and the intensity of the total radiation $I$ is given by equation (6):

$$I = N_s I_s = \tau I_0 \tag{6}$$

with:

$$\tau = \frac{128\pi^4}{3\lambda_0^4} N_s \alpha^2 \pi \tag{7}$$

$\tau$ = turbidity coefficient or turbidity, $N_s$ = number of scattering centres, $\lambda_0$ = wave length in vacuo

The polarizability $\alpha$ of a single particle is related to the dielectric constant $\varepsilon$ and, consequently, to the refractive index $n$.

$$\alpha = \frac{\varepsilon - 1}{4\pi} \frac{1}{N_s} = \frac{n^2 - 1}{4\pi} \frac{1}{N_s} \tag{8}$$

$$\tau = \frac{32}{3} \pi^3 \frac{(n-1)^2}{N_s \lambda_0^4} \tag{9}$$

(when $n \approx 1$; $(n^2 - 1) \approx 2(n-1)$).

When considering scattering by particles in a dilute solution, the intensity of the scattered radiation of the dissolved particles is the difference between the scattered radiation of the solution and that of the solvent. The polarizability $\alpha_B$ of the dissolved molecule is dependent on the properties of the solvent.

$$\alpha_B = \frac{n^2 - n_A^2}{4\pi N_{s,B}} = \frac{2n_A(n - n_A)}{4\pi N_{s,B}} \tag{10}$$

with $N_{s,B} = N_L \cdot \dfrac{c_B}{M_B}$ for $\alpha_B$ get

$$\alpha_B = \dfrac{2n_A M_B (n - n_A)}{4\pi N_L c_B} \tag{11}$$

Consequently, the turbidity $\tau$ is dependent on the molecular weight.

$$\tau = \dfrac{32\pi^3 n_A^2}{3 N_L \lambda_0^4} \left\{ \dfrac{(n - n_A)}{c_B} \right\}^2 c_B M_B \tag{12}$$

$n_A$ = refractive index of the solvent, $n$ = refractive index of the solution, $N_L$ = AVOGADRO number ($6.025 \cdot 10^{23}$ mol$^{-1}$), $N_{s,B}$ = number of the dissolved particles, $(n - n_A)/c_B = (dn/dc_B)$ = refractive index increment, $c_B$ = concentration (g/cm$^3$), $M_B$ = molecular weight (g/mol)

The above considerations are based on a solution diluted to such an extent that no interaction between the dissolved particles occurs. However, light scattering can only be measured at concentrations where the interaction between the particles of the solution cannot be neglected. The properties of the system free from interactions will then be obtained by extrapolation to $c_B = 0$.

Vibrating dipoles radiate energy perpendicularly to the plane of vibration. Fig. 10 shows a dipole excited by polarized light, parallel and perpendicular to the z-axis, and the corresponding intensity distributions.

For non-polarized light ($u$), the scattering intensity is the sum of the vertical and horizontal components:

$$I_u(\vartheta) = I_u(0°) \dfrac{1 + \cos^2 \vartheta}{2} \tag{13}$$

$\vartheta$ = angle between direction of incidence and direction of observation

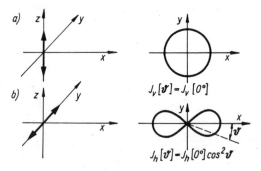

*Fig. 10*: Angular dependence of the intensity of the scattered light for polarized incident light a) vertically polarized light ($I_v$), b) horizontally polarized light ($I_h$)

The reduced scattering intensity $I_{red}$ is defined by equation (14) (RAYLEIGH relation):

$$I_{red}(\vartheta, c_B) = \frac{I(\vartheta, c_B) R^2}{I_0 V_0} \tag{14}$$

$I(\vartheta, c_B)$ = measured intensity of a solution of concentration $c_B$ at angle $\vartheta$, $R$ = distance between the detector and the scattering volume, $V_0$ = scattering volume (amount of solution monitored by the detector)

For the light scattering equation, the following holds:

$$\frac{Kc_B}{I_{red}(\vartheta, c_B)} = \frac{1}{M_w} + 2A_2 c_B \tag{15}$$

with $K$ (optical constant) = $\dfrac{4\pi^2 n_A^2}{N_L \lambda_0^4} \left(\dfrac{dn}{dc_B}\right)^2$

When the particles from which the light is scattered are larger than $\lambda/20$, the scattered light will be weakened by interference between the rays scattered from various positions on the particle.

For describing the angular dependence of the scattered light of large particles, a particle scattering function $P(\vartheta)$ is introduced. This is defined as:

$$P(\vartheta) = \frac{I(\vartheta)}{I(\vartheta = 0°)}; \quad P(\vartheta = 0°) = 1 \tag{16}$$

Great importance is attached to the scattering function in the light scattering of polymeric solutions because it relates the angular dependence of scattered light and the parameters of molecular geometry.

Different scattering functions result from different particle shapes.

The scattering function for a linear chain with radius of gyration $s$ is given by equation (17):

$$P(\vartheta) = (2/x^2) [e^{-x} - (1 - x)]$$
$$x = (16\pi^2/\lambda^2) s^2 \sin^2 \frac{\vartheta}{2} \tag{17}$$

$r^2 = 6s^2$

$r$ = end-to-end distance

Equation (18) is the scattering function of a sphere of diameter $D$ and equation (19) the scattering function of a rod of length $L$.

$$P(\vartheta)_{\text{sphere}} = [(3/x^3)(\sin x - x \cos x)]^2 \tag{18}$$

$$x = (2\pi/\lambda) D \sin \vartheta/2$$

$$s^2 = \frac{3}{20} D^2$$

$$P(\vartheta)_{\text{rod}} = \frac{1}{x} \int_0^{2x} (\sin u/u) \, du - (\sin x/x)^2 \tag{19}$$

$$x = (2\pi/\lambda) L \sin \vartheta/2$$

$$s^2 = \frac{L^2}{12}$$

For molecular weights smaller than $10^5$ g/mol, $P(\vartheta)$ within the limits of error is equal to unity for all angles so that no deductions with respect to particle size can be made. For small $x (M < 10^6$ g/mol), equations (17) to (19) can be expanded into series which result in corresponding expressions when terminated after the 1st term. In this case, the particle shape cannot be ascertained from light scattering measurements.

When taking into account the polydispersity of macromolecules, the following holds for the reciprocal scattering function $P(\vartheta)^{-1}$:

$$P(\vartheta)^{-1} = 1 + \frac{16\pi^2}{3\lambda^2} \langle s^2 \rangle \sin^2 \frac{\vartheta}{2} \tag{20}$$

Hence, the light scattering equation for polydisperse macromolecules is:

$$\frac{Kc_B}{I_{\text{red}}(\vartheta, c_B)} = \frac{1}{M_w} \left[ 1 + \frac{16\pi^2}{3\lambda^2} \langle s^2 \rangle_z \sin^2 \frac{\vartheta}{2} \right] + 2A_2 c_B \tag{21}$$

$\lambda$ = wave length in the medium of refractive index $n$ ($\lambda = \lambda_0/n$)

When extrapolating $Kc_B/I_{\text{red}}$ to $c_B = 0$ (system free from interaction), the size of the scattering particles can be calculated from the slope of the extrapolated curve. The slope of the extrapolated curve $\vartheta = 0$ (interference-free system) is proportional to $A_2$. The weight-average molecular weight is obtained from $Kc_B/I_{\text{red}}$ for $c_B = 0$ and $\vartheta = 0$.

From a consideration of the asymptotic behaviour of $P(\vartheta)$, i.e. when $x$ and, thus, $\vartheta$ are very great, it follows that the slope of the asymptote is a measure of the number-average end-to-end distance. The intercept of the asymptote at $\sin^2 \frac{\vartheta}{2} = 0$

gives $1/(2M_n)$. (These calculations apply to molecules with GAUSSIAN density distribution.)

$$\langle r^2 \rangle_n = \frac{3\lambda^2 \text{ (slope of the asymptote)}}{8\pi^2 \text{ (intercept of the asymptote)}} \qquad (22)$$

This shows that polydispersity affects the angular dependence of the scattering intensity.

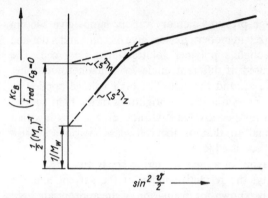

*Fig. 11*: Dependence of $(Kc_B/I_{red})_{c_B \to 0}$ on $\sin^2 \vartheta/2$ for a polydisperse sample showing asymptotic behaviour

The macromolecules are stretched out in thermodynamically good solvents. Such molecules show deviations from the behaviour of a GAUSSIAN chain. These deviations are described by the parameter $\varepsilon'$ which can be calculated from the exponent a of the KUHN-MARK-HOUWINK equation.

$$\varepsilon' = \frac{2a - 1}{3} \qquad (23)$$

Then, the relationship between radius of gyration and end-to-end distance is:

$$\langle s^2 \rangle = \frac{\langle r^2 \rangle}{(2 + \varepsilon')(3 + \varepsilon')} \qquad (24)$$

The excluded volume effect ($\varepsilon' > 0$) affects the scattering function in the same way as the polydispersity so that the two influences on the angular dependence of the scattering intensity can hardly be separated.

When large particles are in the solution (e.g. microgels or associates), the scattering function shows pronounced curvature at small angles.

The angular dependence of scattering intensity furnishes important information about size, shape and polydispersity of the scattering particles. Unfortunately, the effects determining the behaviour of the scattering function superimpose on each other so that in general only $M_w$, $\langle s^2 \rangle_z$ and $A_2$ can be determined with satisfactory accuracy.

## 1.5.3. Experimental Background

Light of known wave length (high pressure mercury vapour lamps give 546 nm or 436 nm and He-Ne-lasers, utilized in modern devices give 633 nm) and a defined polarization state is passed through a polymer solution in a glass cell. The laterally scattered light is measured at different angles with a photomultiplier (PM) which is moved in an arc around the cell. The high voltage applied to the PM is varied in order to balance variations in brightness of the light source. The glass is arranged in a temperature-controlled environment, e.g. a bath with a liquid with refractive index equal to that of the cell glass. Various set-ups described in the literature have been used [5, 6].

Fig. 12 shows the schematic layout of a light scattering apparatus.

Besides the scattering intensity, the refractive index of the solvent and the refractive index increment must be known for evaluation of the appropriate light scattering equation. Since the two aforesaid quantities determine the absolute value of the molecular weight, they should be measured as accurately as possible. $dn/dc_B$ is calculated from the change in the refractive index $\Delta n$ with polymer concentration. For this purpose differences of the refractive indices $\Delta n$ of order of magnitude of $10^{-6}$ must be measurable. Differential refractometers are thus used in preference to interferometers.

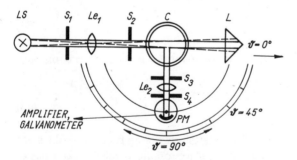

*Fig. 12*: Schematic layout of a light scattering apparatus
LS: Light source, S 1—S 4: Slits 1—4, C — Cell, L — Light trap for incident beam, PM — Photomultiplier, LE — 1 Lens 1, LE — 2 Lens 2

# Determination of Weight-average Molecular Weight

The size of the refractive index increment determines not only the concentration and weight-average molecular weight but also the intensity of the scattered light. It should not be less than 0.1 g/cm³. The concentration (in g/cm³ of solution) of the polymer solution required for light scattering measurements is about

$$c_B > \frac{1}{\left(\dfrac{dn}{dc_B}\right)^2 M_w} \tag{25}$$

$\dfrac{dn}{dc_B}$ in cm³/g, $M_w$ in g/mol

Purification of the solutions and the solvents calls for particular attention and proper experimental procedures. Dust particles distort the scattering intensities especially at small angles of observation, cause misinterpretations and impede the extrapolations necessary for evaluation. Two methods are available for removing dust from the solution and solvent.

a) Centrifugation
The solutions are centrifuged for one hour at about 25 000 rpm. Using a dustfree pipette, the upper part is withdrawn from each centrifuged solution without stirring. Pipetting mus be terminated not later than 10 min after concluding centrifuging.

b) Filtration
The solutions are forced several times with nitrogen through type G 5 glass frit (or millipore filters 0.45) into the purified and dustfree measuring cells.

The purification procedure can lead to changes in sample composition and solution concentration because particles of a high molecular weight ($> 10^7$ g/mol), e.g. microgels which are poorly cross-linked products, may be separated. In order to achieve reproducible results with microgel and high-molecular weight samples, it is always advisable to use the same method for sample purification and employ the same purifying equipment. The concentration of the solutions must be determined after purification and corrected, if necessary.

The light scattering equations (15) and/or (21) describe the relationship between the reduced scattering intensity $I_{red}(\vartheta)$ and the characteristic values of the macromolecule. According to equation (14), $I_{red}(\vartheta)$ is the scattering intensity at an angle of observation $\vartheta$ related to a scattering volume of 1 cm³, at an observation distance of 1 cm, and a primary intensity $I_0 = 1$ cm⁻¹. Since these defined conditions can only be observed with sophisticated equipment, commercial devices measure a relative value and not the absolute value of the reduced scattering intensity. The absolute scattering intensities of standard substances (e.g. benzene) or of scattering standards (glass bodies) are known and enable the apparatus to be calibrated. Errors in the absolute determination of the weight-average molecular weights result from the obligatory calibration because the absolute scattering intensities (RAYLEIGH relations) of the standards are not known exactly. Further uncertainties and sources of error occur when extrapolation the measured quantities to an interaction-free and interference-free system. The smallest observation angles of current light scattering equipment are about 20 to 30° so that extrapolation to the observation angle $\vartheta = 0$ is problematical, especially in the case of polydisperse products and those containing high percentages of microgels. Commercial laser-light scattering apparatus [also employed as detectors in gel permeation chromatography (*GPC*) (Chapter 2.5.)] permit measurements at angles down to 2°. Extrapolation to $c_B = 0$ is made more difficult by the occurrence of entanglement networks in large molecules at concentrations below 1 g/100 cm³. Their scattering behaviour differs from that of individual molecules. Association of macromolecules also exerts a considerable influence on the weight-average molecular weight because it increases the high-molecular weight content of the sample. In the case of polymers capable of association molecular weights must be measured and compared at different temperatures and in different solvents.

The light scattering method is applicable to molecules with molecular weights between $5 \cdot 10^3$ and $10^7$ g/mol. Copolymers must be characterized by measurements in several solvents using special methods [3] since, in contrast to $M_n$, $M_w$ is not additive.

## 1.5.4. Examples

When familiarizing oneself with the light scattering method it is advisable to select a system where no particular problems are involved in measurement (e.g., polystyrene standard in benzene) and to determine the weight-average molecular weight, the second virial coefficient, and the radius of gyration.

Solution preparation is of particular importance for correct results especially for polymers forming associates. For example, differences between the measurements of individually weighed solutions and those of a dilution series may occur.

Slightly soluble polymers are dissolved by shaking and heating for several hours. Care should be taken to avoid thermal degradation. To prevent transitory associate formation, the solutions can be quenched and then measured.

In the selected example, the measurements are carried out in five polystyrene-benzene solutions with concentrations of $1 \cdot 10^{-3}$ to $5 \cdot 10^{-3}$ g/cm³. Solutions, solvents and cells are carefully purified. The measuring vessel of the light scattering equipment is thermostatted to the selected measuring temperature, e.g. 25 °C; and the wave length (e.g. 546 nm) and polarization state of the light (e.g. non-polarized). The readout of the apparatus is calibrated with a benzene

Determination of Weight-average Molecular Weight

standard whose scattering intensity is measured in arbitrary units at an observation angle of $\vartheta = 90°$.

In non-polarized light, the reduced scattering intensity

$$I_{red}(\vartheta, c_B) = \frac{I_{red}^{st}(90°)}{I_{meas}^{st}(90°)} I(\vartheta, c_B) \sin \vartheta \frac{2}{1 + \cos^2 \vartheta} \qquad (26)$$

$I(\vartheta, c_B)$ = difference between the scattering intensities of the polymeric solution of concentration $c_B$ at $\vartheta$ and the scattering intensity of the pure solvent at $\vartheta$ in arbitrary units

$I_{red}^{st}(90°)$ = absolute reduced scattering intensity of the benzene standard at $\vartheta = 90°$, and 25 °C, and using $\lambda_0 = 546$ nm is $I_{red}^{st} = 16.4 \cdot 10^{-6}$ cm$^{-1}$

$I_{meas}^{st}(90°)$ = measured scattering intensity of the standard at $\vartheta = 90°$ in arbitrary units

The quantity $\sin \vartheta$ corrects the angle-dependent scattering volume $V_0$. This correction is necessary because the scattering intensity of the standard is measured at $\vartheta = 90°$ and $V_0(\vartheta = 90°) = V_0(\vartheta) \sin \vartheta$. The factor $\left(\frac{1 + \cos^2 \vartheta}{2}\right)$ takes into account the influence of the polarization state on the angular dependence (equation (13)). The numerical values for the expression $\sin \vartheta/(1 + \cos^2 \vartheta)$ are listed in Table 9.

For evaluation, the reduced scattering intensity of the standard, its relative scattering intensity and the factor 2 are included in the optical constant.

Table 9: Auxiliary quantities required for evaluation of the light scattering measurements

| Angle (°) | $\sin^2 \frac{\vartheta}{2}$ | $\frac{\sin \vartheta}{1 + \cos^2 \vartheta}$ |
|---|---|---|
| 30    | 0.067 | 0.286 |
| 37.5  | 0.104 | 0.374 |
| 45    | 0.146 | 0.471 |
| 60    | 0.250 | 0.693 |
| 75    | 0.370 | 0.905 |
| 90    | 0.500 | 1.000 |
| 105   | 0.630 | 0.905 |
| 120   | 0.750 | 0.693 |
| 135   | 0.854 | 0.471 |
| 142.5 | 0.895 | 0.374 |
| 150   | 0.933 | 0.286 |

To simplify evaluation, the quantity $\dfrac{Kc_B}{I_{red}(c_B, \vartheta)}$ is replaced by the identical quantity $\dfrac{K'c_B}{I_{corr}(c_B, \vartheta)}$

$$K' = \dfrac{2\pi^2 n_A^2 \left(\dfrac{dn}{dc_B}\right)^2 I_{meas}^{st}(90°)}{I_{red}^{st}(90°) N_L \lambda_0^4} f' \tag{27}$$

$$I_{corr}(\vartheta, c_B) = [I_{sol}(\vartheta, c_B) - I_{solv}(\vartheta)] \dfrac{\sin \vartheta}{1 + \cos^2 \vartheta} \tag{28}$$

(Equation (27) applies to non-polarized light)

$I_{sol}(\vartheta, c_B)$ = measured intensity of a solution of concentration $c_B$ at angle $\vartheta$
$I_{solv}(\vartheta)$ = measured intensity of the solvent at angle $\vartheta$
$f'$ = CABANNES-factor which takes into consideration the depolarization:

$$f' = \dfrac{6 + 6\sigma_p}{6 - 7\sigma_p}$$

$$\sigma_p = \dfrac{I_{sol,h}(90°) - I_{solv,h}(90°)}{I_{sol,v}(90°) - I_{solv,v}(90°)}$$

subscript $h$: horizontal component of the scattering intensity, subcript $v$: vertical component of the scattering intensity.
(The CABANNES-factor is equal to unity for polymers of $M > 5 \cdot 10^4$ g/mol.)

The calibrated light scattering device is now used to measure the scattering intensity.

The cells are placed in the measuring vessel of the light scattering apparatus and, after thermostatting (at 25 °C for about 20 min), the scattered light is measured. For this purpose, the angles between 30° and 150° are adjusted successively at intervals of 15° and the respective scale divisions on the galvanometer are read. After measuring a value of $I(\vartheta)$ for each angle $\vartheta$, the cell is turned through about 90° in the measuring vessel and the measurement is repeated. Provided the values measured at the corresponding angles in the two measured series do not differ from one another, a third run may be dispensed with. If a third run is necessary, the cell must be turned through 90° once more prior to thus series of measurements. This procedure is repeated for all weighed samples and for the solvent.

The data can be evaluated by the ZIMM method. This requires extrapolation to $\vartheta = 0$ and $c_B = 0$ in a diagram (other evaluation methods are possible, see Table 12). In the ZIMM plot $\dfrac{K'c_B}{I_{corr}(\vartheta, c_B)}$ is plotted against $\sin^2 \dfrac{\vartheta}{2} + kc_B$ ($k$ must be chosen so sufficiently large for $kc_{B,max}$ to lie between 0.5 to 1; usually $k$ is

100 for g/cm³ units of concentration). After calculating the optical constant $K$ (assuming that $n_A$ and $dn/dc_B$ have already been measured), the quantities required for plotting can be tabulated.

$\lambda_0 = 546$ nm; $\quad I^{st}_{red} = 16.4 \cdot 10^{-6}$ cm$^{-1}$, $\quad \dfrac{dn}{dc_B} = 0.106 \dfrac{cm^3}{g}$

$T = 298$ K; $\quad I^{st}_{meas} = 100$ scd div. $\quad n_A = 1.5020$

$k = 100$

$K = 0.570 \dfrac{\text{mol cm}^3}{g^2}$

After plotting $\dfrac{K'c_B}{I_{corr}}$ against $\left(\sin^2 \dfrac{\vartheta}{2} + kc_B\right)$ in the ZIMM plot (Fig. 13), points measured at the same concentration or angle are joined up. The connection curves, straight lines in the ideal case, correspond to the angular dependence of the reciprocal scattering intensity at constant concentration or the concentration dependence at constant scattering angle. The extrapolated curve to $c_B = 0$ is obtained by extending the constant angle curves past the lowest concentration point (at $\vartheta = 30°$, $c_B = 1 \cdot 10^{-3}$ g/cm³, $\dfrac{K'c_B}{I_{corr}} = 0.443 \cdot 10^{-5}$ g/mol).

The value of $\dfrac{K'c_B}{I_{corr}(30°, 0)}$ corresponding to $c_B = 0$ and $\vartheta = 30°$ should be obtained from the extension at $\sin^2 \dfrac{30}{2} + k \cdot 0 = 0.067$

$\left(\dfrac{K'c_B}{I_{corr}(30°, 0)} = 0.315 \cdot 10^{-5} \text{ mol/g}\right)$.

Values of $K'c_B/I_{corr}(\vartheta, 0)$ are obtained analogously for all angles (marked by crosses in Fig. 13) and joined up and extrapolated to $c_B = 0$. When the curves of the same concentration are extended beyond the smallest observation angle (30°), then the $\dfrac{K'c_B}{I_{corr}(0, c_B)}$ — value corresponding to $\vartheta = 0$ and $c_B$ must be taken from the abscissa value $kc_B \left(\vartheta = 0 \rightarrow \sin^2 \dfrac{\vartheta}{2} = 0\right)$.

For $c_{B1} = 1 \cdot 10^{-3}$ g/cm³, at $kc_B = 0.1$ g/cm³, $\dfrac{K'c_B}{I_{corr}(0, c_{B1})} = 0.43 \cdot 10^{-5}$ mol/g is read off.

Connecting all points of $\dfrac{K'c_B}{I_{corr}(0, c_B)}$ obtained analogously (indicated by triangles in Fig. 13) gives a curve which can be extrapolated to $\vartheta = 0$.

*Table 10*: Scattering intensities and derived quantities required for plotting a ZIMM plot

| Concentration $c_B$ (g cm$^{-3}$) | calculated quantities | angle $\vartheta$ | | | |
|---|---|---|---|---|---|
| | | 30 | 37,5 | 45 | 60 |
| 0 | $I_{solv}/scd$*) | 270 | 210 | 170 | 130 |
| $1 \cdot 10^{-3}$ | $I_{sol}/scd$ | 720 | 540 | 425 | 292 |
| | $I_{sol} - I_{solv}$ | 450 | 330 | 255 | 162 |
| | $I_{corr}$ | 128.7 | 123.4 | 120.1 | 112.3 |
| | $\dfrac{K'c_B}{I_{corr}} 10^{-5}$ mol g$^{-1}$ | 0.443 | 0.462 | 0.475 | 0.508 |
| | $\sin^2 \vartheta/2 + 100 c_B$ | 0.167 | 0.204 | 0.246 | 0.350 |
| $2 \cdot 10^{-3}$ | $I_{sol}/scd$ | 995 | 741 | 590 | 402 |
| | $I_{corr}$ | 207.4 | 198.6 | 197.8 | 188.5 |
| | $\dfrac{K'c_B}{I_{corr}} 10^{-5}$ mol g$^{-1}$ | 0.550 | 0.574 | 0.576 | 0.605 |
| | $\sin^2 \vartheta/2 + 100 c_B$ | 0.267 | 0.304 | 0.346 | 0.450 |
| $3 \cdot 10^{-3}$ | $I_{sol}/scd$ | 1160 | 880 | 696 | 466 |
| | $I_{corr}$ | 254.5 | 250.6 | 247.7 | 232.8 |
| | $\dfrac{K'c_B}{I_{corr}} 10^{-5}$ mol g$^{-1}$ | 0.672 | 0.682 | 0.690 | 0.734 |
| | $\sin^2 \vartheta/2 + 100 c_B$ | 0.367 | 0.404 | 0.446 | 0.550 |
| $4 \cdot 10^{-3}$ | $I_{sol}/scd$ | 1280 | 970 | 765 | 515 |
| | $I_{corr}$ | 288.9 | 284.2 | 280.2 | 266.8 |
| | $\dfrac{K'c_B}{I_{corr}} 10^{-5}$ mol g$^{-1}$ | 0.789 | 0.802 | 0.814 | 0.855 |
| | $\sin^2 \vartheta/2 + 100 c_B$ | 0.467 | 0.504 | 0.546 | 0.650 |
| $5 \cdot 10^{-3}$ | $I_{sol}/scd$ | 1323 | 1005 | 797 | 540 |
| | $I_{corr}$ | 301.2 | 297.3 | 295.3 | 284.1 |
| | $\dfrac{K'c_B}{I_{corr}} 10^{-5}$ mol g$^{-1}$ | 0.946 | 0.959 | 0.965 | 1.003 |
| | $\sin^2 \vartheta/2 + 100 c_B$ | 0.567 | 0.604 | 0.646 | 0.750 |

*) scd = scale division

| 75 | 90 | 105 | 120 | 135 | 142,5 | 150 |
|---|---|---|---|---|---|---|
| 108 | 100 | 108 | 130 | 170 | 208 | 272 |
| 220 | 200 | 210 | 255 | 348 | 429 | 556 |
| 112 | 100 | 102 | 125 | 178 | 221 | 284 |
| 101.4 | 100 | 92.3 | 86.6 | 83.4 | 82.7 | 81.2 |
| 0.562 | 0.57 | 0.617 | 0.658 | 0.680 | 0.690 | 0.702 |
| 0.470 | 0.600 | 0.730 | 0.850 | 0.954 | 0.995 | 1.033 |
| 300 | 272 | 279 | 350 | 476 | 588 | 764 |
| 173.8 | 172 | 154.8 | 152.5 | 144.1 | 142.1 | 140.7 |
| 0.656 | 0.663 | 0.737 | 0.748 | 0.791 | 0.802 | 0.810 |
| 0.570 | 0.700 | 0.830 | 0.950 | 1.054 | 1.095 | 1.133 |
| 350 | 312 | 335 | 417 | 573 | 710 | 915 |
| 219.0 | 212 | 205.4 | 198.9 | 189.8 | 187.7 | 183.9 |
| 0.781 | 0.807 | 0.832 | 0.860 | 0.901 | 0.911 | 0.930 |
| 0.670 | 0.800 | 0.930 | 1.050 | 1.154 | 1.195 | 1.233 |
| 390 | 345 | 370 | 468 | 647 | 800 | 1038 |
| 255.2 | 245 | 237.1 | 234.2 | 224.7 | 221.4 | 219.1 |
| 0.893 | 0.931 | 0.962 | 0.973 | 1.015 | 1.030 | 1.041 |
| 0.770 | 0.900 | 1.030 | 1.150 | 1.254 | 1.295 | 1.333 |
| 418 | 365 | 395 | 490 | 687 | 854 | 1105 |
| 280.55 | 265 | 259.7 | 249.5 | 243.5 | 241.6 | 238.2 |
| 1.016 | 1.075 | 1.097 | 1.142 | 1.170 | 1.180 | 1.196 |
| 0.870 | 1.00 | 1.130 | 1.250 | 1.354 | 1.395 | 1.433 |

From the light scattering curve:

$$\lim_{\substack{c_B \to 0 \\ \vartheta \to 0}} \frac{K' c_B}{I_{\text{corr}}(\vartheta, c_B)} = \frac{1}{M_w} \tag{29}$$

This shows that the intersection of the two extrapolation curves with the ordinate $\dfrac{K' \cdot c_B}{I_{\text{corr}}(\vartheta, c_B)}$ gives $1/M_w$ as the intercept.

Extrapolation to $c_B = 0$ (system free from interaction) results in:

$$\lim_{c_B \to 0} \frac{K' c_B}{I_{\text{corr}}(\vartheta, c_B)} = \frac{1}{M_w} + \frac{1}{M_w}\left(\frac{16}{3}\pi^2 \cdot \frac{\langle s^2 \rangle_z}{\lambda^2} \cdot \sin^2 \frac{\vartheta}{2}\right) \tag{30}$$

From the slope of the curve of $\dfrac{K' \cdot c_B}{I_{\text{corr}}(\vartheta, c_B)}$ extrapolated to $c_B = 0$, the radius of gyration can be calculated.

$$\tan \alpha = \frac{d\left(\lim_{c_B \to 0} \dfrac{K' c_B}{I_{\text{corr}}(\vartheta, c_B)}\right)}{d \sin^2 \dfrac{\vartheta}{2}} = \frac{16\pi^2}{3\lambda^2 M_w}\langle s^2 \rangle_z \tag{31}$$

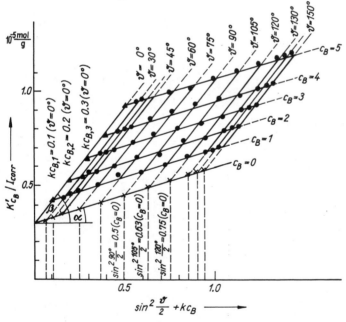

*Fig. 13*: ZIMM plot for a polystyrene sample in benzene

Extrapolating to $\vartheta = 0$ (i.e. excluding intermolecular interference), gives equation (32):

$$\lim_{\vartheta \to 0} \frac{K' c_B}{I_{\text{corr}}(\vartheta, c_B)} = \frac{1}{M_w} + 2A_2 c_B \tag{32}$$

and, thus, the second virial coefficient:

$$\tan \beta = \frac{d\left(\lim\limits_{\vartheta \to 0} \dfrac{K' c_B}{I_{\text{corr}}(\vartheta, c_B)}\right)}{dc_B} = 2A_2 \tag{33}$$

In the ZIMM plot in Fig. 13, the intercept is

$$\lim_{\substack{c_B \to 0 \\ \vartheta \to 0}} \frac{K' c_B}{I_{\text{corr}}(\vartheta, c_B)} = 0.30 \cdot 10^{-5} \text{ mol/g}$$

The average molecular weight of the polystyrene sample is $M_w = 3.3 \cdot 10^5$ g/mol.

From the slopes of $\tan \alpha = 0.30 \cdot 10^{-5}$ mol/g and $\tan \beta = 12.6 \cdot 10^{-4}$ mol cm³/g², the 2nd virial coefficient

$$A_2 = 6{,}3 \cdot 10^{-4} \frac{\text{cm}^3 \text{ mol}}{\text{g}^2}$$

and the radius of gyration $\langle s^2 \rangle_z = 25 \cdot 10^{-12}$ cm² are calculated. The average molecular weight can be determined rapidly by the dissymmetry method based on the evaluation of measurements of scattering intensity at $\vartheta = 45°$, $90°$ and $135°$.

The dissymmetry $z$ is calculated with equation (34):

$$z = \frac{I_{\text{red}}(45°, 0)}{I_{\text{red}}(135°, 0)} = \frac{P(45°)}{P(135°)} \tag{34}$$

When the type of scattering function is known, the scattering function at $\vartheta = 90°$, $P(90°)$ and $D/\lambda$ can be obtained from Tables for a particular value of $z$. $D$ is a characteristic particle size which corresponds to $\langle r^2 \rangle_z^{1/2}$ for chain molecules. Also:

$$M_w = P^{-1}(90°) \cdot [Kc_B/I_{\text{red}}(90°)]_{c_B=0}^{-1} \tag{35}$$

In the example, $z = 90/48 = 1.875$. Using Table 11, $P^{-1}(90°) = 1.7$ can be obtained for polydisperse chains by linear interpolation.

Substituting in equation (35):

$$M_w = (1.7/0.49) \cdot 10^5 \text{ g/mol} = 3.5 \cdot 10^5 \text{ g/mol}$$

*Table 11*: Dissymmetry $z = \dfrac{I_{red}(\vartheta = 45°, 0)}{I_{red}(\vartheta = 135°, 0)}$ and $P^{-1}$ (90°) as a function of $D/\lambda$ (scattering function for polydisperse chains with a SCHULZ-ZIMM-distribution)

| $D/\lambda$ | Rod | | Monodisperse chains | | Polydisperse chains | | Spheres | |
|---|---|---|---|---|---|---|---|---|
| | $z$ | $P^{-1}$ (90°) | $z$ | $P^{-1}$ (90°) | $z$ | $P^{-1}$ (90°) | $z$ | $P^{-1}$ (90°) |
| 0.05 | 1.006 | 1.006 | 1.014 | 1.012 | 1.016 | 1.016 | 1.011 | 1.010 |
| 0.10 | 1.032 | 1.023 | 1.065 | 1.065 | 1.094 | 1.066 | 1.061 | 1.041 |
| 0.15 | 1.070 | 1.050 | 1.135 | 1.103 | 1.200 | 1.148 | 1.136 | 1.090 |
| 0.20 | 1.127 | 1.089 | 1.257 | 1.183 | 1.340 | 1.236 | 1.255 | 1.171 |
| 0.25 | 1.210 | 1.144 | 1.410 | 1.290 | 1.519 | 1.391 | 1.441 | 1.280 |
| 0.30 | 1.279 | 1.207 | 1.585 | 1.434 | 1.715 | 1.520 | 1.695 | 1.439 |
| 0.35 | 1.372 | 1.288 | 1.790 | 1.612 | 1.924 | 1.760 | 2.657 | 1.622 |
| 0.40 | 1.495 | 1.377 | 2.020 | 1.809 | 2.151 | 2.051 | 1.657 | 1.930 |
| 0.45 | 1.620 | 1.486 | 2.283 | 2.049 | 2.360 | 2.330 | 3.692 | 2.341 |
| 0.50 | 1.753 | 1.608 | 2.534 | 2.320 | 2.569 | 2.642 | 5.810 | 2.780 |
| 0.55 | 1.895 | 1.744 | 2.796 | 2.660 | 2.778 | 2.987 | | |
| 0.60 | 1.971 | 1.860 | 3.060 | 2.982 | 2.980 | 3.365 | | |
| 0.65 | 2.058 | 2.010 | 3.303 | 3.413 | 3.169 | 3.776 | | |
| 0.70 | 2.106 | 2.193 | 3.521 | 3.814 | 3.354 | 4.218 | | |
| 0.75 | 2.160 | 2.361 | 3.745 | 4.348 | 3.523 | 4.695 | | |
| 0.80 | 2.200 | 2.500 | 3.915 | 4.776 | 3.681 | 5.205 | | |

With $D/\lambda = 0.34$ (linearly interpolated),

$$\langle s^2 \rangle_z = \frac{\lambda^2 (D/\lambda)^2}{6} = 26 \cdot 10^{-12} \text{ cm}^2$$

in good agreement with the results of the ZIMM method.

Apart from the two extrapolation methods discussed above, similar methods for evaluating light scattering results can be utilized for polydisperse, high-molecular weight and special polymers. A large number of publications deals with the computer evaluation of light scattering measurements [7].

Errors in molecular weight determination are dependent on the two extrapolation curves and, thus, on the accuracy of extrapolation, other errors lie in calibration and the determination of the refractive index increment and the refractive index. In general, the total error is about 5 to 10%.

# Determination of Weight-average Molecular Weight

*Table 12*: Extrapolation methods for the evaluation of light scattering results

| Method | Graphic Plotting | Remarks | References |
|---|---|---|---|
| YANG | $\left(\dfrac{Kc_B}{I_{red}}\right)_{c_B \to 0} \Big/ \sin^2 \vartheta/2$ versus $(\sin^2 \vartheta/2)^{-1}$ | separate extrapolation to $c_B = 0$; very large molecular weights | J. Pol. Sci. **26** (1957), 305 |
| GUINIER | $\log (I_{red}/c_B)\, c_B \to 0$ versus $\sin^2 \vartheta/2$ | separate extrapolation to $c_B = 0$, particularly suitable for polydisperse microgel-containing samples (particle size distribution) | Ann. Physik **12** (1939), 161 |
| LANGE | $\left(\dfrac{I_{red}}{Kc_B}\right)_{c_B \to 0} \Big/ \sin^2 \vartheta/2$ versus $\sin^2 \vartheta/2$ | separate extrapolation to $c_B = 0$, determination of the microgel content | Kolloid Z. Z. Polymere **240** (1970), 747 |
| WIJK/STAVERMAN | $\dfrac{Kc_B}{I_{red}}$ versus $\sin^2 \vartheta/2 - kc_B$ | particularly suitable for samples with curved scattering characteristic at small $\vartheta$ | J. Pol. Sci. A **24** (1966), 1011 |
| BERRY | $\left(\dfrac{Kc_B}{I_{red}}\right)^{1/2}_{c_B \to 0}$ versus $\sin^2 \vartheta/2$ | separate extrapolation to $c_B = 0$, special polymers | J. chem. Physics **44** (1966), 4550 |
| separate extrapolation | $\left(\dfrac{Kc_B}{I_{red}}\right)_{c_B \to 0}$ versus $\sin^2 \vartheta/2$ | separate extrapolation to $c_B = 0$ | |

# Bibliography

[1] HUGLIN, M. B., „*Light Scattering from Polymer Solutions*", Academic Press London and New York 1972
[2] ESKIN, W. E., „*Light Scattering from Polymer Solutions*" (Russian), Nauk House Publ., Moscow 1974
[3] BUSHUK, W. and H. BENOIT, Canad. J. Chem. **36** (1958), 1616
[4] BERNE, B. J. and R. PECORA, „*Dynamic Light Scattering*", Wiley-Interscience, New York 1976
[5] CANTOW, H. J. and G. V. SCHULZ, Z. physik. Chemie, Neue Folge, 1 (1954), 365
[6] HACK, H. and G. MEYERHOFF, Makromol. Chem. **179** (1978), 2475
[7] KAJIWARA, K. and S. B. ROSS-MURPHY, Europ. Polymer J. **11** (1975), 365

## 1.6. Sedimentation and Diffusion

### 1.6.1. Introduction

The study of transport processes in solution forms the basis of a number of molecular weight determination methods. Mass transport can be characterized by diffusion coefficient, sedimentation coefficient and viscosity. All three depend on size and shape of the dissolved particles.

When, for example, a lower layer is formed by addition to dilute solutions of different concentrations of a polymer, a concentration equilibration process is initiated by diffusion. The mass flow of thus process equals the product of concentration gradient and diffusion coefficient. If the relationship between the diffusion coefficient (diffusion constant), extrapolated to zero concentration, and the molecular weight are known, the molecular weight can be determined from a diffusion measurement alone. The molecular weight distribution and/or a molecular inhomogeneity parameter can also be estimated. Diffusion coefficient measurements of sufficient accuracy are however among the technically most difficult in physical chemistry so that they are mainly used for determining molecular weight only in association with the sedimentation coefficient or the intrinsic viscosity.

A universally applicable method for polymer characterization is ultracentrifugation in which the motion dissolved particles in a gravitational field is observed [1]. The method can be used to determine different averages of molecular weight ($M_w$, $M_z$ and exponential averages depending on the methods of evaluation and experiment), molecular weight distribution and particle size distributions of dis-

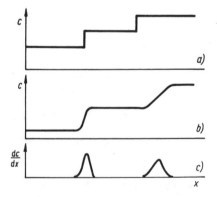

*Fig. 14*: Concentration variation in the measuring cell ($x$ = position coordinate, distance from the axis of rotation)
a) ideally uniform fractions neglecting diffusion, b) ideally uniform fractions including diffusion, c) differential of the concentration variation shown in b)

# Sedimentation and Diffusion

persions. In the case of copolymers (especially graft copolymers) chemical heterogeneity distributions can also be measured by density gradient ultracentrifugation.

## 1.6.2. Principles

In a gravitational field, the particles in a solution will migrate. If the density of the solute is greater than that of the solvent, they will sediment, that is to say, they move in the direction of the gravitational field to the bottom of the measuring cell. Otherwise, they will float and migrate to the meniscus of the solution. If all molecules in a solution are of the same size, weight and shape, and if the gravitational field of an ultracentrifuge were the only effect, a sharp interface would be observed between sedimenting particles and solvent. Actually, diffusion (BROWNian motion) counteracts migration in the gravitational field and blurs the interface. Polymolecularity must also be taken into account where polymers are concerned.

Basically, two different measuring methods are possible in ultracentrifugation: rate measurements and equilibrium measurements.

In the former method, the particle size, particle shape, and distribution width of the dissolved polymer molecules are determined from measurements of sedimentation rate. This method can be utilized for molecules having molecular weights above $2 \cdot 10^4$ g mol$^{-1}$.

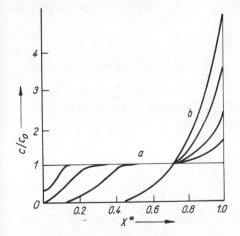

*Fig. 15*: Schematic representation of the concentration in the measuring cell of the ultracentrifuge at different times
a) initial concentration, b) establishment of equilibrium ($c/c_0$: ratio of concentration, $x^*$ relative coordinate)

If the gravitational field of the ultracentrifuge is allowed to act long enough a dynamic equilibrium will be established at the bottom of the measuring cell. Similarly to the variation of atmospheric density with altitude, a particle distribution will occur from which the particle weight and possibly the distribution curve can be calculated. Equilibrium ultracentrifugation is restricted to molecules with molecular weights of $10^3$ to $10^6$ g mol$^{-1}$, otherwise extremely long measuring times would be required for equilibrium to be established.

If the rotational speed of the ultracentrifuge is so high that diffusion is negligible compared to sedimentation, the force acting on the sedimenting particle is determined only by the effective mass of the latter and the centrifugal acceleration.

$$K_w = m_{eff} \omega^2 x \tag{1}$$

$\omega$ = angular velocity ($= 2\pi$ rotational speed), $x$ = distance between particle and centre of rotation

The effective mass is the mass of the particle corrected for buoyancy.

$$m_{eff} = \frac{M_B}{N_L}(1 - \varrho_A v_B) \tag{2}$$

$\varrho_A$ = density of the solvent, $v_B$ = partial specific volume, $M_B$ = molecular weight, $N_L$ = AVOGADRO number ($6.025 \cdot 10^{23}$ mol$^{-1}$).

During migration, the particle is impeded by a velocity-dependent frictional force $K_r$.

$$K_r = f\, dx/dt \tag{3}$$

$f$ = coefficient of friction

When the two forces are in equilibrium, the particle moves without being accelerated, that is to say, at constant speed. Thus:

$$K_w = K_r \tag{4}$$

$$\omega^2 x \frac{M_B}{N_L}(1 - \varrho_A v_B) = f \frac{dx}{dt}$$

Introducing the sedimentation coefficient $s$, relating migration velocity to centrifugal acceleration (SVEDBERG):

$$s = \frac{dx}{dt} \bigg/ \omega^2 x \tag{5}$$

(the SVEDBERG = $10^{-13}s$ is frequently used as a unit of measurement for $s$), we obtain at equilibrium

$$M_B = \frac{f \cdot s N_L}{1 - \varrho_A v_B} \tag{6}$$

Equation (6) includes the coefficient of friction of sedimentation $f$ which can be replaced by expressions containing other measured values. For example, STOKES calculated the coefficient of friction for non-solvated spherical particles of radius $R_k$ in a medium of viscosity $\eta_A$ as:

$$f_k = 6\pi\eta_A R_k \tag{7}$$

For non-spherical particles, a coefficient of friction $f_D$ can be calculated from measurements of the diffusion coefficient:

$$f_D = \frac{kT}{D} \tag{8}$$

$D$ = diffusion coefficient, $k$ = BOLTZMANN constant

Experiments have shown that the friction coefficient of diffusion is equal to that of sedimentation. Substituting equation (8) in equation (6), gives the SVEDBERG-equation

$$M_B = \frac{RTs}{(1 - \varrho_A v_B) D} \tag{9}$$

$R$ = universal gas constant ($= kN_L$), $R = 8.3147 \, \dfrac{\text{J}}{\text{mol deg}}$

The friction coefficient can also be determined from viscosity measurements. Now

$$f_\eta = f_A 6\pi\eta_A \langle s^2 \rangle^{0.5} \tag{10}$$

$f_A$ = asymmetry factor describing the relationship between radius and radius of gyration $\langle s^2 \rangle^{0.5}$ as well as deviations from the friction coefficient of a non-solvated sphere

The radius of gyration can be substituted by the FOX-FLORY-equation (equation (7) of section 1.7.)

$$[\eta] = \Phi (\langle s^2 \rangle / M_B)^{3/2} M_B^{0.5}$$

$\Phi$ = FLORY universal constant

$$f_\eta = f_A 6\pi\eta_A \left(\frac{[\eta]}{\Phi}\right)^{1/3} M_B^{1/3} \tag{11}$$

Substitution of equation (11) into equation (6) leads to the MANDELKERN-FLORY-SCHERAGA-equation.

$$M_B = \left( \frac{N_L \eta_A}{(6\pi f_A)^{-1} (1 - \varrho_A v_B) \, \Phi^{1/3}} \right)^{3/2} [\eta]^{1/2} \, s^{3/2} \tag{12}$$

The literature frequently quotes $6\pi f_A = p$ and $\Phi^{1/3} p^{-1} = \beta$, where $\Phi$, $p$ and $\beta$ have been calculated for different models (sphere, ellipsoid, coil) and can be taken from publications, e.g. [2].

Integration of equation (5) leads to an equation for the sedimentation coefficient $s$.

$$s = \frac{\ln x_1 - \ln x_0}{\omega^2 (t_1 - t_0)} \approx \frac{2(x_1 - x_0)}{(x_1 + x_0) \, \omega^2 (t_1 - t_0)} \quad \text{if} \quad (x_1 - x_2) \ll x_0 \tag{13}$$

It has already been pointed out that diffusion counteracts sedimentation. If we consider a cross-section of area $q$ in the measuring cell, then the mass flow of diffusion through $q$ is:

$$\frac{dm_D}{dt} = -Dq \frac{\partial c_B}{\partial x} \tag{14}$$

If a mass $m_s$ sediments through $q$ in unit time

$$\frac{dm_s}{dt} = c_B q \frac{\partial x}{\partial t} = s c_B \omega^2 x q \tag{15}$$

Hence, the resulting mass flow through $q$ is

$$\frac{dm}{dt} = \frac{dm_s}{dt} + \frac{dm_D}{dt} = \left( s c_B \omega^2 x - D \frac{\partial c_B}{\partial x} \right) q \tag{16}$$

The amount of substance present between $x$ and $x + dx$ from the centre of rotation decreases in a time $dt$ by

$$-\frac{\partial}{\partial x} \left( s c_B \omega^2 q x - Dq \frac{\partial c_B}{\partial x} \right) dx \tag{17}$$

The change in concentration in the volume element $q \, dx$, multiplied by $q \, dx$, is $(\partial c_B / \partial t) \, q \, dx$. Since the measuring cells are sector-shaped, $q$ is not constant.

$$q = \varphi x h \tag{18}$$

$\varphi$ = sector angle of the measuring cell, $h$ = height of the measuring cell

# Sedimentation and Diffusion

The variation of concentration with time at a position $x$ is described by LAMM's differential equation derived as follows:

$$\frac{\partial c_B}{\partial t} \varphi x h \, dx = \frac{\partial}{\partial x}\left[-sc_B\omega^2\varphi x^2 h + D\varphi x h \frac{\partial c_B}{\partial x}\right] dx$$

$$\frac{\partial c_B}{\partial t} = \frac{1}{x}\frac{\partial}{\partial x}\left[-sc_B\omega^2 x^2 + Dx\frac{\partial c_B}{\partial x}\right] \tag{19}$$

If the speed of the ultracentrifuge is adjusted so that $(\partial c_B/\partial t) = 0$, then $\frac{dm_D}{dt} + \frac{dm_s}{dt} = 0$ and $s$ can be calculated from the change in concentration with a change in the space coordinate $x$ (equilibrium centrifuge or sedimentation equilibrium method). When $(\partial c_B/\partial t) = 0$, the expression in brackets in equation (19) must be equal to zero.

$$s\omega^2 c_B x - D\frac{\partial c_B}{\partial x} = 0 \tag{20}$$

Replacing the sedimentation coefficient by the SVEDBERG-equation (9) gives the variation of concentration with position:

$$\frac{1}{c_B}\frac{\partial c_B}{\partial x} = \frac{d \ln c_B}{dx} = \frac{M_B \omega^2 x (1 - \varrho_A v_B)}{RT} \tag{21}$$

(The derivations are correct only when the frictional forces of diffusion and sedimentation are equal and $D$ and $s$ are independent of $c_B$.)

Integrating equation (21) leads to an expression for the concentration distribution over the bottom of the measuring cell:

$$c_B = c_{B,0} \exp\left[\frac{M_B}{2RT}(1 - \varrho_A v_B)\omega^2(x^2 - x_b^2)\right] \tag{22}$$

$c_{B,0}$ = concentration at the bottom of the measuring cell ($x = x_b$)

The molecular weight $M_B$ can be calculated from the concentrations $c_{B,1}$ and $c_{B,2}$, measured at the two positions $x_1$ and $x_2$. At locus $x_1$:

$$\ln c_{B,1} = \ln c_{B,0} + \frac{M_B}{2RT}(1 - \varrho_A v_B)\omega^2(x_1^2 - x_b^2) \tag{23}$$

Similarly, the concentration at $x_2$ is

$$\ln c_{B,2} = \ln c_{B,0} + \frac{M_B}{2RT}(1 - \varrho_A v_B)\omega^2(x_2^2 - x_b^2) \tag{24}$$

Subtracting the natural logarithms,

$$\ln c_{B,2} - \ln c_{B,1} = \ln \frac{c_{B,2}}{c_{B,1}} = \frac{M_B}{2RT}(1 - \varrho_A v_B)\omega^2(x_2^2 - x_1^2) \tag{25}$$

and rearranging gives the equation for the molecular weight:

$$M_B = \frac{2RT \ln \frac{c_{B,2}}{c_{B,1}}}{(1 - \varrho_A v_B)\omega^2(x_2^2 - x_1^2)} \tag{26}$$

The above equations apply only to ideal conditions, that is to say, in interaction-free solutions. The sedimentation coefficient $s$ is, however, dependent on concentration and its reciprocal value can be represented by a power series in concentration.

$$\frac{1}{s} = \frac{1}{s_0} + k'c_B + k''c_B^2 \tag{27}$$

Since the sedimentation coefficient $s_0$ extrapolated to $c_B = 0$ (frequently called sedimentation constant) is included in the evaluation of the ultracentrifuge measurements, a concentration series must be measured, as with all methods of molecular weight determination.

In centrifugation, a polydisperse sample does not show a sharp interface between the sedimenting polymers and the supernatant polymer-free solvent. Since particles of higher molecular weight will sediment more rapidly than those of lower molecular weight, the concentration profile must reflect the molecular weight distribution of the polymer sample. Naturally, diffusion will produce a perturbing effect, particularly over long periods, which will also broaden the concentration profiles. The molecular weight dependence of the sedimentation coefficient extrapolated to $c_B = 0$ can be written analogously to that of the limiting viscosity.

$$s_0 = KM_B^u \tag{28}$$

Molecular weight distribution can be determined from ultracentrifuge measurements providing that:
a) an adequate calibration relation, equation (28), obtained by measuring polymer fractions is available;
b) the distribution of the sedimentation coefficients has been measured;
c) the latter is determined exclusively by the mass fractions of the individual molecular species and not by other factors such as molecular interactions and diffusion effects (see example).

## Sedimentation and Diffusion

The ultracentrifuge can be used to determine different averages of the molecular weight. When the molecular weight is calculated from the relation $s/D$ using the SVEDBERG-equation, different $\bar{M}_{s,D}$ will be obtained for the molecular weight depending on the averaging of $s$ and $D$. Combination of $s_w$ and $D_w$ leads to an average of $\bar{M}_{s_w, D_w}$ (double weight average) which lies between number-average and weight-average molecular weight.

When a polydisperse sample is in sedimentation equilibrium, the concentration gradient (equation (21)) is proportional to the sum of the products of molecular weight and concentration of the individual components.

$$\frac{dc_B}{dx} = \frac{\omega^2 x(1 - \varrho_A v_B)}{RT} \sum_i c_{B,i} M_i \tag{29}$$

At any position $x$ in the measuring cell, the weight-average $M_{w,x}$ or the $z$-average (ultracentrifug-average) $M_{z,x}$ can be calculated.

$$M_{w,x} = \frac{\sum_i c_{B,i}(x) M_i}{\sum_i c_{B,i}(x)} = \frac{dc_B}{dx} \frac{RT}{\omega^2 x(1 - \varrho_A v_B) c_B(x)} \tag{30}$$

$$M_{z,x} = \frac{\sum_i c_{B,i}(x) M_i^2}{\sum_i c_{B,i}(x) M_i} = \frac{d}{dx}\left(\frac{dc_B}{dx}\right) \frac{RT}{\omega^2 x(1 - \varrho_A v_B) \frac{dc_B}{dx}} - \frac{RT}{\omega^2 x^2(1 - \varrho_A v_B)} \tag{31}$$

Integration of equations (30) and (31) over the length of the measuring cell leads to expressions for calculating the weight-average molecular weight [3],

$$M_w = \frac{\int_{x_m}^{x_b} M_{w,x} x c_B(x)\, dx}{\int_{x_m}^{x_b} x c_B(x)\, dx} = \frac{RT[c_B(x_b) - c_B(x_m)]}{(1 - \varrho_A v_B)\omega^2 \int_{x_m}^{x_b} x c_B(x)\, dx} \tag{32}$$

and the ultracentrifug-average of the molecular weight

$$M_z = \frac{\int_{x_m}^{x_b} M_{z,x} \frac{dc_B}{dx}\, dx}{\int_{x_m}^{x_b} \frac{dc_B}{dx}\, dx} = \frac{RT\left[\left(\frac{dc_B}{dx}\right)_b \Big/ x_b - \left(\frac{dc_B}{dx}\right)_m \Big/ x_m\right]}{(1 - \varrho_A v_B)\omega^2 [c_B(x_b) - c_B(x_m)]} \tag{33}$$

$M_n$ results from equation (34) which is valid if the zone of observation is immediately above the bottom of the cell.

$$M_n = \frac{RT \int_{x_m}^{x_b} c_B(x)\,dx}{(1 - \varrho_A v_B)\,\omega^2 x_b \int_{x_m}^{x_b} dx \int_{x_m}^{x_b} c_B(x)\,dx} \qquad (34)$$

$x_m$ = space coordinate of the meniscus, $x_b$ = space coordinate of the cell bottom

## 1.6.3. Experimental Background

Since the introduction of the ultracentrifuge for observating the sedimentation process by THE SVEDBERG at the beginning of the twenties, it has been subject to continuous technical improvement. Fig. 16 shows the basic design of an ultracentrifuge.

At $(60 \ldots 70)\,10^3$ revolutions per minute and a mean distance between the measuring cell and the axis of rotation of $6 \ldots 7$ cm, centrifugal accelerations of $(2 \ldots 4)\,10^5$ g (g = gravitational acceleration) are achieved. The maximum acceleration is limited by the strength of the rotor material which is mainly duraluminium, titanium or chromium-nickel-steel alloys. Two or more holes in the rotor parallel to the rotor axis house the measuring cells. Expensive ultracentrifuges are in operation in which up to five samples can be measured simultaneously. The rotor runs in a steel chamber for protection against accidents. Frictional heat

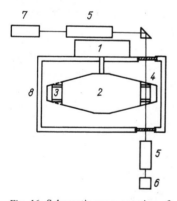

*Fig. 16*: Schematic representation of an ultracentrifuge
1 — Motor and gear, 2 — Rotor, 3 — Measuring cell, 4 — Quartz window, 5 — Optics, 6 — Light source, 7 — Recording device (camera, photomultiplier), 8 — Thermostatic enclosure

is suppressed either by evacuating this chamber using oil-diffusion pumps or by filling it with hydrogen, which is a good heat conductor, to about $10^3$ Pa. During measurement the temperature of the rotor must be kept constant to within $\pm 0.1$ deg. The temperature in commercial ultracentrifuges can be regulated over the range from 20 °C to 150 °C.

Usually, the rotor is suspended on a thin shaft which is elastic in order to balance it and which extends outside the evacuated steel chamber. Different systems are used for driving the rotor. The first ultracentrifuges by SVEDBERG were provided with turbines at the end of the axis of rotation which were usually operated hydraulically. Ultracentrifuges with air-driven turbines are also known. Recently, rotors driven by an electric motor, which is connected to the rotor either directly or by a gear mechanism have been widely used. Speed can be kept constant to within 30 rpm.

The measuring cells which accomodate the samples, have a length between 0.2 and 3 cm, and are segment shaped with sector angle between 2° and 4°. They are provided with quartz or sapphire windows. The processes in the ultracentrifuge are monitored by measuring the concentration of the solution at a point $x$. Refractometric methods (interference optics, Schlieren optics) and absorption measurements (absorption optics) are particularly suitable. All three optical systems can be used for sedimentation and diffusion measurements.

Light absorption can be utilized for the determination of the variation of concentration if the dissolved particles absorb more intensely than the solvent in a certain wavelength range. Numerous proteins and polystyrene exhibit intense absorption bands in the near UV. Absorption optics have the advantage that they are suitable for measuring solutions of low concentration (about $10^{-5}$ g cm$^{-3}$). A disadvantage of older equipment is the complicated evaluation

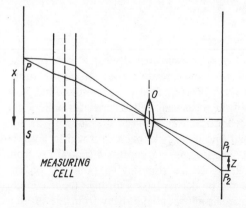

*Fig. 17*: Diagram of the ray path in LAMM's scale method

of photographed density curves that call for calibration. In modern devices, however, this disadvantage is obviated by the use of photomultipliers and dot scanning of the cell.

Refractometric methods are based on the dependence of the refractive index of the solution on the concentration of the dissolved particle. LAMM's scale method (Fig. 17) is the most accurate.

When passing through a cell filled with a homogeneous medium, a light ray emanating from point $P$ on a scale $S$ is refracted, undergoes parallel displacement and is imaged through a lens on a screen at $P_1$. If, however, a concentration gradient is present along the cell the light ray will be bent towards the direction of increasing refractive index and hit the screen at point $P_2$ which is displaced by a distance $Z$ from point $P_1$. $Z$ is proportional to the change in refractive index with position and is thus proportional to the concentration gradient. The scale photographed at different times is then compared with a reference scale (Fig. 18).

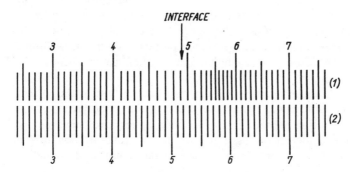

*Fig. 18*: Scale record of a sedimentation test
1 — Start of test, 2 — Reference scale

A method due to PHILPOT and SVENSON (Schlieren optics) is not quite so accurate, but enables the curve of the concentration gradient $dc_B/dx$ to be plotted directly. A horizontal slit diaphragm is imaged on a photographic plate by a Schlieren lens located in front of the measuring cell. There is an oblique slit in front of the camera lens and a cylindrical lens between the camera lens and plate. The ray represented by a solid line in Fig. 19 passes through a layer of constant refractive index in the cell. The oblique slit permits a central ray from the slit diaphragm to pass through its centre resulting in point $P_1$ on the photographic plate. The ray represented by dashes passes through the maximum refractive index gradient. The oblique slit allows a lateral ray from the slit diaphragm to reach the plate where it produces point $P_2$, the maximum of the concentration gradient. The oblique slit allows only that portion of the rays, emanating from the slit diaphragm to pass corresponding to a certain $dn/dx$. The ordinate of the curve of the concentration gradient can be altered by changing the inclination of the oblique slit.

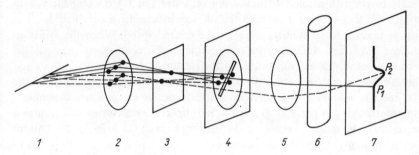

*Fig. 19*: Schematic representation of the ray path for Schlieren optics (PHILPOT and SVENSON)
1 — Slit diaphragm, 2 — Schlieren lens, 3 — Measuring cell, 4 — Oblique slit, 5 — Lens, 6 — Cylindrical lens, 7 — Photographic plate

Interference optics record the shift of interference lines whose number is proportional to the refractive index difference and, hence, to the concentration difference.

The procedures for ultracentrifuge methods are based on the principles and derivations present in Section 1.6.2. The following techniques are used:

## Measurement of sedimentation velocity

The speed of the ultracentrifuge is set sufficiently high that the sedimentation velocity is considerably greater than the diffusion velocity. The movement of the concentration gradient is measured and $s$ is calculated either from equation (5) or the integrated form, equation (13). Owing to the molecular weight distribution and the slight diffusion effects, the boundary between solution and solvent is diffuse and because of this, the sedimentation maximum is followed. Its position is measured at times $t_1$ and $t_0$. If the curve of the concentration gradient is asymmetric, a weight average of the sedimentation coefficient $s_w$ is determined from an empirical statistical approximation:

$$s_w = \frac{3s_m - s_{max}}{2} \qquad (35)$$

Where $s_m$ is the sedimentation coefficient resulting from the movement of the median value of the $dc_B/dx$-curve; and $s_{max}$ is the sedimentation coefficient resulting from the movement of the maximum of the $dc_B/dx$-curve. Polymer solutions of different concentrations (in the case of Schlieren optics $(1 \ldots 10) \times 10^{-3}$ g cm$^{-3}$)

must be measured in order to be able to extrapolate to $c_B = 0$. Solution purification need not be as through as for light scattering measurements because coarse impurities sediment rapidly during initial centrifugation. In order to obtain a concentration gradient and, thus, a defined space coordinate for the meniscus at the start of centrifugation the measuring cells are not filled completely with solution. The temperature is set and the ultracentrifuge drive switched on. The centrifuge takes 5 to 10 minutes to reach the desired speed. When the diffusion coefficient is known, the diffusion and sedimentation coefficients extrapolated to zero concentrations are substituted in the SVEDBERG-equation (9).

The denominator of equation (9) includes the density of the solvent and the partial specific volume of the polymer. Both quantities are pressure dependent. During centrifuging, a pressure gradient builds up in the measuring cell requiring a correction to be made (pressure at the meniscus about 0.1 MPa, at the bottom of the cell up to 50 MPa). The molecular weight distribution and diffusion lead to a broadening of the gradient curve proportional to $t$ and to $t^{0.5}$, respectively. Broadening due to diffusion is corrected by extrapolating the experimental data to $t = \infty$.

When the $[\eta]$-$M$-relationship of the product is known, the weight-average molecular weight can be calculated from the MANDELKERN-FLORY-SCHERAGA-equation (12) even if $D$ is unknown.

### Sedimentation equilibrium

At low rotor speeds and long measuring times equilibrium sets in between sedimentation and diffusion. The change in concentration with time at a position $x$ is equal to zero, and LAMM's differential equation reduces to equation (20). Applying the SVEDBERG equation leads to equation (21) which can be rearranged to give:

$$M_B = \frac{RT}{(1 - \varrho_A v_B)\omega^2 c_B x} \frac{\partial c_B}{\partial x} \tag{36}$$

Equation (26) which can also be used to determine molecular weight can be derived by integrating (21). These two equations are strictly true only in theta solvents otherwise an apparent molecular weight is obtained. The true molecular weight is obtained by plotting $(M_B)^{-1}$ versus $c_B$ and extrapolating to $c_B = 0$. The averages of the molecular weight of polydisperse samples are determined using equation (32) to (34).

### ARCHIBALD's transition state method

In order to avoid the long times required for the appearance of sedimentation equilibrium without, however, determining the diffusion coefficient, it is possible to measure the concentration at the meniscus or at the bottom of the measuring cell shortly after the start of sedimentation (ARCHIBALD [4, 5]). Since the dissolved substance cannot migrate into or out of the measuring cell, the condition $dm/dt = 0$ must also apply at these points. An apparent weight-average molecular weight $M_{w,\text{app}}$ can be determined with the help of equation (36) from the polymer concentration at the meniscus, $c_{B,m}$ and $(\partial c_B/\partial x)_m$. At the meniscus, the proportion of molecules of high molecular weight decreases because they sediment more rapidly. The true molecular weight is obtained by measuring $M_{w,\text{app}}$ at intervals and extrapolation to $t = 0$. The extrapolated molecular weights are concentration-dependent so that $(M_{w,\text{app}}(t = 0))^{-1}$ must be plotted against $c_B$ and extrapolated to $c_B = 0$. In the case of very high molecular weight components, e.g. aggregated molecules, the molecular weight of the aggregated product can be determined at the meniscus. The molecular weight of the total sample is determined by measuring the rotational speed dependence of the apparent molecular weight and extrapolating to $\omega = 0$.

## TRAUTMAN's transition state method [6]

In this method, higher rotor speeds are used than those in the previous method. While the centrifuge is running, the polymer concentration gradient at the meniscus $(\partial c_B/\partial x)_m$ and the decrease in polymer concentration at the meniscus $\Delta c_{B,m}$ are measured until a pure solvent layer forms at the meniscus. Then:

$$M_w c_{B,0} - M_Z \Delta c_{B,m} = \frac{(\partial c_B/\partial x)_m RT}{\omega^2 x_m^2 (1 - \varrho_A v_B)} \tag{37}$$

$c_{B,0}$ = polymer concentration of the initial solution.

$(\partial c_B/\partial x)_m$ is plotted against $\Delta c_{B,m}$ and extrapolated to $\Delta c_{B,m} = 0$. The extrapolated value of the concentration gradient and equation (37) are used to calculate an apparent molecular weight which again is extrapolated to $c_B = 0$.

## Density gradient ultracentrifugation

For density gradient ultracentrifugation, the polymer is dissolved in a mixture of two or more solvents of different densities. During operation of the centrifuge, a density gradient occurs in the measuring cell and the dissolved molecules move to the point at which their density coincides with that of the solvent mixture. BROWNian motion causes them to form a zone. For polymers that are uniform chemically, sterically and, with respect to their molecular weight, this zone is GAUSSIAN shaped with a width inversely proportional to molecular weight. The concentration distribution at distance $\bar{x}$ from the rotor axis is:

$$c_B(x) = c_B(\bar{x}) \exp\left(-(x - \bar{x})^2/2\sigma^2\right) \tag{38}$$

$$\sigma = RT \Big/ \left[M_B \omega^2 \bar{x} \left(\frac{d\varrho}{dx}\right)_{\bar{x}} v_B\right] \tag{39}$$

$(d\varrho/dx)_{\bar{x}}$ = density gradient at $\bar{x}$

Density gradient ultracentrifugation is particularly suitable for establishing the composition distribution of copolymers, especially of graft copolymers, with components of different densities [7]. Homopolymer tacticity can also be determined because different stereospecific molecules have different partial specific volumes.

## 1.6.4. Example

Construction of a molecular weight distribution curve from measurements of sedimentation coefficient is demonstrated using polystyrene as an example.

The measurements are taken at the theta point (solvent cyclohexane, $T_\theta = 307 \text{ K}$) where the virial coefficients of the concentration dependency of $s$ and $D$ are equal to zero. Experiments have shown that the second virial coefficients are actually equal to zero [8]. However, because the third virial coefficients have a small but finite value, it is advisable to select a low polymer concentration.

200 mg of the polystyrene sample are weighed out into a 100 cm³ measuring flask which is then filled with cyclohexane. The sample is dissolved at 40 °C while shaking and then placed in the measuring cell of the ultracentrifuge. The temperature of the ultracentrifuge is maintained at 307 K. After about 10 minutes, the desired speed of $50 \cdot 10^3$ revolutions per minute is attained and recordings are made of the sedimentation at intervals. Fig. 20 shows a PHILPOT-SVENSON plot schematically.

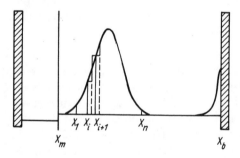

*Fig. 20*: Division of a PHILPOT-SVENSON plot (gradient curve) into equidistant sections

The concentration gradient is plotted as a function of distance from the centre of rotation. The concentration gradient curve is divided into equidistant sections and the heights are measured. The $dc_B/dx$-values must be corrected because of the sector-shape of the cell by multiplying the gradient values at position $x_i$ by $(x_i/x_0)^2$. ($x_0$ is the distance of the initial point of migration from the centre of rotation.) The corrected values are listed with $x$ in Table 13 and plotted in Fig. 21.

The area under the sedimentation peak is measured by planimeter or calculated by, for example, the trapezium rule. It corresponds to the total amount of sedimenting polymers. If the sedimentation peaks are split into a sufficient number of sections containing certain weight fractions of the sample (here, for simplification, 10%, 25%, 50%, 75% and 90%), the migration of the respective weight fractions can be observed. The $x$-values of the sedimentation curves at different times, corresponding to the weight fractions are listed in Table 14. Determination of the concentration gradients at $t = 0$ is problematical because of the finite starting phase of the ultracentrifuge. The time at which the test starts is specified as

$$t_0 = \frac{1}{3} t_{\text{real}} \tag{4}$$

$t_{\text{real}}$ = time until the selected speed is reached

*Table 13*: Measured heights of the gradient curve $\Delta h(x_i)$; $x_i = x_1 + i \Delta x$

| $t$ (min) | 0 | 12 | 37 | 65 | 90 | 120 |
|---|---|---|---|---|---|---|
| $x_1$ (cm) | 5.65 | 5.68 | 5.73 | 5.91 | 6.02 | 6.17 |
| $\Delta x$ (mm) | 0.4 | 0.4 | 0.6 | 0.6 | 0.7 | 0.7 |
| $\Delta h$ for | | | | | | |
| $i = 0$ | 0.4 | 0.3 | 0.25 | 0.25 | 0.25 | 0.25 |
| $i = 1$ | 1.1 | 0.65 | 0.6 | 0.6 | 0.5 | 0.5 |
| $i = 2$ | 2.0 | 1.15 | 1.25 | 1.1 | 1.0 | 1.0 |
| $i = 3$ | 2.8 | 1.9 | 1.9 | 1.75 | 1.7 | 1.5 |
| $i = 4$ | 3.2 | 2.6 | 2.6 | 2.3 | 2.25 | 1.8 |
| $i = 5$ | 3.1 | 2.8 | 2.65 | 2.5 | 2.25 | 2.1 |
| $i = 6$ | 2.7 | 2.8 | 2.15 | 2.1 | 1.8 | 1.75 |
| $i = 7$ | 1.8 | 2.5 | 1.25 | 1.4 | 1.2 | 1.1 |
| $i = 8$ | 1.25 | 2.0 | 0.75 | 0.9 | 0.65 | 0.65 |
| $i = 9$ | 0.7 | 1.3 | 0.3 | 0.6 | 0.4 | 0.35 |
| $i = 10$ | 0.4 | 0.75 | 0.25 | 0.25 | 0.2 | 0.2 |
| $i = 11$ | 0.25 | 0.4 | 0.15 | 0.15 | 0.1 | 0.1 |
| $i = 12$ | | 0.25 | | | | |
| $A$ | 7.88 | 7.78 | 8.46 | 8.34 | 8.61 | 7.91 |

$A$ = area under the gradient curve

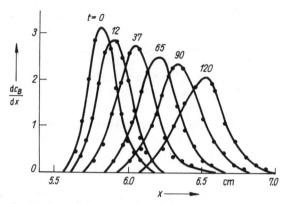

*Fig. 21*: PHILPOT-SVENSON plots of the sedimentation of polystyrene in cyclohexane at different times (in min)

*Table 14*: Space coordinates of integral weight fractions (%) at different times

| $t$ (min) | 10% | | 25% | | 50% | | 75% | | 90% | |
|---|---|---|---|---|---|---|---|---|---|---|
| | $x$ (cm) | $\ln x$ | $x$ (cm) | $\ln x$ | $x$ (cm) | $\ln x$ | $x$ (cm) | $\ln x$ | $x$ (cm) | $\ln x$ |
| 0 | 5.70 | 1.741 | 5.75 | 1.749 | 5.82 | 1.761 | 5.88 | 1.771 | 5.93 | 1.780 |
| 12 | 5.75 | 1.749 | 5.82 | 1.761 | 5.90 | 1.775 | 5.96 | 1.785 | 6.04 | 1.798 |
| 37 | 5.87 | 1.770 | 5.94 | 1.782 | 6.05 | 1.800 | 6.11 | 1.810 | 6.18 | 1.821 |
| 65 | 6.01 | 1.793 | 6.07 | 1.803 | 6.20 | 1.825 | 6.28 | 1.838 | 6.37 | 1.852 |
| 90 | 6.12 | 1.812 | 6.21 | 1.825 | 6.34 | 1.847 | 6.43 | 1.861 | 6.54 | 1.878 |
| 120 | 6.26 | 1.834 | 6.37 | 1.852 | 6.53 | 1.876 | 6.64 | 1.893 | | |

According to equation (13), the sedimentation coefficient $s$ can be calculated from the slope of the plot of the natural logarithm of the distance of the space coordinates of the selected weight fractions against time (Fig. 22).

$$s = \frac{\Delta \ln x}{\omega^2 \Delta t} \tag{41}$$

The relationship between sedimentation coefficient and molecular weight for the polystyrene-cyclohexane system at 307 K is taken from the literature [9, p. 272].

$$s = 1.55 \; 10^{-2} \; M_w^{0.5} \tag{42}$$

where $M_w$ in gmol$^{-1}$ and $s$ in $10^{-13}$ s.

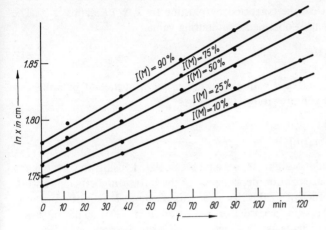

*Fig. 22*: Plot of the natural logarithm of the space coordinate against time for different integral mass fractions for use in equation (41)

Average molecular weights can now be assigned to the weight fractions using the sedimentation coefficient (Table 15). The resultant integral weight distribution is shown in Fig. 23.

*Table 15*: Gradients of the $\ln x - t$ plot (Fig. 22), and sedimentation coefficients, calculated using equation (41), from which $M_w$ is calculated with equation (42)

| $I(m)/\%$ | $\tan \alpha / 10^{-4}$ min$^{-1}$ | $s/10^{-13}$ s | $M_w/10^3$ g mol$^{-1}$ |
| --- | --- | --- | --- |
| 10 | 7.75 | 4.72 | 93 |
| 25 | 8.33 | 5.07 | 107 |
| 50 | 9.58 | 5.83 | 142 |
| 75 | 10.33 | 6.29 | 165 |
| 90 | 11.0 | 6.69 | 186 |

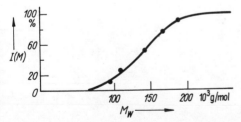

*Fig. 23*: Integral molecular weight distribution of a polystyrene sample

The diffusion effect can be corrected by plotting $(\ln x_i/x_m)/t$ against $t^{-1}$. The sedimentation coefficient is obtained as a limiting value

$$s = \frac{1}{\omega^2} \lim_{t^{-1} \to 0} (\ln x_i/x_m)/t \tag{43}$$

If the appropriate numerical values are known, the influence of pressure can be corrected using the compressibility factor $k$.

$$k = \frac{\eta_A(p)\,[1 - \varrho_A(p_0)\,v_B(p_0)]}{\eta_A(p_0)\,[1 - \varrho_A(p)\,v_B(p)]} \tag{44}$$

The average molecular weight $M_{s,D}$ can be calculated, using the SVEDBERG-equation, from the concentration dependence of the diffusion coefficient and of the sedimentation coefficient (Table 16). For this purpose, $D$ is plotted against $c_B$ and $s^{-1}$ against $c_B$ (in accordance with equation (27)). The diffusion and sedimentation coefficients extrapolated to $c_B = 0$ are then obtained from the graph. $D/s$ can, however, be extrapolated directly to $c_B = 0$. Both methods give the correct result.

Table 16: Concentration dependence of the sedimentation coefficient and diffusion coefficient of polystyrene in toluene at 293 K [10].

| $c_B$ $(10^{-3}$ g cm$^{-3})$ | 1 | 2 | 3 | 4 |
|---|---|---|---|---|
| $s/10^{-13}$ s | 4.09 | 3.72 | 3.43 | 3.20 |
| $s^{-1}/10^{13}$ s$^{-1}$ | 0.24 | 0.27 | 0.29 | 0.31 |
| $D/10^{-7}$ cm$^2$ s$^{-1}$ | 5.35 | 5.47 | 5.50 | 5.70 |
| $\dfrac{D}{s}\Big/10^6$ cm$^2$ s$^{-1}$ | 1.31 | 1.47 | 1.60 | 1.78 |

$v_B \varrho_A = 0.795$

From Fig. 24:

$\lim\limits_{c_B \to 0} s^{-1} = 0{,}224 \cdot 10^{13}$ s$^{-1}$

$s_{c_B \to 0} = 4{,}46 \cdot 10^{-13}$ s

$\lim\limits_{c_B \to 0} D = 5{,}22 \cdot 10^{-7}$ cm$^2$ s$^{-1}$

and

$\lim\limits_{c_B \to 0} (D/s) = 1{,}15 \cdot 10^6$ cm$^2$ s$^{-2}$

Inserting the values into equation (9), both extrapolation methods yield
$M_{s,D} = 102 \cdot 10^3 \text{ g mol}^{-1}$

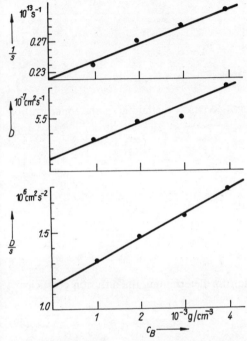

*Fig. 24*: Extrapolation of the reciprocal sedimentation coefficient, of the diffusion coefficient and of the ratio $D/s$

## 1.6.5. Determination of Diffusion Coefficient

Determination of the diffusion coefficient $D$ in equation (9) is based on the relationship between the change in concentration with time $dc_B/dt$ and the concentration gradient formulated in the second FICK law:

$$\frac{dc_B}{dt} = D \frac{\partial^2 c_B}{\partial x^2} \tag{45}$$

($D$ is independent of $c_B$)

The solutions of the partial differential equation (45) depend on the geometry of the apparatus and on the boundary and initial conditions. If the dissolved particles diffuse into their own solvent, then at time $t = 0$, the concentration is equal to $c_{B,0}$ for all negative $x$ and for all positive $x$ it is equal to $c_B = 0$. Assuming that $D$ is independent of concentration and that the total concentration remains unchanged, equation (45) can be integrated to give

$$c_B(x, t) = \frac{c_{B,0}}{2} \left( 1 - \frac{2}{\sqrt{\pi}} \int_0^y e^{-y^2} dy \right) \tag{46}$$

$y = x/2Dt$, $c_{B,0}$ = initial concentration

Measuring the diffusion by the PHILPOT-SVENSON method gives the concentration gradient

$$\frac{dc_B}{dx} = \frac{c_{B,0}}{2\sqrt{\pi Dt}} e^{-x^2/4Dt} \tag{47}$$

with the maximum height $h$ at $x = 0$.

$$h = \frac{c_{B,0}}{2\sqrt{\pi Dt}} \tag{48}$$

Rearrangement results in the equation for determining the diffusion coefficient

$$D = \frac{c_{B,0}^2}{4\pi t h} \tag{49}$$

# Bibliography

[1] ELIAS, H. G., „Theorie und Praxis der Ultrazentrifugentechnik", (Beckman Instruments GmbH, Munich)
[2] SCHERAGA, H. A. and L. MANDELKERN, J. Amer. chem. Soc. **75** (1953), 179
[3] LANSING, W. D. and E. O. KRAEMER, J. Amer. chem. Soc. **57** (1935), 1369
[4] ARCHIBALD, W. J., J. phys. Colloid Chem. **51** (1947), 1204
[5] ELIAS, H. G., Angew. Chem. **73** (1961), 209
[6] TRAUTMAN, R., J. physic. Chem. **60** (1956), 1211
[7] HERMANS, J. J., Ber. Bunsenges. physik. Chem. **70** (1966), 280
[8] CANTOW, H. J., Makromolekulare Chem. **30** (1959), 169
[9] MCCORMICK, H. W., „Sedimentation", Chapter C. 2 in: „Polymer Fractionation", ed. M. J. R. CANTOW, Academic Press, New York, London 1967
[10] MEYERHOFF, G., Z. physik. Chem. Neue Folge, **23** (1960), 100

## 1.7. Determination of the Viscosity-average Molecular Weight $M_\eta$ of Polymers

### 1.7.1. Introduction

Viscometry of polymer solutions is the most frequently applied method for characterizing macromolecules because it is rapid and simple and furnishes comprehensive microstructural information on polymer molecules. This includes information on the average molecular size, gyration radii and end-to-end distances of linear and branched molecules, long- and short-range interaction parameters as used in the excluded volume theory and thus on the quality of solvents. In addition, information is also provided on the molecular structures of concentrated polymer solutions and melts as well as their flow and relaxation characteristics.

Viscometry is not an absolute method of determining molecular weight because the exact relationship between the measured limiting viscosity and the molecular weight is still not known.

### 1.7.2. Principles [1, 2]

Viscometric methods are based on the fact that the viscosity of a liquid to which a particle is added increases proportionally with the volume of the particle. In homologous series of polymers the volume of the macromolecules increases with molecular weight. Viscosity increase must therefore be related to molecular weight. The relative viscosity increase is generally referred to as specific viscosity $\eta_{sp}$.

$$\eta_{sp} = \frac{\eta - \eta_0}{\eta_0} = \frac{\eta}{\eta_0} - 1 = \eta_{rel} - 1 \tag{1}$$

$\eta$ = viscosity of the solution, $\eta_0$ = viscosity of the solvent, $\eta_{rel}$ = viscosity ratio or relative viscosity

$\eta_{sp}$ is dependent of concentration and of interaction forces. In order to characterize the viscosity increase due to the dissolved macromolecules, $\eta_{sp}$ is divided by the concentration and then extrapolated to infinite dilution. Since the viscosity increase is dependent on the shear or velocity gradient $G$, the limiting value at $G = 0$ must be obtained. The quantity obtained is called the limiting viscosity or STAUDINGER-index $[\eta]$.

$$[\eta] = \lim_{\substack{c_B \to 0 \\ G \to 0}} \frac{\eta_{sp}}{c_B} \tag{2}$$

The empirical KUHN-MARK-HOUWINK equation relates limiting viscosity to molecular weight.

$$[\eta] = K\bar{M}^a \tag{3}$$

The quantities $K$ and $a$, which cannot be exactly calculated from theory, must be determined from calibration measurements. They are dependent on the geometric shape (hydrodynamic volume) and, consequently, on the interaction with the solvent. The limiting viscosity of a polydisperse sample is a weight average given by the empirical PHILIPPOFF's rule:

$$[\eta] = \frac{\sum_i w_i [\eta]_i}{\sum_i w_i} \tag{4}$$

The molecular weight calculated from equation (3) is a viscosity-average $M_\eta$. Since it is an exponential average (see 1.1.), it is dependent on the magnitude of the exponent:

$$M_\eta = \left( \frac{\sum_i w_i M_i^a}{\sum_i w_i} \right)^{1/a} \tag{5}$$

For $a = 1$, $M_\eta$ is identical to the weight average and for $a = -1$ to the number average of the molecular weight. A viscosity-average is obtained from equation (3) only if the calibrating samples are either molecularly homogeneous or if their viscosity average is known. In other cases, $K$ is dependent upon the molecular inhomogeneity of the calibration substances. If the samples used for calibration are not molecularly homogeneous, but their molecular weight distribution is known, corrections can be made for the effect of the type and the width of molecular weight distribution.

The required correction factors can be calculated and are tabulated in [3]. We have

$$[\eta] = KM_\eta^a = Kq_w M_w^a = Kq_n M_n^a \tag{6}$$

If the molecular weight distribution is not too wide ($M_w/M_n < 3$), the viscosity average is only a few per cent smaller than the weight average. Then, the constants in equation (3) can be determined from the dependence of the limiting viscosity on the weight-average molecular weight of the calibration samples.

The exponent $a$ and the constant $K$ of the KUHN-MARK-HOUWINK equation can be calculated for a few model systems. Basic viscosity theory for polymer molecules applies to macromolecules with segments exhibiting a GAUSSIAN density distribution about the centre of gravity. In 1949, FOX and FLORY derived a universal viscosity function which associates the limiting viscosity with the radius of gyration:

$$[\eta] = \Phi \frac{\langle s^2 \rangle_n^{3/2}}{M_n} \tag{7}$$

$\Phi$ = FLORY universal constant

The limiting viscosity is proportional to the ratio of the effective coil volume to the molecular weight. The proportionality factor, the FLORY universal constant $\Phi$, is dependent upon the segment distribution about the centre of gravity of the macromolecule and, hence, upon its geometric shape and the forces of interaction inside the solution:

$$\Phi = \Phi_0(1 - 2.63\varepsilon' + 2.86\varepsilon'^2) \tag{8}$$

$\Phi_0$ = FLORY universal constant of a coil in the theta state = $4.18 \cdot 10^{24}$ (mol macromolecule)$^{-1}$; related to the radius of gyration, $\varepsilon'$ = parameter which characterizes the deviation of the real state of the solution from the theta state, $\varepsilon' = \dfrac{2a - 1}{3}$

The unperturbed dimension $\langle s_0^2 \rangle$, achieved in the theta state in the polymer solution is a dimension characteristic of the isolated non-interacting macromolecule. It can be calculated with the help of random walk statistics from segment length and segment number and/or bond number, bond length, valency angle and rotational barrier potential. The deviation of the true molecular dimension is allowed for by introducing an expansion coefficient, $\alpha = (\langle s^2 \rangle / \langle s_0^2 \rangle)^{0.5}$.

$$[\eta] = [\eta]_\Theta \, \alpha^3 \tag{9}$$

$$[\eta]_\Theta = \Phi_0 \frac{\langle s_0^2 \rangle^{3/2}}{M_n}$$

$[\eta]_\Theta$ = limiting viscosity in the theta state

According to the "two parameter theory" or "excluded volume theory" the expansion coefficient is a function of the short-range interaction parameter, which includes the interaction of nearest neighbours and is thus dependent on the bond angle, bond length and rotational barrier potential, and the long-range interaction parameter which is dependent on interactions between segments of different macromolecules and segments of one macromolecule which are separated by a large number of bonds.

The expansion coefficient can be expanded as a series in the excluded volume parameter. This parameter depends on the segment number and the excluded volume per segment, $\beta$. Inserting the respective expressions of the „two-parameter theory" for $\alpha$ into equation (9) gives:

$$\frac{[\eta]}{M_n^{1/2}} = K_0 + \Phi_0 C' B M_n^{1/2}; \qquad K_0 = \Phi_0(\langle s_0^2 \rangle / M_n)^{3/2} \qquad (10)$$

$B = \beta/M_s^2$, where $M_s$ = molecular weight of a segment, $C'$ = numerical factor

On plotting $[\eta]/M_n^{1/2}$ versus $M_n^{1/2}$, $K_0$ is the intercept and the unperturbed dimension can be calculated. The excluded volume $\beta$, associated with the second virial coefficient, is obtained from the slope.

The analytical relationship between $\langle s^2 \rangle$ and $\bar{M}$ determines the exponent of the KUHN-MARK-HOUWINK equation. In the theta state, the square of the radius of gyration is directly proportional to the molecular weight so that in the case of the unperturbed coil, $a = 0.5$. The deviation of the $\langle s^2 \rangle - \bar{M}$ relationship from that in the theta state can be characterized by parameter $\varepsilon'$ of equation (8) which increases as the thermodynamic state of the solution approaches ideality.

$$\langle s^2 \rangle \sim \bar{M}^{1+\varepsilon'} \qquad (11)$$

In the limiting case of the free-draining coil $a = 1$; with good solvents $a$ lies within the range from 0.6 to 0.9.

For non-solvent or equally solvated spheres of constant density, $\langle s^2 \rangle^{3/2} \sim M$, and the limiting viscosity must thus be independent of the molecular weight.

Very high molecular weight macromolecules contract more and more with increasing molecular weight and approach the viscosity behaviour of a sphere. Very small molecules ($M < 5 \cdot 10^3$ g/mol) do not coil and the relationships between their limiting viscosity and molecular weight may be similar to that of rigid rods, $a = 2$. Branched and slightly cross-linked products contract more intensely than linear molecules of the same molecular weight and the relationship between their limiting viscosity and molecular weight are different to those of linear products. Differences in the viscosity behaviour of linear and branched products may be used to determine the branching coefficient.

## 1.7.3. Experimental Background

The viscosity of high-polymer solution is commonly measured with three types of viscosimeter: capillary, rotation and drop viscosimeters. The limiting viscosity is most frequently determined with capillary viscometers. According to the HAGEN-POISEUILLE law,

$$\eta = \frac{\pi \Delta p \, r^4}{8 l v} t \qquad (12)$$

$\Delta p = \varrho g h$ (pressure difference), $t$ = flow time (time of passage), $l$ = capillary length, $r$ = capillary radius, $h$ = rise in manometer level

The dynamic viscosity is proportional to the product of the flow time $t$ through the capillaries and the density of the solution $\varrho$ or the solvent $\varrho_0$ at constant flow volume $v$ and constant equipment dimensions. Therefore, it is not necessary to determine absolute viscosity values in order to obtain relative and specific viscosities. Since the densities of solvent and solution differ only slightly, $\varrho \approx \varrho_0$, they can be neglected and only the times of flow of the solution $t$ and those of the solvent $t_0$ need to be measured. Viscosities determined in this way, however, are distorted by various effects. On entering the capillary, the liquid is accelerated and loses potential energy which is converted into frictional heat. The HAGEN-POISEUILLE law, applicable only to infinitely long capillaries, must be modified by the HAGENBACH correction which depends on viscometer design.

$$\eta = \frac{\pi \, \Delta p \, r^4}{8lv} t - \frac{c'v\varrho}{8\pi l t} \tag{13}$$

$c' =$ constant ($\approx 1.12$)

The correction term decreases with increasing time of flow and may thus be neglected provided the viscometer capillaries are sufficiently narrow and long. Surface tension effects can be reduced to negligible levels by the use of viscometers with suspended levels.

The two types of capillary viscometer are the OSTWALD and UBBELOHDE designs (Fig. 26).

Specially modified UBBELOHDE viscometers are generally utilized for characterizing polymers. The viscometer capillaries have a length of about 10 to 20 cm

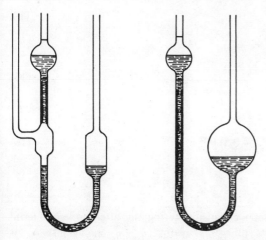

*Fig. 25*: Diagram showing an UBBELOHDE Viscometer (left) and an OSTWALD Viscometer (right)

and an internal diameter of anything between 0.3 and 0.4 mm. A vessel with calibration marks is arranged above the capillary.

A reservoir in which the polymer solution can be dilute with solvent can be connected to the end of the capillary by means of a U-tube. The end of the capillary is vented by an additional tube so that the hydrostatic pressure is independent of the amount of liquid in the reservoir (suspended level). Since viscosity checks are frequently routine, automatically recording viscometers are widely used. The flow time of the calibrated volume is measured by photoelectric cells in such instruments.

The capillary diameter of a viscometer is selected so that the relative viscosities are between 1.2 and 2. The concentration of the polymer solution should be 0.1 to 1 g/100 cm³. If the concentrations are too small, the viscosity values will exhibit unexplained anomalies probably due to adsorption and weighing errors. Solutions of higher concentrations may exhibit structural viscosity and complicate extrapolation to $c_B = 0$. Extrapolation to $G = 0$ is not required for the alsmost universally utilized capillary viscometer and low polymer concentrations provided that $[\eta] < 500$ g/cm³. In the case of high molecular weight samples ($\bar{M} > 5 \cdot 10^5$ g mol$^{-1}$), the relative viscosity decreases with increasing shear gradient (structural viscosity). It can be effectively reduced by long coiled capillaries so that large limiting viscosities can be measured correctly.

The extrapolation to $c_B = 0$, for the determination of the limiting viscosity, is usually carried out graphically. Plots resulting from the SCHULZ-BLASCHKE, HUGGINS or KRAEMER equations are frequently used.

SCHULZ-BLASCHKE equation

$$\frac{\eta_{sp}}{c_B} = [\eta] + k_{SB}[\eta]\,\eta_{sp} \tag{14}$$

HUGGINS equation

$$\frac{\eta_{sp}}{c_B} = [\eta] + k_H[\eta]^2\,c_B \tag{15}$$

KRAEMER equation

$$\frac{\ln \eta_{rel}}{c_B} = \frac{\ln(\eta_{sp}+1)}{c_B} = [\eta] + (k_K - 0.5)[\eta]^2\,c_B \tag{16}$$

$\frac{\ln \eta_{rel}}{c_B}$ = inherent viscosity

All three equations are power series expansions terminated after the second term. They are interchangeable. Plots of $\eta_{sp}/c_B$ against $\eta_{sp}$, $\eta_{sp}/c_B$ against $c_B$ or $\ln \eta_{rel}/c_B$ against $c_B$, are straight lines with the limiting viscosity as the intercept

## Determination of the Viscosity-average Molecular Weight

providing that the polymer concentrations are not too high. The respective constants, $k_{SB}$, $k_H$ or $k_K$, can be calculated from the slope. They depend on the solution state, temperature, polymer structure, etc. For thermodynamically good solvents, the SCHULZ-BLASCHKE constant lies between 0.28 to 0.30 and the HUGGINS constant between 0.3 to 0.4. In thermodynamically poor solvents (theta solvents), $k_H$ is frequently 0.5. When the constant of one of equations (14—16) has been measured for a system under defined conditions, $[\eta]$ can be determined by a single-point measurement.

In production testing, viscosity is usually measured at a fixed polymer concentration, typically 0.5 or 1 g/100 cm³. The $K_F$ or FIKENTSCHER value is then calculated from equation (17).

$$\log \eta_{\text{rel}} = \left( \frac{75k^2 c_B}{1 + 1{,}5kc_B} \right) + kc_B \tag{17}$$

$K_F = 1000k$

The relationship between $\log \eta_{\text{rel}}/c_B$ or $\eta_{\text{rel}}$ and $K_F$ is tabulated. The FIKENTSCHER equation is valid only for good solvents and has little theoretical basis.

Viscometric measurements require constant temperatures because changes of 0.1 deg. will cause errors of 1% in the $\eta_{\text{rel}}$-values. Further sources of error are: viscometers suspended out of true in the thermostat bath resulting in differences in measurements of the hydrostatic pressure of the solvent and solution; parallax errors; concentration changes due to evaporation of solvents; association, polymer cross-linking or degradation; microgel content and insufficient dissolution of the polymer; possible structural viscosity in the case of high molecular weights.

### 1.7.4. Example

#### 1.7.4.1. Molecular Weight Determination

The procedure for determining molecular weights by viscometry is illustrated below using a linear polyethylene (low pressure polyethylene) as an example. The choice of solvent (n-decalin) and temperature (135 °C) is governed by the $[\eta]$-$\bar{M}$-relationship on which evaluation is based. The measurements are carried out in a BERGER-DECKERT UBBELOHDE viscometer (Fig. 26) which is particularly suitable for high-temperature measurements. The capillary is of the K 5 type.

The BERGER-DECKERT dilution viscometer enables a dilution series to be prepared from a stock solution. There may be differences, however, between a dilution series and solutions obtained from individually weighed samples because of the dif-

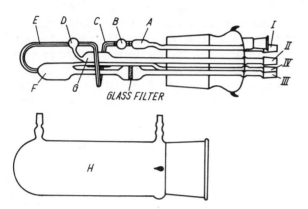

*Fig. 26*: High-temperature BERGER-DECKERT dilution viscometer
I — Viscometer tube, II — Air tube (aeration and pressure equalization), III — Riser and filling tube, IV — Gas feed or discharge pipe, A — Reservoir, B — Measuring volume (0.5 to 0.8 cm$^3$), C — Capillary tube (about 200 mm long), D — Ball for forming the suspended level, E — Capillary to the solution reservoir, F — Reservoir (30 cm$^3$), G — Dilution chamber (15 cm$^3$), H — Thermostatting vessel

ference in history. Viscosity is dependent upon the arrangement of the macromolecules in the solution. In the most highly concentrated sample solutions, for example, entanglements and coiling may occur which can be identified by dependence of the flow time on the number of passes through the capillary. They do not necessarily unravel on dilution. Measurements on solutions obtained from individually weighed samples are discussed below.

Five solutions with concentrations ranging from 0.1 to 0.5 g/100 cm$^3$ are prepared by weighing the appropriate amount of sample into the measuring flask. The flask is two thirds filled with n-decalin and about 0.1% (related to the amount of the sample) of n-phenyl-$\beta$-naphthylamine is added to prevent oxidative degradation of the sample. The flasks are thermostated and then closed and shaken in the oven for about two hours at 135 °C until complete dissolution. The measuring flasks are topped up with n-decalin at 135 °C and shaken once more for about 15 min. Dust particles are removed from solutions and solvents by a glass frit fused into the feed tube of the viscometer. In the case of high molecular weight and microgel-containing products, the concentration may have to be determined after purification.

The viscometer is heated to a temperature of 135 °C. During measurement, care must be taken to ensure that the temperature remains constant because the viscosity of the polyethylene solutions is highly temperature-dependent. As usual in measurements with capillary viscosimeters, the flow times (in seconds) of the pure solvent $t_0$ and the polymer solution are measured. Five individual measure-

## Determination of the Viscosity-average Molecular Weight

ments are carried out at each concentration. The polymer solution is injected with a syringe as quickly as possible to avoid cooling of the solution and precipitation of the polymer. After a measurement, the solution is removed and the next solution of a higher concentration injected into the viscometer. After rinsing the capillary several times, the measurement can be carried out.

The procedure for a dilution series is as follows: 5 ml of the stock solution is fed into the viscometer. After measuring, the next concentration is prepared by adding 1 ml of hot n-decalin and rinsing the measuring bulb. The measurement is taken again. Subsequently 1.5 ml, 2.5 ml and finally 5 ml of n-decalin are added.

The relative and specific viscosities are determined from the flow times $t$ and $t_0$ using equation (18):

$$\eta_{sp} = \eta_{rel} - 1 = \frac{\eta - \eta_0}{\eta_0} = \frac{t - t_0}{t_0} \tag{18}$$

The measured flow times, the specific viscosities calculated from equation (18), and the reduced viscosities are listed in Table 17.

Fig. 27 shows a SCHULZ-BLASCHKE plot.

Table 17: Flow times, specific and reduced viscosities of a polyethylene sample

| $c_B$ ($10^{-3}$ g cm$^{-3}$) | $\bar{t}$ (s) | $\eta_{sp}$ | $\eta_{sp}/c_B$ (cm$^3 \cdot$ g$^{-1}$) |
|---|---|---|---|
| 1.3240 | 138.2 | 0.133 | 100.7 |
| 2.4808 | 153.4 | 0.258 | 104.0 |
| 3.2244 | 163.6 | 0.342 | 105.9 |
| 4.1632 | 177.0 | 0.451 | 108.4 |
| 4.9917 | 189.4 | 0.553 | 110.8 |
| 0 | 121.95 | | |

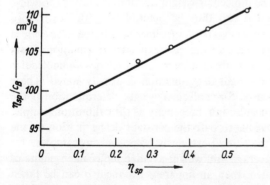

Fif. 27: SCHULZ-BLASCHKE plot of the data in Table 17

The limiting viscosity $[\eta] = 96.8$ cm³/g is the intercept of the plot. The SCHULZ-BLASCHKE constant is $k_{SB} = 0.29$.

When calculating molecular weights from limiting viscosities, the chosen units of limiting viscosity (here cm³/g) must be noted because the units $dl/g = 100$ cm³/g are frequently used. ($[\eta]$ is then often designated as $Z_\eta$.)

The $[\eta]$-$\bar{M}$-relationship [5] is:

$$[\eta]_{\text{n-decalin}}^{135°C} = 6{,}2 \times 10^{-2} \bar{M}^{0,70} \tag{19}$$

Substituting the measured limiting viscosity in equation (19), gives the molecular weight of the sample as:

$M_\eta = 3.6 \times 10^4$ g/mol

The total error in the molecular weight determination is governed by the error in measuring the limiting viscosity, which is about 5 to 10%, and the validity of the $[\eta]$-$\bar{M}$-relationship. It is advisable to state the $[\eta]$-$\bar{M}$-relationship on which the calculation of molecular weight is based.

## 1.7.4.2. Derivation of a $[\eta]$-$\bar{M}$-relation

Calculation of molecular weight from limiting viscosity measurements depends on equation (3) which relates $[\eta]$ to $\bar{M}$. In the case of narrowly-distributed fractions, the logarithm of molecular weight determined by any method can be plotted against the logarithm of the limiting viscosity. The slope of the resulting straight line is the exponent of the KUHN-MARK-HOUWINK relationship $a$ and the intercept is the logarithm of the constant $K$. As previously stated, $M_\eta$ is only a little smaller than $M_w$ if the molecular weight distribution is not too wide. As a first approximation $K$ and $a$ can thus be determined with satisfactory accuracy by plotting $\log [\eta]$ against $\log M_w$. Agreement with the true values improves as $a$ approaches unity, i.e., as the coil expands. If equation (3) is set up on the basis of samples that are not very homogeneous in molecular weight, the method of molecular weight determination is usually stated. These $[\eta]$-$\bar{M}$-relationships can only be used for molecular weight determination if the test sample exhibits the same molecular inhomogeneity as the calibration sample. Since this rarely occurs it is advisable to base the constant determination on the number-average weight only in the case of extremely homogeneous samples.

If distribution type, number-average and weight-average molecular weights of the calibration samples are known, then an apparent exponent $a$ can be taken from the plot of $\log [\eta]$ versus $\log M_w$. The relationship between $M_n$, $M_w$ and $M_\eta$ is known for various types of molecular weight distribution. Thus, for the logarithmic normal distribution:

$$M_\eta = M_w \left(\frac{M_w}{M_n}\right)^{(a-1)/2} \tag{20a}$$

or

$$M_\eta = M_n(M_w/M_n)^{(a+1)/2} \tag{20b}$$

For a SCHULZ-FLORY distribution with the degree of coupling $k$, $k = (M_w/M_n - 1)^{-1}$:

$$M_\eta = \frac{M_w}{(k+1)} \left(\frac{\Gamma(k+a+1)}{\Gamma(k+1)}\right)^{1/a} \tag{21a}$$

$$M_\eta = \frac{M_n}{k} \left(\frac{\Gamma(k+a+1)}{\Gamma(k+1)}\right)^{1/a} \tag{21b}$$

$\Gamma(x)$ = gamma function

The factors $q_n$ and $q_w$ of equation (6) are the ratios

$$\frac{M_\eta}{M_n} = (q_n)^{1/a}; \quad \frac{M_\eta}{M_w} = (q_w)^{1/a} \tag{22}$$

(Example: For $M_w/M_n = 3$ and $a = 0.7$, $q_n = 2.02$ and $q_w = 0.94$ for the SCHULZ-FLORY distribution.)

Substituting the apparent $a$ in equation (20) or (21) leads to a preliminary $M_\eta$. The log log plot of this and $[\eta]$ provides a value of $a$ which can be used to calculate an $M_\eta$ from equation (20) or (21). This procedure is repeated until the exponent $a$ remains constant and has thus assumed its true value.

The validity of the experimentally established $[\eta]$-$\bar{M}$-relationship must not be extended a priori beyond the molecular weight range of the calibration samples. As explained in 1.7.2. macromolecules of molecular weight less than $10^4$ g mol$^{-1}$ and high molecular weight samples do not necessarily exhibit a linear variation of log $[\eta]$ against log $\bar{M}$. A $[\eta]$-$\bar{M}$-relationship, which is also valid for low molecular weights, can be established for a few polymers by inserting a correction term $b_\eta$ in the limiting viscosity. On plotting $[\eta]$ versus $\bar{M}$, $[\eta] = b_\eta$ is obtained at $\bar{M} = 0$. In most cases, $|b_\eta|$ is about 1 cm$^3$/g. If log ($[\eta] + b_\eta$) is plotted as a function of log $\bar{M}$, a straight line is obtained even for low molecular weights. From this $K$ and $a$ can be determined. For high values of the limiting viscosity, $b_\eta$ is so small that it can be neglected and the $[\eta]$-$\bar{M}$-relationship applicable to high molecular weights is obtained.

Table 18 shows the limiting viscosities of polystyrene fractions of different molecular weights.

*Table 18*: Limiting viscosities of polystyrene fractions in benzene ($T = 293$ K)

| $[\eta]$ (cm³ g⁻¹) | log $[\eta]$ | $\bar{M}$ (g mol⁻¹) | log $\bar{M}$ | $\{[\eta] - b_\eta\}$ (cm³ g⁻¹) | log $\{[\eta] - b_\eta\}$ |
|---|---|---|---|---|---|
| 2.90 | 0.46 | 1180 | 3.07 | 1.8 | 0.26 |
| 3.80 | 0.58 | 2170 | 3.34 | 2.7 | 0.43 |
| 6.40 | 0.81 | 5900 | 3.77 | 5.3 | 0.72 |
| 13.7 | 1.14 | 18.5 × 10³ | 4.27 | 12.6 | 1.10 |
| 51.5 | 1.71 | 124 × 10³ | 5.09 | 50.4 | 1.70 |
| 110 | 2.04 | 313 × 10³ | 5.50 | 108.9 | 2.04 |
| 162 | 2.21 | 520 × 10³ | 5.72 | 160.9 | 2.21 |

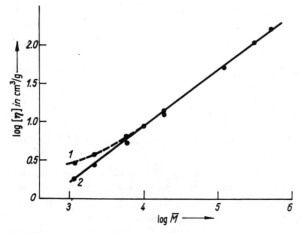

*Fig. 28*: Log-log plot of $[\eta]$ against $\bar{M}$
Curve 1: without correction, Curve 2: with correction

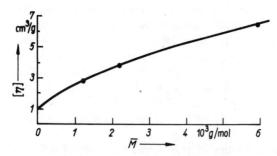

*Fig. 29*: Plot of $[\eta]$ against $\bar{M}$ for the determination of $b_\eta$

Curve 1 in Fig. 28 is obtained by plotting log [$\eta$] versus log $\bar{M}$. It clearly exhibits a curvature at low molecular weights.

Correcting with $b_\eta = 1.1$ cm$^3$/g, taken from Fig. 29, gives curve 2 in Fig. 28 which is linear over the entire range of molecular weight. The slope of the straight line gives $a = 0.72$ and the intercept is $1.2 \cdot 10^{-2}$. The [$\eta$]-$\bar{M}$-relationship for polystyrene in benzene at 293 K is thus:

$$[\eta]_{20°C}^{benzene} = b_\eta + 1.2 \cdot 10^{-2} \bar{M}^{0.72} \tag{23}$$

$b_\eta = 1.1$ cm$^3$/g, [$\eta$] in cm$^3$/g

# Bibliography

[1] SCHURZ, J., „*Viscositätsmessungen an Hochpolymeren*", Berliner Union Kohlhammer, Stuttgart 1972
[2] MEYERHOFF, G., Fortschr. Hochpolymeren-Forsch. **3** (1961), 59
[3] BRANDRUP, J. and E. H. IMMERGUT, ed. „*Polymer Handbook*", 2nd ed. Wiley-Interscience, New York 1975, IV—115
[4] LANGHAMMER, G., BERGER, R. and H. SEIDE, Plaste und Kautschuk **11** (1964), 472
[5] CHIANG, J. R., Polymer Sci. **36** (1959), 91

# 2. Determination of Molecular Weight Distribution and Chemical Heterogeneity

## 2.1. Molecular Weight Distribution

### 2.1.1. Introduction

Throughout the history of macromolecular chemistry, determination of molecular weight distribution has been a central problem of practically the same importance as the molecular weight determination itself. A large number of individual papers and various comprehensive monographs [1—4] and reviews (e.g. [5—8]) have there fore been devoted to this problem. In theoretical and experimental research into structures and properties the important influence of molecular weight distribution on the property spectrum of polymers in the solid state, melts and solutions has been pointed out repeatedly. In kinetically ideal reactions the molecular weight distribution can be predicted theoretically [4]. This is limited, however, in the case of large-scale production because of process and reactor variations and is impossible in the case of processed polymers because variable chemical and mechanochemical operations occur during forming. However, the microstructure and supermolecular structure which are imprinted by the forming process are the principal factors governing the use of the material. Thus, experimental measurement remains the only precise method of determining the molecular weight distribution, at least for industrial products, and, consequently, the most reliable basis for predicting properties. All experimental methods of determining molecular weight distribution ($MWD$) are based on fractionation, i.e., the separation of the substance into fractions that are molecularly homogeneous as far as possible. The distribution functions are then constructed from mass fractions $m_i$ and molecular weights $M_i$. Great importance is also attached to preparative fractionating methods, used for obtaining larger amounts (10 to 100 g) of monodisperse substances required for investigating the relationship between structure and properties.

The therm "fractionation", which was restricted formerly to separation according to molecular size, now serves as a general term for methods of separating macromolecules according to all microstructural parameters, e.g. chemical composition and sequence length of polymers (chemical heterogeneity and sequence length distribution), tacticity, etc.

## 2.1.2. Principles of Separation by Fractionation

The most important separation methods for polymers are based on thermodynamic and kinetic properties or make use of the different hydrodynamic volumes of the dissolved molecules for separating them. Table 19 gives a survey of the separation effects on which the principal methods are based.

*Table 19*: Separation effects and methods for polymer fractionation

| Separation effects | Separation methods |
| --- | --- |
| 1. *Thermodynamic properties* | |
| solubility | precipitation fractionation |
| | fractionated extraction |
| | solubility fractionation with or without supporting material — with solvents/precipitation agents and/or temperature gradients — as single-stage or multi-stage process in separation columns and also with variations of pressure |
| melting behaviour | zone-melting process (restricted to crystallizable polymers) |
| 2. *Effects of interfacial energy* | |
| adsorbability | thin-layer chromatography (preferred for separating copolymers according to chemical heterogeneity) |
| 3. *Kinetic and particle-size effects* | |
| diffusion rate | ultracentrifugation (Section 1.6.) |
| sedimentation rate | |
| migration speed in an electrical field | electrophoresis in the presence of supporting materials, in the form of thin layers or columns |
| gel permeation | gel permeation chromatography on thin layers or in columns |
| gel filtration | with porous inorganic or organic gels with or without the |
| size exclusion | application of pressure |

It is apparent that separation of macromolecules takes place mainly in solution. Several separating effects are usually superimposed in the methods (e.g. solubility — adsorption — diffusion). Frequently, this combination is intentional in order to increase separation efficiency.

Gel permeation chromatography is now the preferred method for analytical fractionation of homopolymers (see Section 2.5.). For fractionation of copolymers and graft polymers, i.e. multi-component systems which differ in polymolecularity, chemical heterogeneity and sequence length distribution, solubility or elution fractionation is of primary importance. It is frequently coupled with gel permeation chromatography. Despite its poor selectivity, solubility fractionation, which depends not only on molecular size but also on chemical composition, stereo-configuration, branching and cross-linking, is still often used. Process parameters such as temperature, pressure, solvent/precipitant combination, etc. can be varied over a wide range making it a convenient experimental technique.

With the increasing availability of commercial liquid chromatographs, more attention will be paid to solubility or elution fractionation in the future.

### 2.1.3. Principles of Solubility Fractionation

All theoretical descriptions of solubility fractionations based on thermodynamics start from the precept of phase separation of the original homogeneous solutions at the relevant critical conditions. A polymer with degree of polymerization, $P_i$, is distributed between two coexisting liquid phases in equilibrium according to:

$$\ln \frac{\Phi'_i}{\Phi''_i} = -\sigma P_i \tag{1}$$

$\Phi'_i$ = volume fraction of the i-meric in the sol-phase, $\Phi''_i$ is the respective volume fraction in the gel-phase

The distribution coefficient $\sigma$, which determines the position of equilibrium is dependent on the polymer itself, its molecular weight, distribution function, concentration and its interaction with the solvent, can be calculated for quasi-binary systems by equation (2) (FLORY-HUGGINS theory [9]):

$$\sigma = \Phi'_B \left(1 - \frac{1}{P'_n}\right) - \Phi''_B \left(1 - \frac{1}{P''_n}\right) + \chi(\Phi'^2_A - \Phi''^2_A) \tag{2}$$

$\Phi_B = \Sigma \Phi_{i,B}$ of the polymer, $\Phi_A$ = volume fraction of the solvent, $\chi$ = interaction parameter

Using the well-known phase stability criteria

$$\left(\frac{\partial \ln a_A}{\partial \Phi_B}\right)_{p,T} = 0; \quad \left(\frac{\partial^2 \ln a_A}{\partial \Phi_B^2}\right)_{p,T} = 0 \tag{3}$$

$a$ = activity,

KONINGSVELD [12] calculated the critical quantities $\Phi_{B,C}$ and $\chi_C$ as

$$\Phi_{B,C} = \frac{1}{1 + P_w \cdot P_z^{-1/2}} \tag{4}$$

$$\chi_C = \frac{1}{2}\left[(1 + P_w^{-1/2})^2 + \frac{(P_z^{1/2} - P_w^{1/2})^2}{P_w \cdot P_z^{1/2}}\right] \tag{5}$$

$P_w$, $P_z$ = weight-average or z-average of the degree of polymerization

The critical quantities are thus dependent on the degree of polymerization and its distribution function $H(P)$ and, in the case of infinite degrees of polymerization, tend towards 0 or 0.5, respectively.

Equation (1) corresponds to the relationship obtained and confirmed experimentally based on the extended BRÖNSTED equation for polymers (SCHULZ [10]):

$$\ln \frac{m_i'}{m_i''} = \ln \gamma - \sigma P_i \tag{6}$$

Where $m_i'/m_i''$ is the mass ratio of the polymer in the two phases and $\gamma$ is the phase volume ratio. From equation (6) separation efficiency, i.e. the quantitative enrichment of the low molecular weight component in the sol phase and of the higher molecular weight component in the gel phase depends largely on $\gamma$ and $\sigma$. Since concentration, molecular weight and pressure dependence of the $\chi$-parameter are neglected, equations (1) and (6) provide only a qualitative description of the separation processes in solubility fractionation. Better agreement between theory and experiment can be achieved for quasibinary systems of isobaric processes by replacing the constant $\chi$-parameter in equation (2) by the expression (KONINGSVELD [10]):

$$\chi = a + \frac{b_{(0)} + \dfrac{b_{(1)}}{P_n}}{1 - c\Phi_B} \tag{7}$$

In this expression, $a$, $b_{(0)}$, $b_{(1)}$ and $c$ are system-specific constants. $b_{(0)}$ includes the temperature dependence of $\chi$ — which is assumed to be inversely proportional to $T$, while $b_{(1)}$ takes into account the molecular weight dependence and $c$ the effect of dilution on the solvent-polymersegment contact number.

The difficulty in determining these constants experimentally and the fact that numerous other relationships are given in the literature, lead one to the conclusion that it is not possible at present to describe solubility fractionation quantitatively even for quasibinary systems. This is even truer of quasiternary solvent precipitant polydisperse polymer systems which are of even greater practical importance.

In view of this, experimental evidence such as that of KLEIN [7, 8] is of great value in the understanding of solubility fractionation and its optimization. According to this work, the separation process in precipitation or solubility fractionation is considered not as a demixing equilibrium but as a solubility equilibrium in which a „saturated" solution is in equilibrium with the „deposit" (gel). According to the general formula:

$$\gamma^* = f(c_B, M_B, T, S/P_r) \tag{8}$$

$\gamma^*$ = weight fraction of $P_r/(S + P_r)$[1])

---

[1]) $S$ is the abbreviation for solvent, $P_r$ for precipitant.

solubility curves of dilute polymer solutions are dependent on concentration, molecular weight, temperature and the solvent and precipitant pair and can be determined experimentally by turbidimetric titration (see also 2.4.). Turbidity curves only indicate the dependence of turbidity temperature and composition on polymer concentration. They thus generally differ from the coexistence and equilibrium curves obtained by investigating the composition of coexistent phases while a variable is changing. KLEIN's approach does not concern itself with the phase composition with respect to polymer, $S$ and $P_r$ but with the more relevant ratio of the amount of polymer in the two phases. The separation characteristics of the polydisperse system can be plotted from the experimental solubility diagram for individual components (Fig. 30) using a CLAESSON diagram [13] (Fig. 31).

This shows the relationship between the saturation concentrations $c_i$ of the individual polymer species and the molecular weight $M_i$ for specified solution

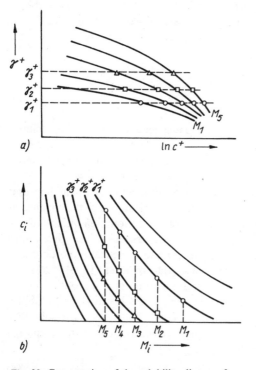

*Fig. 30*: Construction of the solubility diagram from equilibrium curves [7]
a) Equilibrium curves for various polymer species $M_i$ from turbidity titrations (constant $T$), $c^+$ = ideal equilibrium concentration, b) Solubility lines $c_i = f(M_i)$ for constant solution conditions (constant $\gamma$ and $T$)

conditions. The $c_i$-$M_i$-functions are generally called solubility lines. Information on phase separation during fractionation can be obtained from the solubility lines and the molecular weight distribution of the sample.

Phase separation takes place when the saturation concentration is exceeded at the particular molecular weight (see Fig. 31).

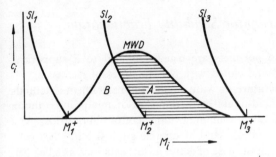

*Fig. 31*: CLAESSON diagram for the graphical treatment of the polymer fractionation
$c_i$ — saturation concentration, $Sl$ — solubility line, $M_i^+$ — critical molecular weight of the polymer solubility, $M_i$ — molecular weight of the component $i$, MWD — molecular weight distribution

At the start of solubility fractionation, the polymer is in the gel phase ($Sl_1$). With decreasing $\gamma^*$ of the $S/P_r$ system or increasing temperature, portion $B$ of the polymer will be dissolved ($Sl_1 \rightarrow Sl_2$). At the start of precipitation fractionation, however, the molecules are present in the dissolved form ($Sl_3$) and portion $A$ is precipitated as gel when the solubility conditions of $Sl_3$ approach $Sl_2$. The comments made so far assume an ideal mixing ratio in which the solubilities of individual components do not influence each other. In real polymer mixtures, however, the equilibrium concentration of the mixture ($c_i$, mixt.) deviates from the ideal equilibrium concentration $c_i^+$ of the individual component. Solubility is reduced by adding a particular component to a polydisperse dilute polymer solution. This reduction is described by an „activity coefficient" $f$ defined through:

$$c_i, \text{mixt.} \cdot f = c_i^+ \qquad (9)$$

In studies of precipitation fractionation, $f$-values between 1.5 ... 2.0 were found in high molecular weight fractions and between 1.3 ... 1.5 in medium $MWD$ [7]. In the case of low molecular weight fractions, the activity coefficients are less than 1.3. Only in highly dilute solutions, obtained by precise control of the solubility fractionation process, is the separation process unaffected by the activity coefficients. Solubility fractionation is thus a more efficient separation technique than precipitation for analytical purposes. The following section, therefore deals only with solubility fractionation.

## 2.1.4. Experimental Conditions for Solubility Fractionation

### 2.1.4.1. Experimental Set-up for Solubility Fractionation

In solubility fractionation, the polymers are usually applied to a supporting material contained in columns and eluted with $S/P_r$-mixtures of decreasing $\gamma^*$-values either at constant temperature or with a temperature gradient. Methods without a support, such as liquid-liquid extraction, are of minor importance. Support materials include perforated aluminium films, cotton fabricstrips, glass spheres with a diameter of 0.1 mm, sea-sand of suitable grain size or silica gels. They must be well purified and, in the case of active adsorbents such as silica gel, conditioned by treatment with dimethyldichlorosilane solutions in n-hexane, drying and then heating to 130 °C to saturate active centres. The use of perforated aluminium film or cotton fabric strips as support can only be recommended if elution is carried out by a shaker-extractor. The best known example of this arrangement is that reported by FUCHS [14]. Where columns are used, glass spheres and sea-sand are preferred as supporting material. The dimensions of the columns depend mainly on the sensitivity of the detectors used for determining the concentration in the fractions. In the classical method with gravimetric determination of concentration following sample precipitation, purification and drying, the dimensions of the columns are:

length: 1.00 ... 1.20 m
internal diameter: 1.0 ... 4.0 cm

Test methods in which the sample is distributed to the entire supporting material are known as solubility or elution fractionations. Methods in which the dissolving zone is limited and adjacent to the migration path are termed elution chromatography. Elution chromatography with temperature gradient and $S/P_r$-gradient introduced by BAKER-WILLIAMS is known as dissolution-precipitation-chromatography [15] because of the continuously repeated dissolution-precipitation-process. The latter can also be carried out on a micro-scale thanks to the high sensitivity of the concentration detectors now available. In spite of the use of columns, elution chromatography is a single stage method because multi stage methods require a phase change repeated several times. The BAKER-WILLIAMS fractionation is a continous solubility fractionation process, which takes place at the column end temperature; the temperature gradient does not influence separation efficiency, although it may improve attainment of equilibrium.

## 2.1.4.2. Optimum Process Parameters for Solubility Fractionation

The separation efficiency of solubility fractionation, i.e. the resolution into fractions that are, as far as possible, molecularly-homogeneous, can be influenced by the following process parameters:
— solvent/precipitant pair
— $\gamma^*$-gradient of the elution mixture
— $\Delta\gamma^*$, step-width of fractionation
— charging of the supporting material
— contact time
— mode of operation (continuous or discontinuous).

### 2.1.4.2.1. Criteria for Selection of Optimum Solvent/Precipitant-Systems ($S/P_r$-Systems)

Suitable $S/P_r$-combinations can be selected according to thermodynamic parameters or turbidimetric titration data. The interaction parameter $\chi$ (equations (2), (5)) is used for thermodynamic classification of solvents. The value for the solvent should be a little less than 0.5 and for the precipitant a little greater than 0.5. Good separation is thus achieved in fractionations close to the $\Theta$-range (see Chapter 4.1.). According to studies by KONINGSVELD [5] best results are obtained when the solvent/polymer $\chi$-parameter is 0.45 and the polymer/precipitant $\chi$-parameter is 0.60. The largest miscibility gap occurs under these conditions. Separation efficiency is also favoured by a positive concentration dependence of the $\chi$-parameter, $\partial\chi/\partial\Phi_B$.

If values of the $\chi$-parameters are not available, the solvent power can be estimated roughly via the solubility parameters using a simplified equation derived from lattice theory

$$\chi_{AB} = \frac{(\delta_A - \delta_B)^2 V_A}{RT} \tag{10}$$

(see Sections 4.2.2. and 4.2.3.). Substituting $\chi_C = 0.5$ for infinitely large molecules at the phase boundary in (10) gives the condition for solubility to occur:

$$(\delta_A - \delta_B)^2 = \frac{RT_c}{2V_A} \tag{11}$$

where $T_c$ = critical temperature, $V_A$ = molecular volume of the solvent.

$\chi$-parameters and solubility parameters can also be taken from Tables [16].

If such information is unavailable, solvents and precipitants may be selected on the basis of turbidimetric titration data. Turbidity curves may be of various shapes depending essentially on molecular weight distribution and the $S/P_r$-combination. These factors are illustrated schematically in Fig. 32 (GRUBER [17]), which shows that the suitability of an $S/P_r$-combination for solubility fractionation increases the flatter the turbidity curve and the steeper the solubility line $Sl$ determined from it.

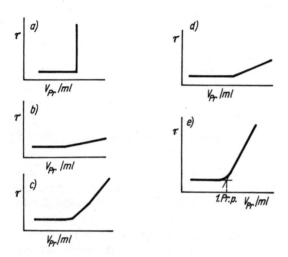

*Fig. 32*: Schematic representation of the behaviour of turbidity curves in case of different molecular weight distribution and polymer — solvent/precipitant-interaction
a — narrow molecular weight distribution, strong precipitant, b — wide molecular weight distribution, weak precipitant, c — wide molecular weight distribution, strong precipitant, d — a kink appears only in case of a relatively high turbidity, e — ideal case, $\tau$ — turbidity

## 2.1.4.2.2. Optimum $\gamma^*$-Gradient of the Elution Mixture

The $\gamma^*$-gradient of the elution mixture must be tailored exactly to the solubility conditions; this can be achieved via the turbidimetric titration curve. Since the amount of polymer should, as far as possible, be distributed equally over all fractions, a logarithmic $\gamma^*$-gradient is unsuitable, although recommended in the literature. The logarithmic $\gamma^*$-gradient is calculated from:

$$c_{P_r} = c_{P_r,\infty} + (c_{P_r,0} - c_{P_r,\infty}) \exp\left(-\frac{v_t}{V}\right) \tag{12}$$

$c_{P_r}$ = precipitant concentration at time $t$, $c_{P_r,\infty}$ = precipitant concentration in the storage vessel, $c_{P_r,0}$ = precipitant concentration in the mixing vessel, $v_t$ = flow rate, $V$ = mix volume

Alogarithmic $\gamma^*$-gradient will not result in good separation in the high molecular weight range because most of the polymer is concentrated in this range (see Fig. 33).

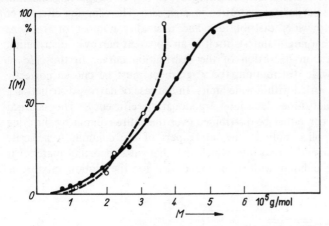

*Fig. 33*: Comparison of the solution fractionation of PS with logarithmic (○) or adapted (●) $\gamma^*$-gradient

## 2.1.4.2.3. Fractionation Step-width

The step-width $\Delta\gamma^*$ is defined as the change of $\gamma^*$ per fraction step. $\Delta\gamma^*$ affects the fractionation separation efficiency greatly. The homogeneity of the fractions increases with decreasing step-width. However, it tends towards a limit at a given concentration. In the main range of separation, $\Delta\gamma^*$ must lie between 0.001 ... 0.002. The number of fractions should range from about 10 to a maximum of 15.

## 2.1.4.2.4. Charging the Supporting Material

Sample loading and charging partially determine the separation efficiency of fractionation. The sample is loaded outside the separation column. For this purpose, a polymer solution of about 1% in a good solvent is mixed with the supporting material and the solvent is slowly evaporated while stirring. During evaporation, pre-fractionation takes place, because as solvent content decreases, the large molecules precipitate first on the surface of the support. Charging is as important to solubility fractionation as the initial concentration is to precipitation fractionation. It must thus be as low as possible. It depends on the distribution width of the *MWD* and must not exceed 2 mg/g of supporting material for narrowly distributed products ($M_w/M_n < 1.5$); in the case of widely distributed products, 4 to 8 mg of sample material can be charged per 1 g of supporting material because of the lower partial concentration in each case. The charging effect is of minor importance to wider distributions and a smaller number of fractions suffices. Excessive charging manifests itself in an apparent narrower distribution or, in extreme cases, in distortion of the distribution curve. In the case of substances with a wide distribution, the $\gamma^*$-gradient must be chosen correctly but charging can be varied within wide limits. In the case of narrow distributions, charging and contact time determine fractionation efficiency. The charged supporting material can either be distributed over the entire column by slurrying in precipitant or applied only to the upper part of the column. A separate column head is frequently used into which the charged supporting material is filled. A separating column with a mixing device for the elution mixture is shown in Fig. 34.

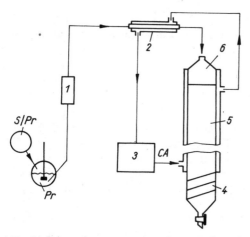

*Fig. 34*: Schematic representation of a separation column for elution fractionation of polymers
*1* — micro-dosing pump, *2* — pre-heater, *3* — thermostat, *4* — sump heating, *5* — column, *6* — sample charging zone, *CA* — circulating agent, $P_r$ — precipitant, *S* — solvent

## 2.1.4.2.5. Contact Time and Mode of Operation

The contact time for establishing solubility equilibrium is dependent on charging. According to Fuchs about 20 min is correct for intermittent operation [14]. According to systematic investigations by Klein, not less than 2 h are needed for fractionating narrowly distributed products because, with an average charge of 1.8 mg/cm² of surface, there is a depth of some 40 to 50 layers of polymers. For the continuous methods now used exclusively in industry, flow rates of 100 to 120 ml/h at normal pressure are optimum although, in general, decreases in separation efficiency have to be accepted for continuous processes. They can only be balanced by a painstaking selection of operating conditions which have to be revised for every modified product. While excessive column flow rates may lead to turbulence, insufficient flow rates reduce separation efficiency because of back-diffusion. Besides the single stage isothermal mode of operation with $S/P_r$-gradient considered so far, the non-isothermal mode with constant $\gamma^*$-value is also of importance [18]. In the latter mode, elution is continued at constant temperature with an $S/P_r$-mixture of constant $\gamma^*$-value until polymer is absent from the eluate. The temperature is then increased and elution is restarted with the same elution mixture until polymer is absent. Sharper fractions are obtained by this multi-step method.

## 2.1.4.3. Processing and Characterization of Fractions

For plotting the distribution functions, the mass fractions $m_i$ and the relevant molecular weights $M_i$ must be determined. As already indicated in Section 2.1.4.1., $m_i$ can be determined gravimetrically after precipitating with excess methanol, purifying and drying or by other methods.

$M_i$ is determined by the methods described in Chapter 1. In the gravimetric determination of $m_i$, polymer solvent retention caused by hindered diffusion at temperatures below the glass point must be taken into account. Solvent retention is dependent on interaction with the polymer and on temperature and layer thickness of the sample. Only polar extracting agents with a high vapour pressure can be used for displacing retained polar solvents. Methanol and acetone with retentions in PAN of less than 0.02% after drying for 30 h at 60 °C and 13.3 Pa are most suitable for this purpose. Classical processing methods are being increasingly replaced by automatic techniques. The RI detector ($RI$ = refractive index) has proved useful for determining the $m_i$ of homopolymers because the refractive index of the mixture is the sum of the indices of the components and is independent of them in the case of high molecular weights. The $S/P_r$-composition in elution fractionation usually changes continuously and must be taken into account by comparison measurement. If absorption coefficients are sufficiently high, UV or IR detectors may be used to determine $m_i$. In our experience, the use of a viscosity detector is recommended for the determination of $M_i$, and the statements in Sections 2.4. apply to the determination of $M_i$ by turbidimetric titration. Use of a scattered light detector for the continuous determination of $M_w$ is described in Section 2.5.

## 2.1.4.4. Separation Efficiency Investigations

Investigations of the separation efficiency, or selectivity of separation, are indispensable for the determination of distribution functions of molecular weight. The simplest and quickest way to obtain information about separation efficiency is to compare turbidimetric titrations of the fractions with the turbidity curve of the non-fractionated product. A steep turbidity curve is indicative of a fraction of a relatively narrow distribution. Absolute values of the molecular inhomogeneity of the fractions cannot be derived from these investigations, however. Such values can only be calculated from values of $M_w$ and $M_n$ which were measured for the fractions. A fraction is well fractionated if $M_w/M_n < 1.05$. Errors in the molecular weight determinations render this criterion rather uncertain for fractionation. Further, we can check whether the relationship

$$[\eta] = \frac{\Sigma\, m_i[\eta_i]}{\Sigma\, m_i} \tag{13}$$

or the HOSEMANN equation [20]:

$$U = U' + \bar{U}_{Fr}(1 + U') \tag{14}$$

are satisfied.

$U^1$) = molecular inhomogeneity $\dfrac{M_w}{M_n} - 1$ of the initial polymer, $\bar{U}_{Fr}$ = mean molecular inhomogeneity of the fractions, $U'$ = molecular inhomogeneity of the initial polymers calculated from the fractions.

For good fractionation $U$ must equal $U'$. A more sophisticated and reliable method for checking separation efficiency is refractionation principally by *GPC*.

## 2.1.4.5. Constructing the Distribution Curve

For plotting *MWD*, the sum of the $m_i$-values is first normalized to unity, i.e., each $m_i$-value is divided by the sum of all weighed amounts or all analogous concentration-proportional quantities. The *MWD* is plotted from available pairs

---

[1]) $U$ is related to the German "Uneinheitlichkeit", which means the variance of the $h(P)$-distribution, divided by $P_n^2$ (see Table 20).

# Molecular Weight Distribution

of values $m_i(M_i)$. Taking into consideration the overlap of individual fractions (see Fig. 35) leads to the integral $MWD$ $I(M)$ according (SCHULZ [10])

$$I_m(M_i) = \frac{m_i}{2} + \sum_{j=1}^{j=i-1} m_j \tag{15}$$

This evaluation method is based on the assumption that the fractions are symmetrically distributed. Another evaluation method, which also takes into account the molecular inhomogeneity of the fractions and leads directly to the differential mass distribution of the molecular weight $H(M)$, was proposed by BEALL [21]. The differential molecular weight distribution can be determined from the integral molecular weight distribution calculated from equation (15) by differentiation (for the mathematically correct designation of $I(M)$ and $H(M)$-functions see Table 20):

$$H(M) = \frac{dI_m(M)}{dM} \tag{16}$$

The latter can be effected graphically or by numerical differentiation [22].

The integral distribution obtained from equation (15) are, strictly speaking step functions. Since $M > M_o$ (molecular weight of the monomer), we can integrate instead of summing, despite the discrete values of $\Delta M$.

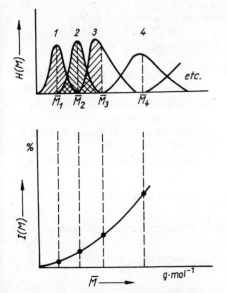

*Fig. 35*: Construction of the integral distribution curve of the molecular weight $I(M_i)$ according to SCHULZ [10], taking into consideration the overlapping of the individual fractions

## 2.1.4.6. Determination of Molecular Inhomogeneity and Polydispersity

The width of the distribution curves, is usually characterized for polymers by the molecular inhomogeneity $U$ (see subscript on page 118):

$$U = \frac{M_w}{M_n} - 1 \quad \text{or} \quad U = \frac{P_w}{P_n} - 1 \tag{17}$$

$P$ = degree of polymerization

Molecular inhomogeneity refers only to the density function of the numerical frequency distribution $h(M)$ and/or $h(P)$. It is related to $H(M)$ by

$$h(M) = \frac{H(M)}{M} \quad \text{or} \quad h(P) = \frac{H(P)}{P} \tag{18}$$

For $H(M)$, a quantity, polydispersity $g_w^2$ analogous to the molecular inhomogeneity, is introduced:

$$g_w^2 = \frac{M_z}{M_w} - 1 \quad \text{or} \quad g_w^2 = \frac{P_z}{P_w} - 1 \tag{19}$$

Table 20: Quantities for evaluating distribution functions of the molecular weight

| Notation | Mathematical expression | Explanation |
|---|---|---|
| $x$ | | random quantity which can adopt the values of $x$ |
| $F(x)$ | $F(x) = \int_0^x f(t)\,dt$ | value of the distribution function of a continuous random quantity (probability that the event $X < x$ will occur) |
| $f(x)$ | $f(x) = dF(x)/dx$ | probability density or density function of $x$ |
| $\mu^{(m)}(x)$ | $\mu^{(m)}(x) = \int_{-\infty}^{+\infty} x^m f(x)\,dx$ | moments of the density function $f(x)$ |
| $\mu^{(0)}(x)$ | $\mu(x) = \int_{-\infty}^{+\infty} \mu^0 f(x)\,dx = 1$ | $0^{th}$ moment, normalizing condition |
| $MX = \mu^{(1)}(x)$ | $\mu^{(1)}(x) = \int_{-\infty}^{+\infty} x f(x)\,dx = \bar{x}$ | predicted value, average value |
| $DX$ | $DX = \int_{-\infty}^{+\infty} (x - \bar{x})^2 f(x)\,dx$ $= \mu^{(2)}(x) - [\mu^{(1)}(x)]^2$ | variance |

Table 20 (continued)

| Notation | Mathematical expression | Explanation |
|---|---|---|
| $\sigma$ | $\sigma = \sqrt{DX}$ | scattering or standard deviation |
| $h(P), h(M)$ | $h(P) = h(M)/M_0$ | density function of the numerical frequency distribution, frequently termed numerical differential molecular weight distribution |
| $H(P), H(M)$ | $H(P) = Ph(P)$ <br> $H(M) = Ph(M)$ | density function of the frequency distribution in terms of mass, frequently termed differential molecular weight distribution |
| $I(P), I(M)$ | $I(P) = \int\limits_0^P H(P)\,dP$ <br> $I(M) = \int\limits_0^M H(M)\,dM$ <br> $H(P) = dI(P)/dP$ | distribution function of the degree of polymerization or of the molecular weight, frequently termed integral molecular weight distribution |
| $\mu_n^{(1)}(P)$ | $\mu_n^{(1)}(P) = \int\limits_0^\infty Ph(P)\,dP = P_n$ | holds only when $\mu_n^{(0)}(P) = 1$, otherwise: <br> $P_n = \mu_n^{(1)}(P)/\mu_n^{(0)}(P)$ |
| $\mu_w^{(1)}(P)$ | $\mu_w^{(1)}(P) = \int\limits_0^\infty PH(P)\,dP = P_w$ | holds only when $\mu_w^{(0)}(P) = 1$, otherwise: <br> $P_w = \mu_w^{(1)}(P)/\mu_w^{(0)}(P)$ <br> N.B.: since $H(P) = Ph(P)$, it is also usual to normalize $\mu_n^{(1)}(P)$ to unity |
| $\mu_n^{(m)}(P)$ <br> $\mu_w^{(m)}(P)$ | $\mu_n^{(m)}(P) = \mu_w^{(m-1)}(P)$ <br> $\mu_w^{(m)}(P) = \mu_n^{(m+1)}(P)$ <br> $\mu_n^{(1)}(P)/\mu_n^{(0)}(P) = P_n$ <br> $\mu_w^{(1)}(P)/\mu_w^{(0)}(P) = P_w$ <br> $\mu_n^{(2)}(P)/\mu_n^{(0)}(P) = P_w P_n$ <br> $\mu_n^{(3)}(P)/\mu_n^{(0)}(P) = P_z P_w P_n$ | general formulae for the averages of molecular weight and relationships between the moments of the density functions $h(P)$ and $H(P)$ |
| $\sigma_n^2$ | $\sigma_n^2 = \mu_n^{(2)}(P) - \{\mu_n^{(1)}(P)\}^2$ <br> $= P_w P_n - P_n^2$ | variance of $h(P)$, <br> $\mu_n^{(0)}(P) = 1$ |
| $U$ | $U = \sigma_n^2/P_n^2 = P_w/P_n - 1$ | molecular inhomogeneity |
| $\sigma_w^2$ | $\sigma_w^2 = \mu_w^{(2)}(P) - \{\mu_w^{(1)}(P)\}^2$ <br> $= P_z P_w - P_w^2$ | variance of $H(P)$, <br> $\mu_w^{(0)}(P) = 1$ |
| $g_w^2$ | $g_w^2 = \sigma_w^2/P_w^2 = P_z/P_w - 1$ | polydispersity |

These two quantities can be calculated easily from fractionation data using the definitions of $M_n$, $M_w$, $M_z$ (see Chapter 1.1.). All quantities for describing and evaluating distribution functions are summarized in Table 20.

## 2.1.5. Distribution Functions of Molecular Weight and Their Mathematical Treatment

The possibility of calculating molecular weight distributions from kinetic data was pointed out in the introduction. Numerous examples of polymerization and polycondensation reactions were compiled by PEEBLES [4] and other authors. Empirical distribution functions are of technological and scientific importance and are frequently the sole way of describing the molecular weight distribution of complex reactions. A summary of the principal distribution functions and their mathematical treatment is given by BERGER [23].

A distinction is made between
— GAUSSIAN distribution in terms of logarithms and
— exponential distributions.

The former are frequently also called logarithmic normal distributions. They can be represented as a general 3-parameter equation:

$$H(P) = \frac{1}{\sigma(2\pi)^{0.5}} \frac{P^z}{P_0^{z+1} \exp\{(z+1)\sigma^2/2\}} \exp\left[-\frac{(\ln P - \ln P_0)^2}{2\sigma^2}\right] \quad (20)$$

$\sigma$ = standard deviation of $\ln P$; i.e. $\sigma^2 = \int_0^\infty (\ln P - \ln P_0)^2 H(P)\,dP$,
$P_0$ = degree of polymerization at $I(P) = 0.5$

With $z = -1$, the WESSLAU distribution is obtained:

$$H(P) = \frac{1}{\sigma(2\pi)^{0.5}} P^{-1} \exp\left[-\frac{(\ln P - \ln P_0)^2}{2\sigma^2}\right] \quad (21)$$

or

$$h(P) = \frac{1}{\sigma(2\pi)^{0.5}} P^{-2} \exp\left[-\frac{(\ln P - \ln P_0)^2}{2\sigma^2}\right] \quad (22)$$

From the moments of the density function, the average of the degree of polymerization are:

$$P_n = P_0 \exp(-\sigma^2/2) \quad (23\,\text{a})$$
$$P_w = P_0 \exp(\sigma^2/2) \quad (23\,\text{b})$$
$$P_z = P_0 \exp(3\sigma^2/2) \quad (23\,\text{c})$$

which lead to:

$$P_0 = (P_n \cdot P_w)^{1/2} \tag{24}$$

and

$$P_w/P_n = P_z/P_w = \exp \sigma^2 \tag{25}$$

The maximum of the $H(P)$-function is about $P_n^{3/2}/P_w^{1/2}$ and that of the $h(P)$-function about $P_n^{5/2}/P_w^{3/2}$. Fig. 36 shows the $h(P)$-function and Fig. 37 the $H(P)$-function of the logarithmic normal distribution for $P_n = 100$ and variable $P_w$-values as a function of $P$.

The $P_0$-values and $\sigma$-values of the logarithmic normal distributions are determined by plotting the fractionation data $I(P)$ against $P$ on a logarithmic sum probability grid. On this grid, the ordinate is divided according to the GAUSSIAN integral and the abscissa logarithmically. The intersection of the straight line of the distribution curve with the straight line for $I(P) = 0.5$ gives the value of $\ln P_0$. At $I(P) = 0.500 \pm 0.341$, $(\ln P - \ln P_0) = \pm \sigma$.

$\sigma$ can also be obtained from the gradient of the $I(P) - P$-curve at $P = P_0$:

$$\sigma = \cfrac{1}{(2\pi)^{0.5} \left( \cfrac{dI(P)}{dP} \right)_{P=P_0}}$$

The WESSLAU distribution is an empirical function suitable for describing broad distributions. It is frequently used for evaluating fractionation data of high-pressure PE (LDPE).

A relationship (equation 26) analogous to equation (21) can be derived for the intrinsic viscosity $[\eta]$ provid the viscosity law $[\eta] = K \cdot M^a$ is satisfied (MUSSA [24]). Then

$$H([\eta]) = \frac{1}{a\sigma(2\pi)^{0.5}} \frac{1}{[\eta]} \exp\left[ -\frac{\ln^2 \frac{[\eta]}{[\eta]_0}}{2(a\sigma)^2} \right] \tag{26}$$

$a$ = exponent of the KUHN-MARK-HOUWINK equation

Again $[\eta]_0$ is obtained from the intersection of the distribution curve with the straight line at $I([\eta]) = 0.5$ and $\sigma$ from the slope of the straight line.

$$\left( \frac{dI([\eta])}{d[\eta]} \right)_{[\eta]=[\eta]_0} = H([\eta]_0) \tag{27}$$

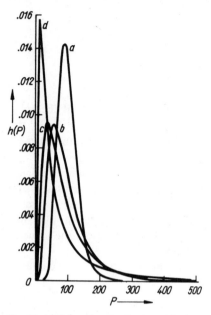

*Fig. 36*: $h(P)$ function for the logarithmic normal distribution as a function of $P$ for $P_n = 100$. $P_w = 110$ (a); 150 (b); 200 (c); 500 (d)

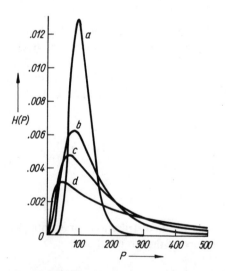

*Fig. 37*: $H(P)$ function for the logarithmic normal distribution as a function of $P$ for $P_n = 100$ and the same $P_w$ values as given in Fig. 36

and

$$a\sigma = \frac{1}{H([\eta]_0)\sqrt{2\pi}\,[\eta]_0} \qquad (28)$$

Setting $z = 0$ in equation (20), the LANSING-KRÄMER distribution is obtained:

$$H(P) = \frac{1}{\sigma(2\pi)^{0.5}} \frac{1}{P_0 \exp\left(\frac{\sigma^2}{2}\right)} \exp\left[-\frac{\ln^2\frac{P}{P_0}}{2\sigma^2}\right] \qquad (29)$$

The most important exponential distribution can be represented in a generalized form by the KOTLIAR function [25]:

$$H(P) = K'P^k e^{-yP^b} \qquad (30\text{a})$$

or

$$h(P) = K'P^{k-1} e^{-yP^b} \qquad (30\text{b})$$

with

$$K' = \frac{(k+1)\,y^{\frac{(k+1)}{b}}}{\Gamma\left[\frac{k+1}{b}+1\right]} \quad \text{(for } \mu_w^{(0)} = 1\text{)} \qquad (30\text{c})$$

and

$$y = \frac{k}{P_{\max}}$$

$k$ = coupling constant, $P_{\max} = P_n$

Equations (30a) and (30b),

for $b = 1;\ k > 0$ give the SCHULZ-ZIMM distribution

and for $b = k + 1;\ k > 0$ give the TUNG distribution.

The following generally holds for the moments of distribution:

$$\mu_n^{(m)}(P) = K' \int_0^\infty P^{k+m-1} \exp(-yP^b)\,dP = \frac{(k+1)\cdot\Gamma\left(\frac{k+m}{b}+1\right)}{(k+m)\cdot y^{\frac{m-1}{b}}\cdot\Gamma\left(\frac{k+1}{b}+1\right)} \qquad (31)$$

$\Gamma$ = gamma function; for $k$ a positive integer $\Gamma(k+1) = k! = k\Gamma(k);\ \Gamma(k) = (k-1)!$

$\mu_n^{(m)}$ reduces to the integral

$$\int_0^\infty x^n e^{-ax} dx = \frac{\Gamma(n+1)}{a^{n+1}} = \frac{n!}{a^{n+1}} \tag{32}$$

With $b = 1$, corresponding to a SCHULZ-ZIMM distribution, a linear function is obtained from equation (30a) by taking logs:

$$\log H(P) = \log K' + k (\log P - P \cdot 0.4343/P_{max}) \tag{33}$$

By plotting the experimentally determined $\log H(P)$-values against $\log P - P \cdot 0.4343/P_{max}$ we obtain, in the ideal case, straight lines with slope $k$, the coupling constant. $k = 1$ corresponds to termination of radical polymerization by disproportionation, $k = 2$ to a recombination termination. $k \approx 1.5$ is frequently found in the case of transfer reactions. The type of chain termination also determines the width of the molecular weight distribution because the molecular inhomogeneity $U$ for the SCHULZ-ZIMM distribution is inversely proportional to $k$ as shown below:

With:

$$h(P) = \frac{\left(\dfrac{k}{P_n}\right)^{k+1}}{k!} P^{k-1} \exp[-kP/P_n] \tag{34}$$

and

$$H(P) = \frac{\left(\dfrac{k}{P_n}\right)^{k+1}}{k!} P^k \exp[-kP/P_n] \tag{35}$$

and from equations (31) and (32):

$$\mu_n^{(m)}(P) = \left(\frac{P_n}{k}\right)^{m-1} \frac{(k+m-1)!}{k!} \tag{36a}$$

$$\mu_w^{(m)}(P) = \left(\frac{P_n}{k}\right)^m \frac{(k+m)!}{k!} \tag{36b}$$

Defining $P_n = \mu_n^{(1)}(P)/\mu_n^{(0)}(P)$ and $P_w = \mu_w^{(1)}(P)/\mu_w^{(0)}(P)$ or $\mu_n^{(2)}(P)/\mu_n^{(1)}(P)$

$$P_n = \frac{P_n \cdot k!}{(k-1)! \, k}; \quad (37a) \quad P_w = P_n \frac{(k+1)}{k}, \tag{37b}$$

and hence
$$U = \frac{P_w}{P_n} - 1 = \frac{k+1}{k} - 1 = \frac{1}{k} \tag{38}$$

Analogously:
$$P_z = P_n \frac{(k+2)}{k} \qquad g_w^2 = \frac{1}{k+1} \tag{39}$$

and
$$P_n : P_w : P_z = 1 : (1+U) : (1+2U) \tag{40}$$

According to SCHULZ, HENRICI-OLIVÉ and OLIVÉ [26]:
$$P_\eta \approx P_n \left(1 + \frac{1+a}{2k}\right) \tag{41}$$

where $a$ = exponent of the KUHN-MARK-HOUWINK equation or more exactly
$$P_\eta = \frac{P_n}{k} \left[\frac{\Gamma(1+k+a)}{\Gamma(1+k)}\right]^{1/a} \tag{42}$$

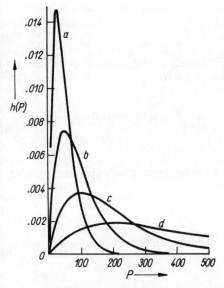

Fig. 38: $h(P)$ function of the SCHULZ-ZIMM distribution as a function of $P$ and with $k = 2$

|       | (a) | (b) | (c) | (d) |
|-------|-----|-----|-----|-----|
| $P_n$ | 50  | 100 | 200 | 400 |
| $P_w$ | 75  | 150 | 300 | 600 |

Figs. 38 and 39 show the $h(P)$ and $H(P)$ functions of the SCHULZ-ZIMM distributions as a function of $P$ with $k = 2$.

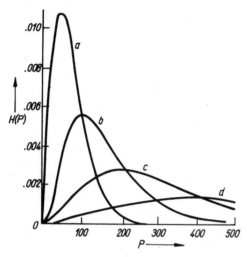

*Fig. 39*: $H(P)$ function of the SCHULZ-ZIMM distribution as a function of $P$ and with $k = 2$; the $P_w$ and $P_n$ values are the same as given in Fig. 38

The SCHULZ-ZIMM distributions are frequently called „normal distributions" because the apply to many polyreactions. Substituting $k = 1$ in equation (34) gives

$$h(P) = \left(\frac{1}{P_n^2}\right) \exp\left(-P/P_n\right) \approx \left(\frac{1}{P_n^2}\right) \cdot \left(1 - \frac{1}{P_n}\right) \tag{43}$$

Replacing $\exp(-k/P_n)$ by $q$ in equation (35) and with $k = 1$ gives the equation derived by SCHULZ:

$$H(P) = (-\ln q)^2 \, Pq^P \tag{44}$$

If $(-\ln q)$ can be approximated as $(1 - q)$ this becomes the 1-parameter SCHULZ-FLORY equation

$$H(P) = Pq^P(1 - q)^2 \tag{45}$$

The latter was derived specifically for polycondensation reactions. In these, $q$ corresponds to the amount of conversion.

POISSON distributions apply mainly to ionically polymerized products:

$$H(P) = \frac{\exp(-P_n) P_n^P P}{P_n P!} \tag{46}$$

or

$$h(P) = \frac{\exp(-P_n) P_n^P}{P_n P!} \tag{47}$$

Summations are required for calculating the moments with the summation index $P$ running from 0 to $\infty$:

$$\mu_n^{(0)}(P) = P_n^{-1} \tag{48a}$$

$$\mu_n^{(1)}(P) = 1 \tag{48b}$$

$$\mu_n^{(2)}(P) = 1 + P_n \tag{48c}$$

and from them

$$P_w = 1 + P_n \tag{49a}$$

as well as

$$U = P_n^{-1} \tag{49b}$$

Since the smallest degree of polymerization is unity, $P$ in equations (46) and (47) strictly expresses the number of addition steps. When the initiator molecule is included in $P$, then summations from $P = 1$ to $\infty$ have to be carried out and $h(P)$ must be replaced by equation (50a) [4].

$$h(P) = \frac{\exp(-P_n) P_n^{P-1}}{P_n(P-1)!} \tag{50a}$$

When calculating the moments and averages, other expressions are obtained,

$$\mu_n^{(1)}(P) = 1 + P_n^{-1} \tag{50b}$$

$$\mu_n^{(2)}(P) = 3 + P_n^{-1} + P_n, \tag{50c}$$

and equations (49a) and (49b) are approximately valid.

In the case of these narrow distributions, there is not normally a great difference between the $h(P)$ function and the $H(P)$ function; the variance $\sigma_n^2$ increases together with increasing $P_n$-value (Fig. 40).

The WEIBULL-TUNG distribution is frequently preferred in practice because of its relatively simple evaluation. It is usually represented in the following form:

$$H(P) = ybP^{b-1} \exp(-yP^b) \tag{51}$$

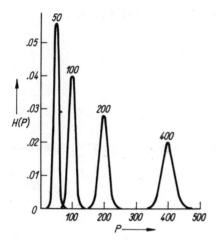

*Fig. 40*: POISSON distributions as a function of $P$ with increasing $P_n$-values

or

$$I(P) = 1 - \exp(-yP^b) \tag{52}$$

The values for $y$ and $b$ are obtained from the slope and intercept, respectively of the plot of

$\ln P$ against $\ln \left[ \ln \dfrac{1}{1 - I(P)} \right]$.

The following expressions also apply to the WEILBULL-TUNG distribution:

$$P_{max} = \left(\frac{1}{y}\right)^{1/b} \left(1 - \frac{1}{b}\right)^{1/b} \tag{53}$$

$$P_n = \left(\frac{1}{y}\right)^{1/b} \left[\Gamma\left(1 - \frac{1}{b}\right)\right]^{-1} \tag{54a}$$

$$P_w = \frac{\Gamma\left(1 + \dfrac{1}{b}\right)}{y^{1/b}}, \tag{54b} \qquad P_z = \frac{\Gamma\left(1 + \dfrac{2}{b}\right)}{y^{1/b} \Gamma\left(1 + \dfrac{1}{b}\right)} \tag{54c}$$

Fig. 41 shows a graphical representation of a comparison of the WEILBULL-TUNG and SCHULZ distribution with the logarithmic normal distribution for variable $P_w$-values.

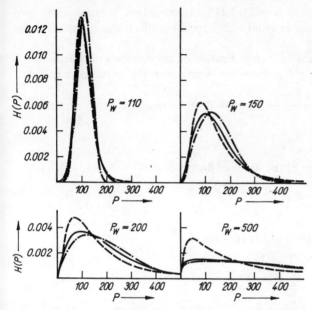

*Fig. 41*: Comparison of the WEIBULL-TUNG (————) and SCHULZ (———) $H(P)$ functions and logarithmic normal distribution (—·—·—) as a function of $P$ and with different $P_w$-values ($P_n = 100$)

## 2.2. Chemical Heterogeneity

### 2.2.1. Introduction

The distribution of the chemical composition of copolymers is called chemical heterogeneity. It is dependent on reactivity parameters and the copolymerization conversion and can be calculated approximately if these are available [9, 27]. The complexity of the non-ideal copolymerization process makes an experimental check of the results indispensable, especially when assessing properties. The properties of copolymers depend on molecular weight, chemical composition and their distributions as well as on the distribution of the sequence lengths. The characterization of copolymers, is therefore, rather complicated. Chemical heterogeneity can be determined by fractionation and subsequent analysis of the fractions, thin-layer chromatography, turbidimetric titration, light-scattering measurements in different solutions, sedimentation analysis using an ultracentrifuge, and NMR spectroscopy. Frequently, the use of the methods is dependent on the polarity of the comonomers or is considerably restricted by the average composition. The most widely utilized method is still solubility fractionation. It is frequently performed, after *GPC* in order to avoid superimposition effects due to the molecular weight distribution.

Three parameters based on a consideration of the moments of distribution of the chemical composition $E$ are used to characterize the constitutional heterogeneity. These are:

the first moment of the mass distribution of the component $i$:

$$\bar{E} \equiv \mu_m^{(1)}(E) = \frac{\Sigma m_i E_i}{\Sigma m_i} \tag{55}$$

the variance as a measure of the width of distribution:

$$\sigma_m^2 \equiv \mu_m^{(2)}(E) - [\mu_m^{(1)}(E)]^2 = \Sigma m_i (E_i - \bar{E})^2 / \Sigma m_i \tag{56}$$

$\bar{E}$ = arithmetic mean

and the skew $\Omega$ as a measure of the symmetry:

$$\Omega_m \equiv \frac{v_m^{(3)}}{2(v_m^{(2)})^{3/2}} = \frac{\dfrac{\Sigma m_i (E_i - \bar{E})^3}{\Sigma m_i}}{2\left\{\dfrac{\Sigma m_i (E_i - \bar{E})^2}{\Sigma m_i}\right\}^{3/2}} \tag{57}$$

When $\Omega_m = 0$, distribution is symmetric. In equations (55) to (57), the $\mu_m$ correspond to the moments of the component distribution and the $v_m$ to the moments of the product distribution. The moment of the product distribution is given by:

$$v_m^{(q)}(E) = \frac{\Sigma m_i (E_i - \bar{E})^q}{\Sigma m_i} \tag{58}$$

$U_1$ and $V_1$ (the division ratio) are further quantities for characterizing the chemical heterogeneity

$$U_1 = U_1^+ + U_1^-$$

$$U_1^+ = \frac{\Sigma m_{i,1}^+ \cdot \Delta E_{i,1}^+}{\Sigma m_{i,1}^+} ; \quad U_1^- = \frac{\Sigma m_{i,1}^- \cdot \Delta E_{i,1}^-}{\Sigma m_{i,1}^-} \tag{59}$$

and

$$V_1 = U_1^+ / U_1^- \tag{60}$$

$V_1$ is again a measure of the asymmetry of the distribution. For calculating $U_1^+$ and $U_1^-$, the distribution function must be divided into strips of equal width.

## 2.2.2. Principles of Solubility Fractionation

For copolymers, the position of the sol-gel equilibrium depends on the degree of polymerization, the specific interactions and on the chemical composition $\alpha$. It is described by the following equation (61) (TOPCHIEV et al. [28]) which is an extension of equation (1):

$$\frac{\Phi'_{B,P,\alpha}}{\Phi''_{B,P,\alpha}} = \exp(-P(\sigma + K\alpha)) \tag{61}$$

$\alpha$ = volume fraction of the monomer $A$ in the copolymer $A - B$ (index $B$), $\sigma$ = distribution coefficient

For a quasibinary copolymer/solvent system $K$ is given by:

$$K = (\Phi''_B - \Phi'_B)(\chi_A - \chi_B) \tag{62}$$

$\chi_{A,B}$ = interaction parameter between the solvent and the component $A$ and/or $B$ of the copolymer.

For the more important (in practice) quasiternary system of copolymer ($B$), solvents 1 and 2:

$$K = (\Phi''_1 - \Phi'_1)(\chi_{1A} - \chi_{1B}) + (\Phi''_2 - \Phi'_2)(\chi_{2A} - \chi_{2B}) \tag{63}$$

A true molecular weight distribution is obtained for systems where $K = 0$ because the chemical composition does not affect the position of equilibrium. To obtain a true distribution of the chemical composition, two different $S/P_r$-systems are generally investigated: firstly a system with a positive $K$-value and then a system with a negative $K$-value. This "two-dimensional" mode of operation is called cross fractionation. It is superior to the one-dimensional technique in which only one $S/P_r$-pair is used to determine heterogeneity [29]. Because of the lack of data on interaction parameters, it is practically impossible to calculate the fractionation of copolymers in advance. The correct solvent must therefore be chosen mainly on the basis of preliminary tests.

## 2.2.3. Selection of $S/P_r$-Systems

Fractionation of copolymers according to molecular size requires the use of $S/P_r$-pairs which interact almost equally with the two co-components. In contrast determination of chemical heterogeneity requires $S/P_r$-pairs which interact to different degrees with the two monomer units. They are usually selected with the help of turbidimetric titration after preliminary consideration of the solubility or

$\chi$-parameters or of the second virial coefficients. The solubility parameters of copolymers are additive like those of solvent mixtures. For copolymers:

$$\delta_{AB} = w_A \delta_A + w_B \delta_B \tag{64}$$

$w$ = weight fraction

and for solvent mixtures:

$$\delta_M = \sum_i \Phi_{S,i} \delta_{S,i} \tag{65}$$

Expanding equations (10) and (11) to take into account the molecular weight dependence and the entropy term of the FLORY-HUGGINS $\chi$-parameter $\chi_{AS}$, the difference in the solubility parameters between copolymer ($B$) and solvent ($S$) at which the polymer will just be dissolved in a solvent is:

$$|\delta_B - \delta_S|_{max} = \left( [0{,}5(1 + P^{-0,5})^2 - \chi_{AS}] \frac{RT}{V_S} \right)^{0,5} \tag{66}$$

For polymers $\chi_{AS}$ is usually between 0.1 and 0.4.

According to equation (66), the difference $|\delta_B - \delta_S|_{max}$ increases with increasing temperature, decreasing degree of polymerization of the polymer and decreasing molar volume of the solvent and/or the mixture. Table 21 contains the $|\delta_B - \delta_S|_{max}$ values calculated by KUHN for a temperature of 25 °C and a solvent volume of 100 ml at different degrees of polymerization and $\chi_{AS}$-values [31]. Table 21 lists the degrees of polymerization corresponding approximately to various $V_B/V_S$ values.

On the other hand, $\Phi_c$ values of the respective homopolymers can be determined by the ELIAS method [30] or turbidimetric titrations can be carried out in molecularly homogeneous fractions of the respective copolymers of different chemical composition.

Table 21: $|\delta_B - \delta_S|_{max}$ values at $T = 25$ °C, $V_S = 100$ cm³ and variable $P$ and $\chi_{AS}$ [31]

| $\chi_{AS}$<br>$P$ | 0.0 | 0.1 | 0.2 | 0.4 |
|---|---|---|---|---|
| 1 | 7.04 | 6.85 | 6.67 | 6.30 |
| $10^1$ | 4.64 | 4.40 | 4.05 | 3.40 |
| $10^2$ | 3.87 | 3.56 | 3.17 | 2.25 |
| $10^3$ | 3.64 | 3.27 | 2.86 | 1.80 |
| $10^4$ | 3.56 | 3.19 | 2.78 | 1.66 |
| $10^5$ | 3.52 | 3.15 | 2.73 | 1.58 |

The $\Phi_c$-values are characteristic of the precipitating power of a precipitant. According to ELIAS [30], the $\Phi_c$-value is related logarithmically to the precipitation point $\Phi_{P_r}$ and the polymer concentration $[P]$ see also 4.1.3. and 4.1.4.2.):

$$\log \Phi_{P_r} = \log \Phi_c - K \log [P] \qquad (67)$$

Suitable $S/P_r$-combinations can be found from the difference of the $\Phi_c$-values of the homopolymers. For large differences between the $\Phi_c$-values, separation depends on the chemical composition. For small $\Delta\Phi_c$-values, separation depends mainly on molecular size.

If copolymers of graduated composition are available, these are dissolved in various solvents and titrated with different precipitants at constant temperature to the first precipitation point (see Fig. 32). The systems with the largest differences in precipitability are selected from the turbidity curves.

## 2.2.4. Experimental Foundations

Fractionation methods for determining chemical heterogeneity correspond to those described in Section 2.1. for determining molecular weight distribution. Besides precipitation fractionation, solubility fractionation (FUCHS [14]) and elution chromatography with appropriate $S/P_r$-systems can be used. When using the cross fractionation method mentioned above, the copolymer solution of about 0.5% is first separated into about 4 fractions by successive additions of precipitant. These are then separated into 4 ... 5 fractions by means of a $S/P_r$-pair producing an opposite effect. For the first $S/P_r$-pair $\chi_A$ must be greater than $\chi_B$ and for the second pair $\chi_A$ must be less than $\chi_B$. Fractionation with demixing liquids [31] is a further method of separating polymer mixtures, such as graft copolymers and copolymers with a wide distribution of chemical composition. Separation is effected by demixing a homogeneous solvent mixture (positive heat of mixing) by lowering the temperature. Providing suitable solvents have been selected on the basis of solubility parameters (equation (66)), second virial coefficients or exponents of the KUHN-MARK-HOUWINK equation (see Section 1.7.2.), the chemically different polymers accumulate in the separated phases. The solvent mixtures must be chosen so that one solvent component is active with respect to the first polymer and inactive towards the other polymer and vice versa. For rapid phase separation the solvent pairs must also have an upper critical miscibility gap and their densities should show a marked difference. These requirements are satisfied, for example, mixtures of N,N-dimethylformamide with cyclohexane, methylcyclohexane, heptane and decalin and also mixtures of methanol with cyclohexane, methylcyclohexane and heptane as well as N,N-dimethyl acetamide heptane, etc. The technique of fractionation with demixing liquids is only suitable for polymers with molecular weights above $10^4$ g/mol. In the case of crystalline polymers, a third phase which can also be separated, is sometimes present.

The mass fractions $m_i$, chemical composition $E_i$ and, if required, the molecular weight $M_i$ of the fractions must be determined. Chemical or spectroscopic methods are the main techniques for determining composition. The mass fractions $m_i$ of the isolated fractions are frequently determined gravimetrically after evaporating the solvent since the detection of mass fractions and composition together involves certain difficulties when using $S/P_r$-gradients which lead to inaccuracies. The $m_i$-values are expressed as mass fractions of 1 and the pairs of value $m_i(E_i)$ are arranged according to increasing $E_i$-values for plotting the integral distribution curve of the chemical composition $I_i(E_i)$. The $I_i$-values are determined in the same way as the molecular weight distributions (equation (15)) and plotted against the $E_i$-values (in percentage by weight).

## 2.2.5. Determination of Chemical Heterogeneity by Thin-Layer Chromatography (TLC)

Another method of determining the heterogeneity of copolymers is thin-layer chromatography because adsorption is far more dependent upon chemical composition than on molecular weight. Generally, thin-layer chromatography (*TLC*) of polymers differs only slightly from that of low molecular weight compounds. The principles and experimental methods are described by STAHL [33]. In polymer characterization, use is made of polarity-controlled adsorption and of solubility-controlled phase-separation as well as of the molecular-sieve effect. Depending on experimental conditions, a more or less intense superposition of the individual effects will occur. Thin-layer chromatographic separation of copolymers requires a solvent with which the polymer interacts at least moderately. Usually a gradient technique with solvent mixtures is used. For separating a slightly polar copolymer, a slightly polar elution mixture and a highly active supporting material should be used and vice versa. A quantitative description of the adsorption equilibrium is given by the SNYDER equation [34]:

$$\log K^0 = \log V_a + \alpha(S^0 - A_s \varepsilon^0) \tag{68}$$

$K^0$ = distribution coefficient between adsorbed and nonadsorbed molecules, $V_a$ = specific constant for the interaction between elution agent and supporting material, $\alpha$ indicates the surface activity of the supporting material, $S^0$ and $A_s \varepsilon^0$ are dimensionless quantities which describe the adsorption energy of the polymer and the eluting agent, $A_s$ corresponds to the surface area occupied by the adsorbed molecules

$K^0$ is related to the well-known $R_f$-value by

$$R_f = [(1 + W/V^0) K^0]^{-1} \tag{69}$$

$W$ = weight of the stationary phase, $V^0$ = volume of the mobile phase

($S^0$, $A_s$ and $\varepsilon^0$ values are quoted by SNYDER as a function of solvents and the structure of low molecular weight substances. They may be used as criteria for selecting experimental conditions for polymers [32].)

For the separation of copolymers whose monomers differ greatly in polarity, elution mixtures of a non-polar and a moderately to highly polar solvent are used. Activated silica gel of 0.25 mm layer thickness is generally used as a supporting material. For polymers with molecular weights below $10^5$ g/mol, microporous silica gel is preferred. The amount of copolymers should not exceed 10 µg in order to minimize the influence of dissolution rate. In order to establish the elution gradient, a solvent or solvent mixture is fed into the *TLC* chamber and, after the eluting agent has reached a certain front, the other eluting agent is continuously added dropwise. Depending on the rate of addition, a linear, concave or convex plot is obtained. Separation of weakly polar copolymers requires the use of weakly polar eluting agents. In addition, they must be selected according to polarity and solubility. Suitable eluting agents for the determination of the heterogeneity of copolymers are listed in Table 22.

*Table 22*: Mixtures of eluting agents for determining the heterogeneity of polymers by TLC

| | |
|---|---|
| Styrene-methyl acrylate | ($CCl_4$ + methyl acetate) + methyl acetate |
| Styrene-methyl methacrylate (block copolymer) | $CCl_4$ + methyl ethyl ketone |
| Styrene-acrylonitrile | tetrachloroethane + ethyl acetate |
| Styrene-butadiene | $CCl_4$ + $CHCl_3$ |
| Cellulose acetate | (methylene chloride + n-butanol or methanol) or acetone + ethyl acetate |
| Cellulose nitrate | $CHCl_3$ + ethyl acetate |

For quantitative evaluation, the *TLC* plates can be sprayed with reagents such methanolic iodine solution to render the spots visible and the areas then measured. Usually a fluorescence spectrometer or other *TLC* scanning-spectrodensitometer is used in which the fluorescence intensities or specific absorptions are measured in ultraviolet or visible light. If the individual comonomers absorb at different wavelengths, calibration with copolymers of known composition is not required since different wavelengths can be utilized. Despite progress in the development of *TLC* evaluation equipment, quantitative *TLC* still involves some errors. Determination of heterogeneity by *TLC* is thus still somewhat uncertain.

The rapidity of the method make it particularly suitable for comparing some copolymers but it is not recommended for partially crystalline polymers.

## 2.3. Examples

### 2.3.1. Determination of the Integral Distribution Function $I(M)$ of PVC by Solubility Fractionation (FUCHS Method) [35]

The product is first freed from polymerization auxiliaries by extraction. The product is dissolved in a tetrahydrofuran ($THF$)/water mixture, applied to cotton fabric strips and fractionated by intermittent shaking extraction. After precipitating the product from the fractions, $m_i$ is determined gravimetrically. The molecular weights of the fractions are determined viscosimetrically and the $I(M)$-curve is calculated from the values of $m_i(M_i)$ using equation (15). From the fractionation data, the molecular inhomogeneity $U$ and the polydispersity $g_w^2$ are calculated (equations (17) and (19)).

Impurities and polymerization auxiliaries are removed from the polymer by shaking extraction repeated three times (each for 1 h) using a 1:1 methanol-water mixture. The polymer is then dried in vacuo at 70 °C to constant weight. 5.00 g of the material pre-treated in this way is dissolved in 100 ml of $THF$ with continuous shaking at 25 °C. The supporting material, purified cotton fabric pieces (1.5 × 3.0 cm), is soaked in the PVC solution in a 250 ml evaporating dish

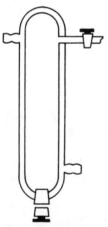

*Fig. 42*: Fractionating vessel (FUCHS [14])

and the *THF* is evaporated slowly. The material is thus pre-fractionated. The slightly moist fabric strips are put into the fractionating vessel (Fig. 42) which is connected to a thermostat and placed on a mechanical shaker.

The PVC on the supporting material is determined from the residual moisture content of a few fabric strips. $S/P_r$-mixtures for fractionation are indicated in Table 23. Fractionating time is 15 min per fraction and the working temperature is $25 \pm 0.05\ °C$. When extraction is complete, the fractions are put into evaporating dishes and mixed with excess methanol. The precipitated polymer is purified by repeated decanting or filtering and washing in methanol. It is then rinsed in weighed evaporating dishes or glass filter crucibles, dried in a vacuum oven at 70 °C to constant weight. The molecular weight is determined by measuring the viscosity of 0.1 per cent solutions in cyclohexanone in an UBBELHODE viscometer with capillary 1 at $25 \pm 0.05\ °C$. The limiting viscosity is calculated from the $\eta_{sp}$ values using a SCHULZ-BLASCHKE constant of 0.28 (see Section 1.7.2.).

$$[\eta] = \frac{\eta_{sp}/c_B}{1 + 0.28\eta_{sp}} \tag{70}$$

Before measurements are taken PVC solutions must be heated to 80 °C to destroy associates. The associates do not reform on cooling.

The molecular weights can be calculated using the BOHDANECKY $[\eta]M_w$ relationship [36]:

$$[\eta] = 1.38 \cdot 10^{-2} M_w^{0.78} \qquad [\eta] \text{ in } cm^3/g \tag{71}$$

In the case of broader fractions, the influence of molecular inhomogeneity upon the intrinsic viscosity must be considered. For this purpose, the correction

*Table 23*: Composition of the $S/P_r$-mixture

| Fraction | Water (ml) | Tetrahydrofuran (ml) |
|---|---|---|
| 1 | 20.0 | 80.0 |
| 2 | 15.0 | 85.0 |
| 3 | 14.0 | 86.0 |
| 4 | 12.0 | 88.0 |
| 5 | 10.0 | 90.0 |
| 6 | 8.5 | 91.5 |
| 7 | 7.5 | 92.5 |
| 8 | 6.0 | 94.0 |
| 9 | 4.0 | 96.0 |
| 10 | 2.0 | 98.0 |
| 11 | 1.0 | 99.0 |
| 12 | 0.0 | 100.0 |

factors for the number-average or weight-average $q_n$ and $q_w$ are calculated from the statistical moments (see Section 1.7.2. [37]). The measured values are tabulated (see Table 24). From data measured for an industrial emulsion PVC the $I(M_w)$

Table 24: Data for the solubility fractionation of industrial emulsion PVC (double determination)

$\dfrac{M_w}{M_n}$ of the initial sample = $\begin{matrix} 83000 \pm 5\% \text{ (g/mol)} \\ 45000 \pm 5\% \text{ (g/mol)} \end{matrix}$

| Fraction No. | $m_i$ (g) | | $m_i$ (%) | | $[\eta]$ (cm³·g⁻¹) | | $M_w \times 10^{-3}$ (gmol⁻¹) | |
|---|---|---|---|---|---|---|---|---|
| | 1. | 2. | 1. | 2. | 1. | 2. | 1. | 2. |
| 1 | 0.078 | 0.092 | 3.5 | 4.2 | 24.2 | 24.1 | 14.4 | 14.3 |
| 2 | 0.176 | 0.169 | 7.9 | 7.7 | 26.0 | 27.0 | 15.8 | 16.6 |
| 3 | 0.112 | 0.100 | 5.1 | 4.6 | 35.2 | 35.3 | 23.3 | 23.4 |
| 4 | 0.140 | 0.162 | 6.3 | 7.4 | 47.0 | 45.0 | 33.8 | 31.9 |
| 5 | 0.250 | 0.227 | 11.2 | 10.4 | 61.0 | 61.0 | 47.2 | 47.2 |
| 6 | 0.393 | 0.397 | 17.7 | 18.1 | 90.0 | 88.0 | 78.0 | 75.4 |
| 7 | 0.328 | 0.315 | 14.8 | 14.4 | 107.0 | 105.0 | 97.0 | 94.6 |
| 8 | 0.257 | 0.249 | 11.6 | 11.4 | 130.0 | 130.0 | 124.0 | 124.0 |
| 9 | 0.200 | 0.198 | 9.0 | 9.0 | 132.0 | 136.0 | 127.0 | 138.0 |
| 10 | 0.184 | 0.180 | 8.3 | 8.2 | 135.0 | 135.0 | 131.0 | 133.0 |
| 11 | 0.103 | 0.100 | 4.6 | 4.6 | 139.0 | 139.0 | 136.0 | 136.0 |

$\sum_{i=1}^{i=11} m_i M_i \cdot 10^{-6} = \begin{matrix} 8.12 \\ 8.15 \end{matrix}$   $\sum_{i=1}^{i=11} m_i/M_i \cdot 10^4 = \begin{matrix} 20.044 \\ 20.845 \end{matrix}$   $\sum_{i=1}^{i=11} m_i M_i^2 \cdot 10^{-11} = \begin{matrix} 8.473 \\ 8.439 \end{matrix}$

| Fraction No. | $I(M_w)$ (%) | | $m_i M_i \times 10^{-4}$ (g·mol⁻¹) | | $m_i/M_i \times 10^4$ (mol·g⁻¹) | | $m_i M_i^2 \times 10^{-8}$ (g²·mol⁻²) | |
|---|---|---|---|---|---|---|---|---|
| | 1. | 2. | 1. | 2. | 1. | 2. | 1. | 2. |
| 1 | 1.7 | 2.1 | 5.04 | 6.00 | 2.43 | 2.94 | 7.26 | 8.6 |
| 2 | 7.5 | 8.0 | 12.48 | 12.78 | 4.76 | 4.64 | 19.72 | 20.2 |
| 3 | 13.9 | 14.2 | 11.88 | 10.76 | 2.19 | 1.97 | 27.68 | 25.2 |
| 4 | 19.6 | 20.2 | 18.90 | 23.60 | 1.86 | 2.32 | 63.88 | 75.3 |
| 5 | 28.4 | 29.1 | 52.86 | 49.08 | 2.37 | 2.50 | 249.5 | 231.7 |
| 6 | 42.8 | 43.3 | 138.06 | 136.47 | 2.27 | 2.40 | 1076.8 | 1029.2 |
| 7 | 59.1 | 59.6 | 143.56 | 136.22 | 1.55 | 1.55 | 1392.9 | 1288.6 |
| 8 | 72.3 | 72.5 | 143.84 | 141.36 | 0.935 | 0.92 | 1783.3 | 1753.4 |
| 9 | 82.6 | 82.7 | 114.30 | 124.20 | 0.708 | 0.652 | 1577.3 | 1714.0 |
| 10 | 91.2 | 91.3 | 108.73 | 109.06 | 0.633 | 0.616 | 1424.3 | 1451.0 |
| 11 | 97.7 | 97.7 | 62.56 | 62.56 | 0.338 | 0.338 | 850.8 | 850.8 |

curve shown in Fig. 43 was obtained together with the following values:

$M_w$ = 81 200 g/mol   $g_w^2$ = 0.28
     = 81 500 g/mol         0.27
$M_n$ = 49 900 g/mol   $U$ = 0.63
     = 48 000 g/mol          0.70
$M_z$ = 104 000 g/mol
     = 103 500 g/mol

$M_w$ of the initial sample was found to be 83 000 g/mol by light scattering measurement while $M_n$ was determined as 45 000 g/mol by membrane osmosis. The $M_w$-values obtained from fractionation agree well with this but the $M_n$-values show greater deviations. Apart from the errors involved in determining $M_n$, such deviations are frequently caused by difficulties in preparing the low molecular weight fractions which exert a great influence on the $M_n$ value. Sharper fractionation can be achieved by column elution fractionation. The discontinuity of the integral distribution curve in Fig. 43 indicates the presence of a mixture of polymers of different distributions.

## 2.3.2. Determination of Chemical Heterogeneity of VC-Methyl Acrylate (VCMA) Copolymers by Column Elution Fractionation with $S/P_r$-gradients

The heterogeneity distribution of *VCMA* copolymers can be determined by elution chromatography with the cyclohexanone/ethylene glycol $S/P_r$-system using columns filled with ballotini at an operating temperature of 90 ... 100 °C, and subsequent determination of the composition of the fractions. The $S/P_r$-system was chosen by reference to the $\Phi_c$ values [38]; this selection depends not only on the copolymer composition but also on the fractionation temperature.
Working conditions:

| | |
|---|---|
| Column length | 120 cm |
| Internal diameter | 4 cm |
| External diameter | 7 cm |
| Column capacity | 1500 cm³ |
| Column head for feeding the sample internal diameter | 4 cm |

| | |
|---|---|
| Volume of the column head | 300 cm³ |
| Support | ballotini ⌀ 0.1 mm |
| Mixture volume ($S/P_r$) | 1500 cm³ |
| Flow rate | 100 ... 120 cm³/h |
| Heating liquid | ethylene glycol |
| Fractionation temperature | 90 °C |
| at the column end | 95 °C |
| Sample size | 1 ... 1.5 g |

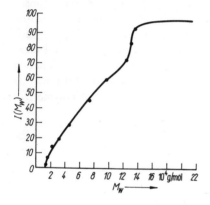

*Fig. 43*: $I(M_w)$-curve of an industrial emulsion PVC determined from solubility fractionation data (see Table 24)

An exponential $S/P_r$-gradient is selected for determining the heterogeneity of the *VCMA* copolymer. The initial $S/P_r$-mixture contains 30 parts of cyclohexanone and 70 parts of ethylene glycol. The polymer sample is dissolved in cyclohexanone to give an approximately 5% solution by shaking at 90 °C. This solution is mixed with the support. After slowly evaporating the solvent it is fed into the column head with the first elution mixture at 50 °C. The head is then connected to the separation column. The column is heated and fractionation is started. About 15 fractions of 100 ml each are taken off and the column is eluted once more with pure cyclohexanone. The fractions are precipitated in methanol (fivefold excess), washed with methanol and dried in a vacuum oven at 50 °C to constant weight. The MA content is determined by saponification. For this purpose, 2.0 ml of 0.1 N ethanolic KOH is added to the sample and left for 3 h at room temperature. The excess KOH is back titrated with 0.1 N $H_2SO_4$. The MA content is calculated from the $H_2SO_4$ consumption $(A-B)$ for the sample $(B)$ and blank $(A)$. Further, saponified chlorine $(C)$ must be determined by argentometric titration and subtracted from $(A-B)$. The content of MA by % weight is then given by

$$\text{MA by \% weight} = \frac{(A-B-C) \cdot 0.86}{\text{sample weight (g)}} \tag{72}$$

Equation (72) is correct for 0.1 N $H_2SO_4$. If necessary, the fractions are arranged in order of increasing MA values and $I(E)$ is calculated by the method for calculating the molecular weight distribution described earlier. The efficiency of separation is checked by turbidimetric titration and refractionation.

Fig. 44 shows the results for the 7th fraction of the *VCMA* copolymer sample. The relevant fractionation data are shown in Table 25 and Fig. 45.

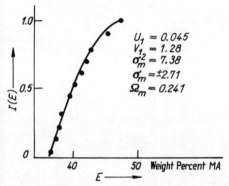

*Fig. 44*: Result from the refractionation of the 7th fraction of a VCMA copolymer

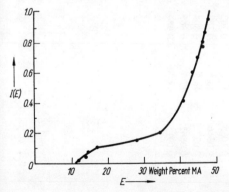

*Fig. 45*: Heterogeneity distribution of a VCMA copolymer with 39.4 weight percent of MA obtained by elution fractionation with cyclohexanone/ethylene glycol (see Table 25)

Table 25: Data for the determination of the chemical heterogeneity of a VCMA copolymer with 39.4 weight % of MA (average from 3 individual measurements)

| Fraction No. | $m_i$ (g) related to 1 g | $E_i$ weight % of MA | $I(E_i)$ | $m_i E_i$ | $m_i(E_i - \bar{E})^2$ | $m_i(E_i - \bar{E})^3$ | $m_{i,1}^+ \Delta E_{i,1}^+ \cdot 10^2$ | $m_{i,1}^- \Delta \bar{E}_{i,1} \cdot 10^2$ |
|---|---|---|---|---|---|---|---|---|
| 1  | 0.0270 | 12.0 | 0.014 | 0.324  | 20.27 | −555.4 |      | 0.74 |
| 2  | 0.0341 | 14.0 | 0.044 | 0.477  | 22.00 | −558.8 |      | 0.87 |
| 3  | 0.0282 | 14.5 | 0.075 | 0.409  | 17.48 | −435.4 |      | 0.70 |
| 4  | 0.0343 | 17.0 | 0.106 | 0.583  | 17.21 | −385.5 |      | 0.77 |
| 5  | 0.0490 | 28.0 | 0.148 | 1.372  | 6.37  | −72.6  |      | 0.56 |
| 6  | 0.0618 | 35.0 | 0.203 | 2.163  | 1.20  | −5.3   |      | 0.27 |
| 7  | 0.3141 | 43.0 | 0.391 | 13.506 | 4.07  | 14.65  |      |      |
| 8  | 0.1015 | 43.5 | 0.599 | 4.415  | 1.71  | 7.00   | 1.13 |      |
| 9  | 0.1035 | 45.0 | 0.702 | 4.658  | 3.25  | 18.2   | 0.42 |      |
| 10 | 0.0316 | 46.5 | 0.769 | 1.469  | 1.59  | 11.3   | 0.58 |      |
| 11 | 0.0372 | 46.5 | 0.804 | 1.730  | 1.87  | 13.3   | 0.22 |      |
| 12 | 0.0862 | 47.0 | 0.865 | 3.882  | 4.98  | 37.8   | 0.26 |      |
| 13 | 0.0915 | 48.0 | 0.954 | 4.392  | 6.77  | 58.2   | 0.66 |      |
|    |        |      |       |        |       |        | 0.79 |      |

From these, the following characteristic chemical heterogeneity parameters are calculated using equations (55)—(57), (59), (60):

$\bar{E} = 39.4\%$
$\sigma_m^2 = 108.8$
$\sigma_m = \pm 10.4$
$\Omega_m = -0.82$
$U = 0.220$
$U_1^+ = 0.053$
$U_1^- = 0.167$
$V_1 = U_1^+/U_1^- = 0.32$

## 2.3.3. Determination of the Distribution Functions of a HDPE from Elution Fractionation Data

The PE samples are fractionated by continuous elution fractionation through a glass column filled with ballotini at 130 °C, using decalin as solvent and 2-ethylhexanol as precipitant. Solvent and precipitant are purified prior to use by vacuum distillation with a VIGREUX column, 1.2 m long.

Specifications for elution fractionation:

| | |
|---|---|
| Column length | 120 cm |
| Column diameter | 34 mm |
| Column capacity | 1.1 $l$ |
| Exclusion volume | 220 cm³ |
| Support | ballotini 0.1 mm in diameter |
| Elution rate | 80 to 100 m$l$/h |
| Fractionation temperature | 130 °C |
| Sample size | 1.5 g |

The column is heated by a thermostat with ethylene gylcol as circulating agent and a resistance heater attached to the column sump. The well insulated separation column maintains the temperature to $\pm 0.2$ °C. The $S/P_r$-mixture is fed by a micro-dosing pump from the stirred mixing vessel (capacity 2$l$) through a reflux condenser, which operates as a preheater, to the top of the column. The solvent is fed into the mixing vessel from a dropping funnel. A 2 per cent solution of the sample is suched in with the heating switched off. Subsequently, the decalin is displaced by the precipitant at room temperature. The column is then heated to 100 °C and then in steps up to 130 °C, thus obtaining the first fractions. The process is continued at 130 °C with an exponential $S/P_r$-gradient.

Table 26: Data for the elution fractionation of HDPE (averages of 3 individual measurements)

| Fraction No. | $m_i$ (g) relative to 1 g of substance | $[\eta]$ (dl/g) | $I([\eta]_i)$ | $\log [\eta]$ | $m_i \cdot [\eta]_i$ (dl) | $\dfrac{m_i}{[\eta]_i} \left(\dfrac{g^2}{dl}\right)$ | $\dfrac{1}{1 - I([\eta])}$ | $\log \log \{1/[1 - I([\eta])]\}$ |
|---|---|---|---|---|---|---|---|---|
| 1 | 0.131 | 0.237 | 0.066 | −0.6255 | 0.0310 | 0.5527 | 1.07 | −1.53 |
| 2 | 0.074 | 0.364 | 0.168 | −0.4389 | 0.0269 | 0.2033 | 1.20 | −1.10 |
| 3 | 0.139 | 0.509 | 0.275 | −0.2932 | 0.0706 | 0.2730 | 1.38 | −0.85 |
| 4 | 0.131 | 0.664 | 0.409 | −0.1778 | 0.0870 | 0.1972 | 1.69 | −0.64 |
| 5 | 0.127 | 0.797 | 0.538 | −0.0985 | 0.1012 | 0.1593 | 2.16 | −0.48 |
| 6 | 0.123 | 0.971 | 0.663 | −0.0130 | 0.1194 | 0.1267 | 2.97 | −0.33 |
| 7 | 0.103 | 1.182 | 0.776 | 0.0726 | 0.1217 | 0.0871 | 4.46 | −0.19 |
| 8 | 0.120 | 1.532 | 0.888 | 0.1852 | 0.1838 | 0.0783 | 8.93 | −0.02 |
| 9 | 0.052 | 2.425 | 0.974 | 0.3512 | 0.1167 | 0.0232 | 38.5 | 0.20 |

# Examples

10 to 12 fractions are taken off. Oxidative degradation is avoided by the use of an argon atmosphere and addition of 0.1% of phenyl-beta-naphthylamine as antioxidant. The polymer is separated by precipitating with acetone, purified, dried and weighed.

Table 26 contains the measured values. Because of the difficulty in determining the absolute molecular weight of PE, only the limiting viscosities are included in the evaluation. They are calculated from measurements in decalin at 135 °C at a concentration of 0.1 g/dl using the SCHULZ-BLASCHKE equation. The molecular inhomogeneity is thus also obtained from the $[\eta]$-averages using:

$$[\eta]_w = \frac{\Sigma m_i [\eta]_i}{\Sigma m_i}, \qquad [\eta]_n = \frac{\Sigma m_i}{\Sigma m_i/[\eta]_i} \tag{73}$$

$$U = [\eta]_w/[\eta]_n - 1 \tag{74}$$

This procedure presupposes the validity of the KUHN-MARK-HOUWINK equation with constant magnitudes of $K$ and $a$ over the whole range of molecular weights.

Using the fractionation data shown in Table 26, equation (73) and (74) lead to:

$[\eta]_w = 0.8583$ (dl/g)
$[\eta]_n = 0.5879$ (dl/g)
$U = 0.46$

Comparison of the $[\eta]_w$ obtained from the fractionation data with the $[\eta]$ value of 0.861 (l/g) determined for the initial sample, leads to the conclusion that the separation efficiency is satisfactory.

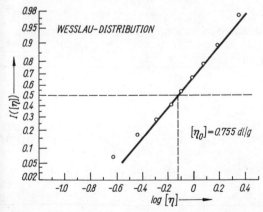

*Fig. 46*: WESSLAU plot of fractionation data from the elution fractionation of HDPE (see Table 26)

In the WESSLAU method of evaluation, $I([\eta]_i)$ is first plotted against $\log [\eta]_i$ on a sum probability grid (Fig. 46) and then against $[\eta]_i$ using normal coordinates (Fig. 47). Fig. 46 shows that the value for $[\eta]_0$ is obtained from the intersection of the straight line of the distribution curve with the straight line $I([\eta]_i) = 0.5$ and is equal to 0.755. $H([\eta]_0)$ is calculated using equation (27) from the slope of the tangent to the fractionation curve at $I([\eta]_i) = 0.5$. The value for a $\sigma$ is obtained from the same slope using equation (28):

$$a\sigma = \frac{1}{0.913\sqrt{2\pi}\,0.755} = 0.58$$

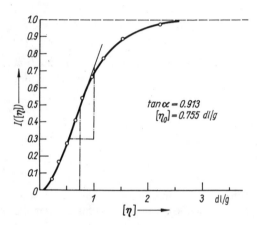

Fig. 47: $I[\eta]$ as a function of $[\eta]$ obtained from elution fractionation data for an HDPE (see Table 26)

The $H(P)$ curve can now be calculated (equation (26)) together with quantities such as $P_n$, $P_w$, $P_z$.

Analogously to equations (23a, b):

$$[\eta]_w = [\eta]_0 \exp\left[\frac{(a\sigma)^2}{2}\right]$$

$$= 0.755 \exp\left[\frac{(0.58)^2}{2}\right] = 0.89 \tag{75}$$

$$[\eta]_n = [\eta]_0 \exp\left[\frac{-(a\sigma)^2}{2}\right]$$

$$= 0.755 \exp\left[\frac{-(0.58)^2}{2}\right] = 0.64 \tag{76}$$

and thus

$$U = \exp(0.58)^2 - 1 = 0.40$$

In order to determine the parameters $y$ and $b$ of the TUNG distribution (equations (51) and (52)), $P$ must again be replaced by $\left[\dfrac{1}{M_0}\left(\dfrac{[\eta]}{K}\right)^{1/a}\right]$ because absolute molecular weights of the fractions are unknown. For this purpose, it is assumed that the KUHN-MARK-HOUWINK equation is valid.

Thus:

$$I[\eta] = 1 - \exp\left\{-\frac{y}{M_0^b}\left(\frac{[\eta]}{K}\right)^{b/a}\right\} \tag{77}$$

and

$$\log\log\frac{1}{[1 - I([\eta])]} = \frac{b}{a}\log\frac{[\eta]}{K} + \log\left(\frac{y}{M_0^b}\log e\right) \tag{78}$$

so that the slope is

$$\tan\alpha' = \frac{b}{a}$$

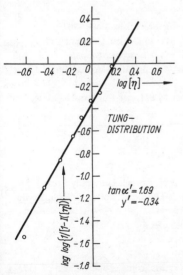

Fig. 48: TUNG plot of the fractionation data of Table 26

and the intercept $y'$ corresponds to

$$y' = -\frac{b}{a} \log K + \log \left( \frac{y}{M_0^b} \log e \right)$$

From Fig. 48:

$\tan \alpha = 1.69$

$y' = -0.34$

and, hence, $a = 0.74$ for

$b = 1.25$

For $K = 3.9 \cdot 10^{-4}$ and $M_0 = 28$

$$\log y = y' + \frac{b}{a} \log K + b \log M_0 - \log \log e$$

$\quad = -0.34 + 1.69 (0.591 - 4) + 1.25 \cdot 1.447 + 0.362$

$y = 1.17 \cdot 10^{-4}$

All further quantities can now be calculated using equations (51), (53—54).

# Bibliography

[1] CANTOW, M. J. R., „Polymer Fractionation", Academic Press New York 1967
[2] TUNG, H. L., „Fractionation of synthetic polymers", Marcel Dekker New York 1977
[3] GLÖCKNER, G., „Polymercharakterisierung durch Flüssigkeitschromatographie", VEB Deutscher Verlag der Wissenschaften Berlin 1980 and Dr. Alfred Hüthig-Verlag-GmbH Heidelberg 1982
[4] PEEBLES, L. H., „Molecular Distributions in Polymers", Polymer Reviews 18 (1971), John Wiley & Sons, Inc.
[5] KONINGSVELD, R., Fortschr. Hochpolymeren-Forsch. 7 (1970—71), 1
[6] WOLF, B. A., Fortschr. Hochpolymeren-Forsch. 10 (1972), 109
[7] KLEIN, J., Angew. Makromolekulare Chem. 10 (1970), 109
[8] KLEIN, J. and G. WEINHOLD, Angew. Makromolekulare Chem. 10 (1970), 49
[9] FLORY, P. J., „Principles of Polymer Chemistry", Ithaca: Cornell University Press 1953
[10] SCHULZ, G. V. in H. A. STUART, „Das Makromolekül in Lösungen", Springer-Verlag Berlin—Göttingen—Heidelberg 1953
[11] KONINGSVELD, R. and L. A. KLEINTJENS, Macromolecules 4 (1971), 637
[12] KONINGSVELD, R., J. Polymer Sci. A 2 6 (1968), 325
[13] CLAESSON, S., J. Polymer Sci. 16 (1955), 193
[14] FUCHS, O., Makromolekulare Chem. 5 (1950), 245; Makromolekulare Chem. 7 (1952), 259

[15] BAKER, G. A. and R. P. J. WILLIAMS, J. chem. Soc. London, (1956), 2352
[16] BRANDRUP, J. and IMMERGUT, E. H., „Polymer HANDBOOK" 2nd edition, Wiley — Interscience, New York 1975 IV — 115
[17] GRUBER, U., Thesis ETH Zürich 1964
[18] KLEIN, J. and M. WERNER, Makromolekulare Chem. **179** (1978), 475
[19] GLÖCKNER, G., KAUFMANN, W., ABDELLATIF, E. and S. RIEDEL, Faserforsch. u. Textiltechn. — J. Polymere **26** (1975), 606
[20] HOSEMANN, R. and W. SCHRAMEK, J. Polymer Sci. **59** (1962), 51
[21] BEALL, G., J. Polymer Sci. **4** (1949), 483
[22] TEICHGRÄBER, M., Faserforsch. u. Textiltechn. J. Polymere **6** (1968), 249
[23] BERGER, R., Plaste und Kautschuk **16** (1969), 326—331, 572—576, 729—731, 822—824
[24] MUSSA, G., J. Polymer Sci. **25** (1957), 441
[25] KOTLIAR, A. M., J. Polymer Sci. A **2** (1964), 4303
[26] SCHULZ, G. V., HENRICI-OLIVÉ, G. and S. OLIVÈ, J. physik. Chem. (Frankfurt/Main) **19** (1959), 125
[27] STOCKMAYER, W. H., J. Chem. Physics **13** (1954), 199
[28] TOPCHIEV, A. V., LITMANOVICH, A. D. and V. Y. SHTERN, Dokl. Akad. Nauk SSSR **147** (1962), 1389
[29] TERAMACHI, S. and Y. KATO, J. Macromol. Sci. — Chem. A **4** (1970), 1785
[30] ELIAS, H. G. and U. GRUBER, Makromolekulare Chem. **78** (1964), 82
[31] KUHN, R., Makromolekulare Chem. **177** (1976), 1525
[32] INAGAKI, H., Fortschr. Hochpolymeren-Forsch. **24** (1977), 189
[33] STAHL, E., „Dünnschichtchromatographie", Springer-Verlag Berlin—Heidelberg—New York 1967 (2nd edition)
[34] SNYDER, L. R., „Principles of adsorption chromatography", Marcel Dekker, New York 1968
[35] SCHRÖDER, E., Plaste und Kautschuk **9** (1962), 395
[36] BOHDANECKY, M., J. Polymer Sci. A2 **5** (1967), 343
[37] FRITZSCHE, P., Faserforsch. u. Textiltechn. — J. Polymere **19** (1968), 505
[38] SCHRÖDER, E. and CH. HANNEMANN, Wiss. Zeitschr. Technische Hochschule Leuna-Merseburg **16** (1974), 364

## 2.4. Determination of Molecular Weight Distribution by Turbidimetric Titration

### 2.4.1. Introduction

In general, turbidimetric titration ($TT$) is defined as the optical monitoring of the behaviour of polymer solutions when a precipitant has been added or the temperature changes. In contrast to precipitation point titration, the entire solubility distribution curve and not only the 1st precipitation point is determined. Turbidimetric titration is a highly valued investigation method for polymer characterization because of the comprehensive information obtained and the low investment in time and material. It is used not only as an aid in polymer fractionation but is particularly employed in the plotting of phase diagrams, determination of solubility parameters and ($\Theta$) temperatures, and in compatibility investigations [1]. Directly coupled with gel permeation chromatography (*GPC*), it has become an indispensable means for the determination of chemical heterogeneity in copolymers [2]. Its use for the determining molecular weight distribution is restricted to homopolymers; systematic uncertainties only permit quantitative evaluation for the comparison of distribution functions of chemically similar species. Therefore, it is less suited for fundamental research but is particularly suitable for quality control and, from the viewpoint of investment costs and the wide range of application, it is superior even to *GPC* [3].

## 2.4.2. Principles

### 2.4.2.1. Fundamentals of Light Scattering in Disperse Systems

For molecular weight determination by *TT*, precipitant is usually added to a very dilute polymer solution at constant temperature with continuous stirring. The amount of the precipitated polymer is determined via the transmittance $I_T/I_0$ or the lateral scattered radiation

$$\tau = \frac{I_{sl}}{I_0} \tag{1}$$

as a function of precipitability

$$\gamma = \frac{V_A}{V_0 + V_{P_r}} \tag{2}$$

$\tau$ = turbidity, $I_0$ = incident light intensity, $I_{sl}$ = scattered light intensity, $I_T$ = transmitted light intensity, $V_{P_r}$ = precipitant volume, $V_0$ = volume of the sample

Because of the continuous changes in dilution $(1 - \gamma)$ during *TT* and the consequent variation in the number of scattering particles, the turbidimetric value $\tau$ must be corrected using equation (3):

$$\tau_{\text{corr}} = \frac{\tau}{1 - \gamma} \tag{3}$$

In empirical methods, precipitant is added continuously. Evaluation of the *TT* presupposes knowledge of the relationship of solubility to molecular weight $M_B = M_B(c_{B,0}, \gamma)$ which must be established for every polymer-solvent-precipitant system by investigations using fractions which should be as uniform as possible.

For intermittent precipitant dosing with measurement of maximum turbidity, the molecular weight of the precipitated polymer particles $M_{w,B}^*$ can be determined directly using the theory of light scattering by disperse systems [6,7]. For spherical particles in which intramolecular and intermolecular interference can be neglected or at small angles where the scattering function $P(\vartheta) \to 1$ (cf. Chapter 1.5.), this theory gives

$$\tau = K \cdot M_{w,B}^* \cdot c_B^* \tag{4}$$

$K$ = optical constant depending on the wavelength of the exciting light, the particle number and the relative refractive index defined as $\tilde{n}_r = \tilde{n}_B/\tilde{n}_A$ ($\tilde{n}_A$ refractive index of the dispersion medium, $\tilde{n}_B$ refractive index of the dispersed polymer), $c_B^*$ = concentration of the dispersed polymer particles

$$c_B^* = x(P) \cdot c_{B,0}(1 - \gamma) \tag{5}$$

$x(P)$ = percentage of the precipitated polymer, $c_{B,0}$ = initial polymer concentration

For larger angles of observation or for molecules where the interference extinction cannot be neglected, equation (4) must be extended by including a particle scattering function (RAYLEIGH-GANS theory).

$$\tau = K \cdot M_{w,B}^* \cdot c_B^* \cdot R(\vartheta) \tag{6}$$

This theory is not universal and can only be used when the conditions $\tilde{n}_r \leq 1.05$ and $(2\pi r)/\lambda \leq 10$ are satisfied ($r$ = particle radius). For larger particles, which are usually present in polymer dispersions, the MIE scattering theory gives the following expression for $\tau$:

$$\tau = n\pi r^2 K \tag{7}$$

$n$ = number of particles per ml, $K$ = MIE scattering coefficient

In contrast to the RAYLEIGH or RAYLEIGH-GANS theory, the MIE theory of scattering in disperse systems is free from limitations with respect to the size and relative refraction numbers of the scattering particles. It takes into account the moments induced by both electrical and magnetic fields. The scattering coefficients $K$ which are very complicate to calculate are tabulated in [8] for various particle diameters and $\tilde{n}_r$-values. For small particles ($D < \lambda/20$), equation (7) simplifies to equation (4) and for systems with $\tilde{n}_r \to 1$ into equation (6). Computer programs are available for calculating the scattering coefficients.

## 2.4.2.2. Coagulation of Precipitated Polymer Particles During Turbidimetric Titration

One of the most important pre-requisites for the determination of molecular weight distribution by $TT$ is the ability to calculate exactly the mass of the precipitated polymer particles $c_B^*$ at each stage of the precipitation process. For this purpose, we usually start by assuming proportionality between $\tau_{corr}$ and $c_B^*$ as in equation (4). This interrelation, however, requires time independence of the coagulation process and constancy of the particle size distribution of the precipitated particles, in contrast to the coagulation theory. According to the latter, supersaturation of the solution causes precipitation of small spherical gel particles which act as nuclei for precipitating polymer particles. With progressive addition of precipitant, the particle size and asymmetry increase and the particle size distribution becomes more uniform [9]. Because of the time dependence of the growth process of the particles, there follows a time dependence of scattered light on turbidimetric values.

In general

$$R = kt^a \tag{8}$$

$R$ = small-angle scattering, $t$ = elapsed time, $k$ = constant

The exponent $a$ is almost unity and, for short times only is independent of the concentration. Taking into consideration the time and concentration dependence of coagulation, we obtain for the particle molecular weight $M_{w,B}^*$

$$M_{w,B}^* = k^* \cdot c_B^* \cdot t^a \tag{9}$$

with $k^* = \dfrac{k'}{Kv^2}$ the coagulation constant ($v^2$ = refractive increment, $k' = \dfrac{R}{c_B^2 t^a}$, $K$ = optical constant).

The coagulation constants continue to be dependent upon the dispersion agent because coagulation is controlled by diffusion and its velocity is inversely proportional to the viscosity $\eta_A$ of the dispersion agent. A dependence of the coagulation velocity of macromolecules on molecular weight, branching and constitution has not been observed [9]. Polymers which tend to associate or crystallize such as PVC, PAN and PMMA show abnormal coagulation behaviour since association and freezing slow down the process. Thus coagulation of polymer particles deviates somewhat from SMOLUCHOWSKI theory and, for this reason, renders exact quantitative evaluation of $TT$ difficult.

## 2.4.3. Empirical Evaluation

### 2.4.3.1. Relation between Solubility and Molecular Weight

Although the complete formalism exists for the determination of molecular weight distribution, difficulties in evaluation mean that semi-empirical methods are still used for comparative determination of molecular weight distributions. These

methods take advantage of the relationship between solubility and molecular weight for the indirect determination of $M^*_{w,\,B}$ and require the following:
1. Exclusive dependence of polymer solubility on molecular weight; attainment of equilibrium at each stage of $TT$.
2. Intensity of the scattered light (and of the transmittance) dependent only on the molecular weight of the precipitated polymer particles.

Requirement 1 is satisfied approximately by very dilute solutions and slow precipitant addition. The solubility of polydisperse systems is however, also influenced by the molecular weight distribution in accordance with thermodynamic theory. The discrepancies with respect to requirement 2 have already been discussed in detail in the preceding Chapter.

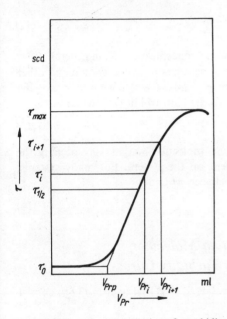

*Fig. 49*: Schematic representation of a turbidity curve
$V_{\text{Prp}}$ — precipitant volume at the first turbidity point, $V_{\text{Pri}}$ — precipitant volume at $\tau_i$, $\tau_0$ — intrinsic turbidity of the solution, $\tau_{1/2}$ — half-value turbidity, $\tau_{\max}$ maximum turbidity

Since the molecular weight $M^*_{w,\,B}$ is a function of $\gamma$ and $c_{B,\,0}$ it is necessary for converting the turbidity curves $\tau = f(V_{P_r})$ (also see Fig. 49) into molecular weight distribution curves, to know the relationship between solubility and concentration $\gamma = f(c_{B,\,0})$ and the relationship between solubility and molecular weight $M^*_{w,\,B}(c_{B,\,0},\,\gamma)$.

According to SCHULZ [10], the following relationships hold:

$$\gamma = a \log c_{B,P_r} + \gamma_0 \tag{10}$$

and

$$\gamma_0 = \gamma_\infty + \frac{b}{M_{w,B}^{*0.5}} \tag{11}$$

$c_{B,P_r}$ = concentration of the polymer solution at the point of turbidity, $\gamma_0$ = precipitability at $c_{B,P_r} = 1$ g/cm³, $\gamma_\infty$ = precipitability in case of infinitely large molecules

Substituting equation (11) into equation (10) we obtain for the solubility equation:

$$\gamma = a \log c_{B,P_r} + \gamma_\infty + \frac{b}{M_{w,B}^{*0.5}} \tag{12}$$

Further relationships between solubility and concentration [3], in general, differ from equation (10) only in the use of other measures of concentration and other specifications for the turbidity points. For example, the following holds for the precipitability $\gamma_{P_r}$, calculated with respect to the halfturbidity (URWIN):

$$\log c_o = a - b\gamma_{P_r} \tag{13}$$

The relationships between solubility and molecular weight — especially the exponents of $M_{w,B}^*$ — are more dependent on the system. For many systems, equation (14) describes the relationship between $M$ and $\gamma$

$$\log M = a - b\gamma \tag{14}$$

## 2.4.3.2. Fundamentals of Calculating Molecular Weight Distribution from Turbidity Curves

All methods for evaluating turbidity curves $\tau_{corr} = f(V_{Pr})$ (e.g. in [3,5]), consist basically of two stages:

1. Conversion of the corrected turbidity curve $\tau_{corr} = f(V_{Pr})$ into a relationship between the mass of the precipitated polymer particles $\Delta m_{i,B}$ and the precipitability according to:

$$\Delta m_{i,B} = \gamma_i - \gamma_{(i-1)} \tag{15}$$

This presupposes quantitative precipitation and, hence, applicability of the relationship

$$\sum_{j=1}^{n} (\Delta m_{i,B})_j = 100 \tag{16}$$

## Molecular Weight Distribution by Turbidimetric Titration

2. Calculation of $\gamma_0$-values from the $\gamma_{P_r}$-values ($\gamma$-value with 50 per cent precipitation) and determination of the molecular weights $M^*_{w,i}$ using equation (11).

The mass of the precipitated polymer particles is obtained from equation (16) using $\tau_i$-values from equation (17):

$$m_{i,B} = \sum_{j=1}^{i} \Delta m_{j,B} = \frac{\tau_{\text{corr},i} - \tau_0}{\tau_{\text{corr,max}} - \tau_0} \tag{17}$$

(For the meaning of the turbidity values $\tau_i, \tau_0, \tau_{\max}$ see Fig. 49.)

From equation (17), we then obtain the „differential" mass fraction of the precipitated particles:

$$\Delta m_{i,B} = \sum_{j=1}^{i} \Delta m_{j,B} - \sum_{j=1}^{i-1} \Delta m_{j,B} = m_{i,B} - m_{(i-1),B} \tag{18}$$

Table 27: Measured values for the determination of the solubility equation of PVC emulsion

| $M_{w,B}$ of the fractions in g/mol | $c_{B,0}$ g/100 cm³ | $\gamma_{P_r} = \gamma_{1/2}$ | $\left[\frac{c_{B,0}}{2}(1-\gamma)\right] \cdot 10^3$ | $\log\left[\frac{c_{B,0}}{2}(1-\gamma)\right]$ | $a$ | $\gamma_0$ calculated |
|---|---|---|---|---|---|---|
| 11 000 | 0.02 | 0.735 | 2.65 | −2.57680 | | |
|        | 0.03 | 0.731 | 4.04 | −2.39362 | −0.030 | 0.6595 |
|        | 0.04 | 0.727 | 5.46 | −2.26281 | | |
|        | 0.06 | 0.720 | 8.40 | −2.07572 | | |
| 35 000 | 0.02 | 0.692 | 3.08 | −2.51145 | | |
|        | 0.03 | 0.685 | 4.73 | −2.32514 | −0.030 | 0.6185 |
|        | 0.04 | 0.680 | 6.40 | −2.19382 | | |
|        | 0.06 | 0.677 | 9.69 | −2.01368 | | |
| 50 000 | 0.02 | 0.685 | 3.15 | −2.50169 | | |
|        | 0.03 | 0.678 | 4.83 | −2.31605 | −0.032 | 0.6100 |
|        | 0.04 | 0.673 | 6.54 | −2.18442 | | |
|        | 0.06 | 0.668 | 9.96 | −2.00174 | | |
| 97 000 | 0.02 | 0.674 | 3.26 | −2.48678 | | |
|        | 0.03 | 0.669 | 4.97 | −2.30364 | −0.032 | 0.6000 |
|        | 0.04 | 0.665 | 6.70 | −2.17393 | | |
|        | 0.06 | 0.657 | 10.29 | −1.99758 | | |

From the $\Delta m_{i,B}$-values of equation (18), the integral mass fractions $I_i(\gamma)$ are calculated using the SCHULZ summation rule.

$$I_i(\gamma) = 100 - \left( \sum_{j=1}^{i-1} \Delta m_{j,B} + \frac{\Delta m_{i,B}}{2} \right) \tag{19}$$

The evaluation is effected according to the scheme in Table 27, 28.

The „differential" $\Delta m_{i,B}$ and the integral $I_i$ mass fractions have been determined at this stage as a function of precipitability.

A useful aid in determining molecular weight distributions is a nomogram suggested by CLAESSON [11]. The $\gamma$-values of calibrating fractions (see example) are plotted on the axes 3 and 4 of Fig. 50 against $\log\left[\dfrac{c_{B,0}}{2}(1-\gamma)\right]$.

axis 1, the $\gamma_0$-values are obtained using equation (10) and are associated with the molecular weight from equation (11). Axis 1 thus denotes a molecular weight division and axis 2 values of $\Delta m_{i,B}$ and $I_i$. Parallels to the solubility lines of the calibrating fractions are drawn through the $\gamma_i$-values from the test.

The $\Delta m_{i,B}$-values and the $I_i$-values are marked off on the solubility line corresponding to the respective $\gamma_i$-value. Joining the $I_i$-values gives the integral distribution $I(M_B)$; joining the $\Delta m_{i,B}$-values gives the differential distribution $\Delta m(M_B)$ approximately. The actual differential distribution must be calculated by differentiation of the integral curve.

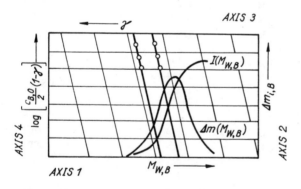

Fig. 50: Schematic CLAESSON nomogram

For interpretation of the distribution curves it should be remembered that the scale of axis 1 is proportional to $M_B^{-0.5}$.

## 2.4.4. Absolute Method for the Evaluation of Turbidity Curves

The main difficulty in evaluating turbidity curves is the exact determination of the mass $m_{i,B}^*$ of the precipitated polymer particles. In empirical methods, it is assumed that precipitation is quantitative and that turbidity is proportional to particle size. The latter assumption is not made in continuous $TT$ methods because of constantly changing particle sizes (see 2.4.2.2.). Direct $m_B^*$-determination is in principle, only possible with a continuous determination of particle size distribution and $\tilde{n}_r$-values. The sources of error already enumerated will decrease considerably if, instead of a continuous measurement, turbidity or scattered light are measured intermittently at maximum turbidity. According to investigations by BEATTIE [6], turbidity at maximum turbidity is almost independent of particle size and proportional to the concentration; this is in agreement with the MIE scattering theory. At maximum turbidity, the particle scattering function also has a maximum value whose position is given by the condition:

$$P = [4\pi r \tilde{n}_A (\tilde{n}_r - 1)/\lambda] = 3 \tag{20}$$

For the concentration of the precipitated polymer $c_B^*$, then, at maximum turbidity and taking into consideration polydispersity and the combination rules for refractive indices, (GLADSTONE-DALE):

$$(\tilde{n}_r - 1)\tilde{n}_A = (n_B - \tilde{n}_A)\Phi_B \tag{21}$$

$n_B$ = refractive index of the polymer, $\Phi_B$ = volume fraction of the polymer in the polymer-rich phase

$$c_B^* = \varrho_B \lambda \tau [3\pi \overline{(K/P)}_w (n_B - \tilde{n}_A)] \tag{22}$$

$\varrho_B$ = density of the polymer, $\overline{(K/P)}_w$ = weight average particle scattering function, $c_B^*$ in g/mol

If equation (20) is assumed to be applicable, $c_B^*$ can be determined absolutely from turbidimetric values for narrow particle size distributions. $P$ need not be known directly because in the case of maximum turbidity, $\overline{(K/P)}_w$ depends insignificantly on $P$ and can be obtained from tables [8]. With wider particle size distributions, $\overline{(K/P)}_w$ can only be determined if the distribution function is known [6]. The condition $P \cong 3$ can also be realized by measurements at different wavelengths.

For experimental determination of $c_B^*$, a series of 0.01 to 0.02 per cent polymer solutions is mixed with various precipitant volumes $V_{P_r}$ while the mixture is constantly stirred and the maximum turbidity measured. The solvent-precipitant system must be strictly isorefractive. The polymer must have a refractive index differing greatly from the latter. Plotting $c_B^*$ against $\gamma$, we obtain the solubility distribution curve. Conversion into the molecular weight distribution curve again calls for knowledge of the solubility equation so that even this improved method of turbidimetric titrations is actually not an absolute method.

### 2.4.5. Experimental Background

Polymer solutions are titrated turbidimetrically according to the principle of „isothermal precipitation" ($IPr$) by adding precipitant or by continuously changing the temperature. Both methods have shortcomings; in the temperature gradient method they are associated with the great temperature difference (up to 100 °C) required for precipitation and in the precipitant method they are associated with the calibration measurements, time, concentration and refractive index corrections required for evaluation. Nevertheless, the precipitant method is usually preferred.

Optimum test conditions:

| | |
|---|---|
| Concentration of polymer solution | 1 ... 10 mg/$l$ |
| Precipitant dosage | 10 ... 20 ml/h |
| Rotational speed of the stirrer | 800 rpm |
| Operating temperature | above the freezing temperature of gel |
| Measuring instrument | light scattering photometer |
| Wavelength of the exciting light | 600 ... 940 nm |
| Measuring angle $\vartheta$ | 45°, 90°, 135° |

Selection of the solvent-precipitant system ($S/P_r$) is of particular importance to $TT$. Because of the requirement that equilibrium be rapidly established during titration, weak solvents and precipitants with $\chi$-values about 0.5 are preferred. When using them, the risk of increased particle aggregation cannot be excluded, however. Optical measurement requires an isorefractive $S/P_r$ pair whose mean refractive index differ from that of the precipitated polymer particles. Because of the constancy of the turbidity, the densities of dispersion agents and dispersed particles should be almost equal; moreover, the coagulation theory calls for a dispersion agent with a viscosity as low as possible.

Turbidimetric titration is carried out at a constant temperature in electrically heated cells with automatic precipitant addition and magnetic stirring. It is advisable to use open or constant volume measuring cells [3]. Turbidimetric measurement can be carried out using scattered-light photometers (see Chapter 1.5.2.) or good quality spectrophotometers with a turbidity attachment see also [1]. The lateral scattering is measured at several fixed angles (45°, 90°, 135°) or at long wavelengths. Alternatively the transmittance can be measured. Devices such as described by STEARNE and URWIN [12] for the combined measurement of scattered light and transmittance have proved successful. For measurements by the temperature gradient method, a double-beam turbidimeter is particularly suitable because it allows simultaneous measurement of sample and standard [13].

To use $TT$ for determining particle weights and molecular weight distributions, light scattering measurements at angles smaller then 20° with light wavelengths above 600 nm are preferred because equation (4) can only be used under these conditions. (Up to $\lambda = 600$ nm, the measured value of the reduced scattering intensity deviates by only a few per cent from that extrapolated to $\vartheta \to 0$.) Up to $\lambda = 940$ nm, light emitting diodes can be used as a light source and Si-photovoltaic cells as a measuring device [9]. $TT$ instrumentation is shown schematically in Fig. 51. Information about the application of a Spekol spectrophotometer (VEB Carl Zeiss Jena) is given in [1].

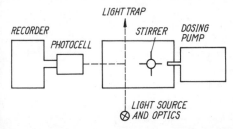

*Fig. 51*: Diagram of apparatus for turbidimetric titration according to [5]

## 2.4.6. Examples

### 2.4.6.1. Determination of the Relationship between Solubility and Molecular Weight of Emulsion PVC

The numerical values for the solubility equation (12) are determined from values of $a$ and $\gamma_0$ found using equation (10).

Solutions of at least four PVC fractions are made up at concentrations between 0.02 and 0.10 g/100 cm³ in cyclohexanone. The $M_w$ values should be known and be as uniform as possible. The samples are dissolved at 80 °C with constant shaking in a thermostat. Having set the temperature of the spectrocolorimeter which possesses a titration attachment (see [1]) 6.00 ml of the prepared solution are put into a 30 ml cell and mixed with 8.00 ml of precipitant. The precipitant is a mixture of nine parts of n-heptane and one part of carbon tetrachloride as a dispersion stabilizer. The titration is carried out while stirring using a microburette, up to the maximum of the turbidity curve $\tau = f(V_{P_r})$ shown by the recorder and the latter is converted into the $\tau_{corr} = f(\gamma)$ curve using equation (2) and (3). In order to reduce errors resulting from the molecular inhomogeneity of the fractions, the precipitability at 50 per cent precipitation $\gamma_{1/2}$ is used in evaluating equation (10). The polymer concentration at $\gamma_{1/2}$ is

$$c^*_{B, 1/2} = \left[\frac{c_{B,O}}{2}(1-\gamma)\right] \quad (c_{B,O} \text{ in g/100 cm}^3).$$

The measured values in Table 27 are represented graphically in Fig. 52. From the slope of the straight line, a value of $-0.03$ is obtained for $a$. The $\gamma_0$ values (obtained graphically or by calculation from the intercepts) are plotted against $M_B^{-0.5}$ and the values for $\gamma_\infty$ and $b$ in the solubility equation (12) are determined (Fig. 53). These are $\gamma_\infty = 0.569$ and $b = 9.7$.

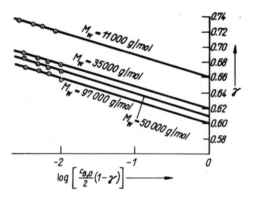

Fig. 52: Determination of solubility-concentration relationships for PVC

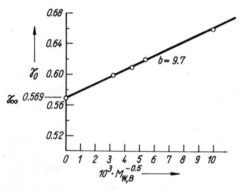

Fig. 53: $\gamma_0$ as a function of $M_{w,B}^{-0.5}$ for PVC

Consequently, the solubility equation for PVC is (see also [14]):

$$\gamma = -0.03 \log c_B + 0.569 + \frac{9.7}{\sqrt{M_{w,B}}} \tag{23}$$

Relationships between solubility and molecular weight for polystyrene are quoted in [3].

*Table 28*: Measured values of the turbidimetric titration of PVC-E

| $i$ | $V_{P_r,i}$ (cm³) | $\gamma$ | $\tau_{corr,i} - \tau_0$ Scd. | $m_{i,B} = \sum_{j=1}^{i} \Delta m_{j,B}$ $= 100 \frac{\tau_{corr,i} - \tau_0}{\tau_{corr,max} - \tau_0} \%$ | $\Delta m_{i,B}$ (%) | $I_i$ (%) |
|---|---|---|---|---|---|---|
| 1  | 9.63  | 0.6161 | —      | —      | —    | 100.0 |
| 2  | 9.96  | 0.6240 | 2.66   | 0.71   | 0.71 | 99.7  |
| 3  | 10.29 | 0.6316 | 8.14   | 2.17   | 1.46 | 98.6  |
| 4  | 10.62 | 0.6390 | 22.16  | 5.90   | 3.73 | 96.0  |
| 5  | 10.95 | 0.6460 | 42.38  | 11.28  | 5.38 | 91.4  |
| 6  | 11.28 | 0.6527 | 67.66  | 18.01  | 6.73 | 85.4  |
| 7  | 11.61 | 0.6593 | 93.95  | 25.01  | 7.00 | 78.5  |
| 8  | 11.94 | 0.6655 | 118.10 | 31.44  | 6.43 | 71.8  |
| 9  | 12.27 | 0.6716 | 143.11 | 38.10  | 6.66 | 65.2  |
| 10 | 12.60 | 0.6774 | 167.40 | 44.57  | 6.47 | 58.7  |
| 11 | 12.93 | 0.6830 | 189.30 | 50.40  | 5.83 | 52.5  |
| 12 | 13.26 | 0.6884 | 208.60 | 55.54  | 5.14 | 47.0  |
| 13 | 13.59 | 0.6937 | 228.50 | 60.84  | 5.30 | 41.8  |
| 14 | 13.92 | 0.6988 | 249.00 | 66.29  | 5.45 | 36.4  |
| 15 | 14.24 | 0.7037 | 266.60 | 70.98  | 4.69 | 31.4  |
| 16 | 14.58 | 0.7084 | 282.90 | 75.32  | 4.34 | 26.9  |
| 17 | 14.91 | 0.7130 | 297.40 | 79.18  | 3.86 | 22.8  |
| 18 | 15.24 | 0.7175 | 311.50 | 82.93  | 3.75 | 19.0  |
| 19 | 15.57 | 0.7218 | 323.50 | 86.13  | 3.20 | 15.5  |
| 20 | 15.90 | 0.7260 | 335.80 | 89.40  | 3.27 | 12.2  |
| 21 | 16.23 | 0.7301 | 346.40 | 92.23  | 2.83 | 9.2   |
| 22 | 16.56 | 0.7340 | 355.20 | 94.57  | 2.34 | 6.6   |
| 23 | 16.88 | 0.7378 | 364.50 | 97.05  | 2.48 | 4.2   |
| 24 | 17.22 | 0.7416 | 373.40 | 99.41  | 2.36 | 1.8   |
| 25 | 17.55 | 0.7452 | 375.40 | 99.95  | 0.54 | 0.3   |
| 26 | 17.88 | 0.7487 | 375.60 | 100.00 | 0.05 | 0     |

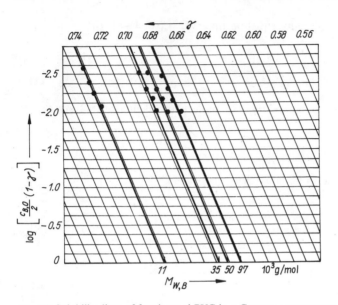

*Fig. 54*: Solubility lines of fractionated PVC in a CLAESSON nomogramm

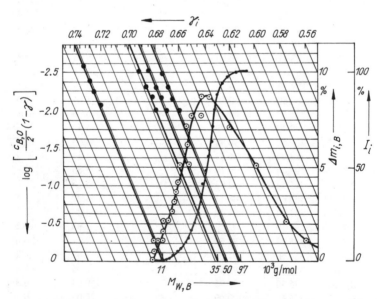

*Fig. 55*: Molecular weight distribution of emulsion PVC obtained from turbidimetric titration

## 2.4.6.2. Determination of the Molecular Weight Distribution of a Commercial Emulsion PVC by Turbidimetric Titration

The product is purified by three cold extractions with a 1:1 methanol-water mixture to remove polymerization aids and impurities, and then dried. For $TT$, a 0.04 per cent solution in AR cyclohexanone is prepared and further treated as described in 2.4.6.1. The turbidimetric values obtained are corrected using equation (3) and converted into the mass fractions and their differential and integral parts according to equations (17) to (19). These are tabulated in Table 28.

Using the solubility lines given in Section 2.4.6.1. and represented in Fig. 52, the CLAESSON nomogram is constructed (Fig. 54) and applied to the $\Delta m_{i,B}$ and $I_i$ values obtained in the test. The various distribution curves $I(M_{w,B})$ are obtained by joining the individual measured points. The distribution curves thus obtained are shown in Fig. 55.

## Bibliography

[1] ARNDT, K.-F. and K.-G. HÄUSLER, Jenaer Rundschau **5** (1978), 238
[2] HOFFMANN, M. and H. URBAN, Makromolekulare Chem. **178** (1977), 2683
[3] URWIN, J. R., in: „Light Scattering from Polymer solution", ed. HUGLIN, M. B., Academic Press New York 1972, p. 391 ff.
[4] HENGSTENBERG, J., Ber. Bunsenges. physik. Chemie **60** (1956), 236
[5] GIESEKUS, H., in: „Polymer Fractionation", ed. CANTOW, M. J. R., Academic Press New York, London 1967, p. 191 ff.
[6] BEATTIE, W. H., J. Polymer Sci. A **3** (1965), 527
[7] BEATTIE, W. H. and H. C. JUNG, J. Colloid Interface Sci. **27** (1968), 581
[8] PANGONIS, W. J., HELLER, W. and A. JACOBSON, „Tables of Scattering Functions for Spherical Particles", Wayne State University Press Detroit 1957
[9] HOFFMANN, M. and H. URBAN, Makromolekulare Chem. **178** (1977), 2661
[10] SCHULZ, G. V., Z. physik. Chem. (B) **46** (1940), 105
[11] CLAESSON, S., J. Polymer Sci. **16** (1955), 193
[12] STEARNE, J. M. and J. R. URWIN, Makromolekulare Chem. **56** (1962), 76
[13] TAYLOR, W. C. and J. P. GRAHAM, J. Polymer Sci. Polymer Letters **2** (1964), 169
[14] SCHRÖDER, E., Plaste und Kautschuk **9** (1962), 525
[15] HOFFMANN, M., KRÖMER, H. and R. KUHN, „Polymeranalytik", Vol. 1, Georg-Thieme-Verlag Stuttgart 1977

## 2.5. Gel Permeation Chromatography

### 2.5.1. Introduction

In spite of its short history, gel permeation chromatography ($GPC$) also called SEC (size exclusion chromatography) is already a standard method of polymer characterization. The main field of application is the separation of polydisperse substances into monodisperse fractions of different molecular weights. Small amounts of substances, relatively short separation times and highly automated equipment are the main advantages of this method over other separation techniques. Because it is a separation method, it can also be employed successfully for the removal and quantitative determination of additives, residual monomers, plasticizers, etc.

$GPC$ is a relative method. Experiment gives an elution chromatogram which is converted into the molecular weight distribution via a calibration curve. The profile of the elution curve is determined by the molecular weight distribution of the sample. When this distribution is stable, the peak maximum correlates with an average molecular weight specific to $GPC$, $\bar{M}_{GPC}$. This quantity is dependent on the type of distribution and lies between the geometric mean of the number-average and weight-average molecular weights $(M_n \cdot M_w)^{1/2}$ and $M_w$ [1].

A characteristic feature of this separation method is independence from test conditions such as temperature, pressure and to some extent from solvent, load and flow rate. The data from such chromatograms are well suited to computer analysis. Thus, the quantities $M_n$, $M_w$, $M_z$, $U$ and $g_w^2$ characterizing a molecular weight distribution can be calculated rapidly.

### 2.5.2. Principles

$GPC$ is a specific form of liquid chromatography. It is based on separation of dissolved macromolecules using porous media. Substance-specific interactions between stationary phase and substance to be eluted must not occur; the choice of the correct solvent is of particular importance. Compared to other chromatography methods, $GPC$ is characterized by three properties [2]:
a) Separation is effected according to molecular size.
b) Larger molecules are eluted before smaller ones.
c) Separation takes place in a volume that is smaller than the total volume of the column.

The concentration in the stationary phase is never larger than that in the mobile phase (DETERMANN [3]). The stationary phase has the same composition as the mobile phase but differs in its lack of mobility.

# Gel Permeation Chromatography

*Table 29*: Elution and separation parameters in GPC

| Parameter | Symbol |
|---|---|
| Gel matrix volume | $V_M$ |
| external volume (interparticle volume) | $V_0$ |
| internal volume | $V_i$ |
| accessible internal volume | $V_{i,\mathrm{acc.}}$ |
| volume of the stationary phase | $V_{St} = V_i + V_M$ |
| total volume | $V_t = V_i + V_M + V_0$ |
| elution volume | $V_e$ |
| distribution coefficient, related to the internal volume | $k_d = \dfrac{V_{i,\mathrm{acc}}}{V_i} = \dfrac{V_e - V_0}{V_i}$ |
| distribution coefficient, related to the volume of the stationary phase | $k_{av} = \dfrac{V_{i,\mathrm{acc}}}{V_{St}} = \dfrac{V_e - V_0}{V_t - V_0}$ |
| width of peak at base | $w$ |

To deal with the separation process theoretically it is necessary to introduce parameters describing the elution process. Table 29 lists the most important elution and separation parameters and their symbols.

The separation process can be interpreted using both equilibrium and non-equilibrium theories [4]. The basis of the exclusion principle is the establishment

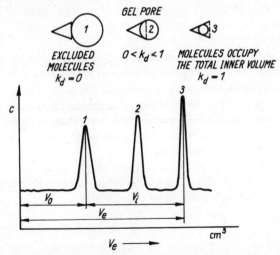

*Fig. 56*: Schematic representation of the separation process of particles of different sizes from porous media

of an equilibrium in sample concentration between the internal and external volume of the gel matrix. The model assumes that the maximum time ($t_m$), available at a defined point in the column for diffusion into the gel phase, must be equal to the time in which half of the substance zone ($t_{0.5}$) passes that point.

The accessibility of the pores by diffusion to different molecules is a function of both molecular and pore size. The smaller the molecules, the deeper they penetrate into the pores of the gel. Smaller molecules are therefore eluted only after larger molecules. Molecules whose average hydrodynamic radius is larger than that of the gel pores cannot be separated. The separation mechanism is represented schematically in Fig. 56.

The maximum retention time of small molecules is related to the sum of the total internal gel volume and the volume between particles ($V_i + V_0$). For retention volumes which are larger than the operating volume of the column, other separating mechanism are superimposed. The sequence of eluted substances in such cases no longer corresponds to that mentioned in a) and b).

Since at any point in the elution process not all molecules diffuse into the available pores, a zone broadening of the elution curve, typical of *GPC*, occurs [5]. This effect interferes with the separation of narrowly distributed products and requires correction. For very broad distributions, this effect can be neglected. The so called skewing effect arising from the change in the elution volume due to the amount of the sample added to the column must also be taken into account. In thermodynamically ideal solvents shifts of the elution volume are measured as a function of sample concentration. This effect increases as the molecular weight increases. In theta solvents, the elution volume remains almost constant [6].

When there is insufficient time for the establishment of equilibrium ($t_m < t_{0.5}$), non-equilibrium effects will occur. This may lead to an excessively high flow rate and dependence of the diffusion coefficient upon the molecular size [2, 4].

None of the propounded theories allows the derivation of a precise correlation between elution volume and molecular size. Therefore, each separating column must be calibrated with polymer standards of known molecular weight. When the solvent is changed, the column must be recalibrated.

An experimentally determined calibration curve is required for every *GPC* analysis. For the entire separation range of the chromatographic system used, the calibration function can be expressed as a polynomial (equation 1):

$$\log M = a_0 + a_1 V_e + a_2 V_e^2 + \ldots + a_n V_e^n \tag{1}$$

$a_0, a_1, \ldots a_n$ = system-specific constants

If the operating range lies only in the linear of this usually *S*-shaped curve, the series can be truncated after the linear term. If calibration fractions are not

# Gel Permeation Chromatography

available, BENOIT [7] has suggested a universal calibration quantity ($[\eta] \cdot M$) related to the hydrodynamic volume:

$$\log([\eta] \cdot M) = a_0 + a_1 V_e + a_2 V_e^2 + \ldots + a_n V_e^n \tag{2}$$

This requires that, at a certain elution volume $V_e$, the product ($[\eta] \cdot M$) is constant. This condition has been confirmed with different polymers, so that:

$$[\eta]_1 \cdot M_1 = [\eta]_x \cdot M_x \quad (V_{e,1} = V_{e,x}) \tag{3}$$

To evaluate the chromatogram only the constants of the $[\eta]$-$M$-relationship of the sample and the calibration substance in the selected solvent must be known along with the calibration function, equation (1). Polystyrene fractions with molecular inhomogeneities <0.1 are usually used as calibration standards.

Table 30 contains some examples of the constants in the KUHN-MARK-HOUWINK equation for the solvents tetrahydrofuran and ortho-dichlorobenzene which are frequently used in *GPC*. Fig. 57 shows a calibration curve of equation (1) obtained using polystyrene standards dissolved in tetrahydrofuran.

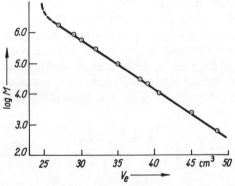

*Fig. 57*: Calibration curve for GPC (3 styragel columns in series, exclusion limits: $1.5 \cdot 10^4$, $1.5 \cdot 10^5$, $10^6$ g/mol, mobile solvent: tetrahydrofuran)

As a measure of the transport processes in the column, the height equivalent to a theoretical plate (HETP) is calculated as in other chromatographic methods from equation (4):

$$h = \frac{L}{n} \tag{4}$$

$L$ = length of column in mm, $n$ = number of theoretical plates

$$n = \left(\frac{4V_e}{w}\right)^2 \tag{5}$$

*Table 30*: [η] — M relationships for various polymers in selected solvents [11]

| Polymer | Solvent | Temperature (°C) | $K \times 10^4$ [η] (d$l \cdot$ g$^{-1}$) | a |
|---|---|---|---|---|
| Polystyrene | THF | 23 | 0.68 | 0.77 |
|  | o-DCB | 135 | 1.38 | 0.70 |
| Polymethyl methacrylate | THF | 23 | 0.93 | 0.72 |
| Polyvinyl chloride | THF | 23 | 1.63 | 0.77 |
| Polyvinyl bromide | THF | 20 | 1.59 | 0.64 |
| Polydioxolan | THF | 25 | 0.94 | 0.87 |
| Polyvinyl acetate | THF | 25 | 3.50 | 0.63 |
| Polyethylene LD | o-DCB | 135 | 5.05 | 0.70 |
| Polypropylene | o-DCB | 135 | 1.30 | 0.78 |
| Polybutadiene (hydrated) | o-DCB | 135 | 2.70 | 0.75 |
| Polydimethylsiloxane | o-DCB | 138 | 3.83 | 0.57 |

Under optimized conditions, heights equivalent to a theoretical plate of 3 to $4 \cdot 10^{-2}$ mm are currently possible in high-resolution GPC.

Equation (5) is frequently utilized for testing the serviceability of the column, using a defined low molecular weight organic compound with large elution volumes. $h$ is only an indirect measure of the separation efficiency only indirectly through its relationship to the width of the peak at the base. For estimating the separation efficiency of a column, the elution volumes and the width of the peak at base of the components 1 and 2 are measured (Fig. 58) and the resolution $R$ calculated from equation (6):

$$R = \frac{2(V_{e,2} - V_{e,1})}{W_1 + W_2} \tag{6}$$

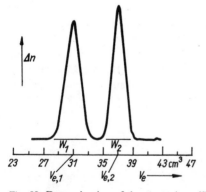

*Fig. 58*: Determination of the separation efficiency of GPC columns

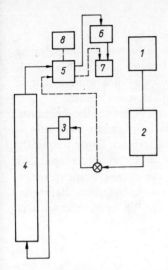

*Fig. 59*: Block diagram of a GPC system
*1* — solvent reservoir, *2* — pump system, *3* — sample charging device, *4* — separating columns, *5* — detector, *6* — siphon fraction collector, *7* — solvent discharge, *8* — recorder
——— sample path, — — — — reference line, —·—·— electrical connection 5 ⋯ 8

Equation (6) shows that the effectiveness of the separation column is determined both by the distance which separates two succeeding peaks and by the width of the two peaks at base. For complete resolution:

$R \geqq 1$.

Introducing the relationship $\log M = f(V_e)$, leads to a resolving power $R_s$ which is specific for *GPC*:

$$R_s = \frac{2(V_{e,2} - V_{e,1})}{(W_1 + W_2) \cdot (\log M_{w,1} - \log M_{w,2})} \quad (7)$$

or a resolution index RI:

$$RI = \left(\frac{M_{w,2}}{M_{w,1}}\right)^z; \quad z = \frac{W_1 + W_2}{2(V_{e,2} - V_{e,1})} \quad 0 < RI < 1 \quad (8)$$

Equations (7) and (8) show that the separation efficiency decreases exponentially with increasing molecular weight. Very high molecular weight products are insufficiently separated.

## 2.5.3. Experimental Background

Generally, *GPC* systems consist of the separation columns, a non-pulsating pressure pump with adjustable flow rate, a sample injector valve with a sample volume loop and a detector. The sample volume loop is filled using a sample injector syringe. Fig. 59 shows the block diagram of the entire system.

The gel types and the packing density of the gel matrix determine the separation efficiency. Table 31 lists some of the packing materials frequently used in high resolution *GPC* and gives brief details of their characteristics. Other frequently used gels are discussed in [4, 9, 10, 15]. The packing of the columns is complicated and depends on the type of gel. For each organic or inorganic gel type there are different packing specifications which depend on the particle diameter. Reference should be made to the relevant technical literature [8, 10]. In most cases it is preferable to use commercially filled columns.

*Table 31*: Selected packing materials for high resolution GPC

| Gel | Particle diameter ($\mu$m) | Material | Average pore diameter ($10^{-8}$ cm) | Exclusion limit; range of application PS-Standards (g mol$^{-1}$) |
|---|---|---|---|---|
| $\mu$-Styragel | 10 | PS-DVB | $10^2$ | $7 \times 10^2$ |
| | | | $5 \times 10^2$ | 0.05 ... 1 ($\times 10^4$) |
| | | | $10^3$ | 0.1 ... 20 ($\times 10^4$) |
| | | | $10^4$ | 1 ... 20 ($\times 10^4$) |
| | | | $10^5$ | 1 ... 20 ($\times 10^5$) |
| | | | $10^6$ | 5 ... >20 ($\times 10^5$) |
| $\mu$-Porasil | 10 | Silica gel | E $1.25 \times 10^2$ | 0.2 ... 5 ($\times 10^4$) |
| | | | E $3 \times 10^2$ | 0.03 ... 1 ($\times 10^5$) |
| | | | E $5 \times 10^2$ | 0.05 ... 5 ($\times 10^5$) |
| | | | E $10^3$ | 5 ... 20 ($\times 10^5$) |
| Li-Chrospher | 10 | Silica gel | $10^2$ | $6 \times 10^4$ |
| | | | $5 \times 10^2$ | $5 \times 10^5$ |
| | | | $10^3$ | $2 \times 10^6$ |
| | | | $4 \times 10^3$ | $> 3.5 \times 10^6$ |
| CPG | 5—10 | Porous glasses | 40 | 1 ... 8 ($\times 10^3$) |
| | | | $10^2$ | 1 ... 30 ($\times 10^3$) |
| | | | $2.5 \times 10^2$ | 2.5 ... 125 ($\times 10^3$) |
| | | | $5.5 \times 10^2$ | 1.1 ... 35 ($\times 10^4$) |
| | | | $1.5 \times 10^3$ | 1 ... 10 ($\times 10^5$) |
| | | | $2.5 \times 10^3$ | 2 ... 15 ($\times 10^5$) |

# Gel Permeation Chromatography

*Table 32*: Column dimensioning for analytical and preparative GPC

| Type of analysis | Internal diameter (mm) | Column length (mm) |
|---|---|---|
| analytical | 2 ... 8 | 250 ... 500 |
| semi-preparative | 12 ... 20 | 250 ... 1000 |
| preparative | 20 ... 625 | 500 ... 2400 |

An appropriate volume of gel must be used. Fig. 60 shows suitable column combinations.

Column dimensions depend on the application (Table 32). The amount of substance separated per passage is dependent on the maximum volume of the polymer solution injected into the column (concentration range from 0.25 to 1 per cent by weight if no other instructions are given by the manufacturer of the separation column). This amount is larger for small molecules than for large molecules. The viscosity of the solution must not be significantly higher than that of the elution agent. The maximum sample volume for the separation columns can be estimated using the VAN-DEEMTER equation.

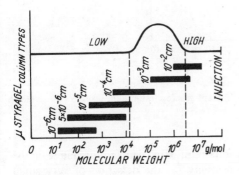

*Fig. 60*: Column selection for the determination of molecular weight distribution of a high-polymeric substance [13]

As a rough guide, 2 to 5 mg of substance can be separated in analytical applications, about 100 mg in semi-preparative work and about 1 g in preparative work.

The choice of eluting agent depends on several factors. When using organic gels, it is particularly important for the sample to be soluble in the same solvent as that in which the gel matrix was swollen. The solvent must not react with the solute or the gel. The eluting agent influences the separation efficiency via the diffusion coefficient. For low eluent viscosities, the number of theoretical separation stages increases. Finally, the solvent must be compatible with the detector. Solvent mixtures cannot be recommended because preferred solvation may occur.

The optimum flow rate depends on the separation column and must be determined experimentally. If the flow rates are too low, back mixing will occur and, if they are too high, the mechanism of separation is affected (see 2.5.2.). In both cases, the separation efficiency will be reduced. A flow rate of 1 m$l$ per min is considered to be the optimum for analytical *GPC* (Fig. 61).

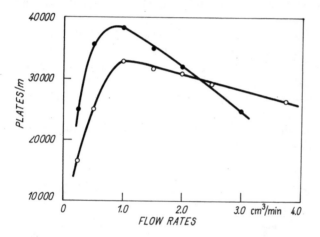

*Fig. 61*: Column efficiencies at various flow rates
○ The column packing is composed of four different gels (exclusion limit up to $5 \times 10^7$ g/mol),
● Gel with exclusion limit $5 \times 10^3$ g/mol

The choice of detector is particularly important. High demands are made on the performance of these devices: high sensitivity for the detection of very small amounts of substance, high linearity of response, a small dead volume in the measuring cell and a short time constant. Usually, RI detectors, which measure the difference of the refractive index between solution and pure solvent, UV filter detectors with fixed wavelengths (254 and 280 nm) and UV spectrophotometers operating in the wavelength range from 200 to 600 nm are used.

For special cases, IR spectrophotometers operating in the wave number range from 4000 to 625 cm$^{-1}$ are employed.

All devices record a physically measured quantity proportional to concentration. This is either converted into the absolute concentration or is used in the calculation as an equivalent quantity.

Recent developments in *GPC* detectors enable concentration and, additionally, the molecular weight to be determined in the eluate. This quantity is measured continuously as a function of the elution volume, thus changing the basis character of classic *GPC*. The chromatographic effect is now utilized only for the separation of the macromolecules and not for the determination of molecular weight [14].

Of the methods for determining molecular weight, already discussed in Chapter 1, those utilizing light scattering and viscosity measurement are particularly suitable because of their compatibility with the flow technique.

For viscosity measurement it is impractical to make a measurement based on time for the flow of a known volume through a capillary in an OSTWALD or UBBELOHDE viscometer. It is far better tu utilize the pressure difference, $\Delta p$, across the capillary (HAGEN-POISEUILLE equation) which builds up continuously with $V_e$. The signal is evaluated analogously to the method described in Section 1.7. Since the concentrations in the eluate are very low, $[\eta]$ is given approximately by:

$$[\eta]_i = \frac{\Delta p_i - \Delta p_0}{\Delta p_0} \cdot \frac{1}{c_i} \tag{9}$$

Low-angle laser light scattering photometry (*LALLS*) enables the continuous determination of $M_i$ in the *GPC* eluate. The *LALLS* technique is distinguished by the fact that, due to the high intensity of the laser light, very low scattering intensities can still be measured from small scattering volumes ($\leq 0.1$ $\mu l$). Further, measurements can be taken at a specified angle between 2 and $10°$ so that extrapolation to the scattering angle $0°$ is unnecessary. These characteristics and a flow cell with a volume of $\leq 10$ $\mu l$ make *LALLS* photometry a suitable molecular weight detector for *GPC*. In contrast to continuous viscosity measurement, the *LALLS* detector has the additional advantage that it does not require a stable non-pulsating flow of solvent through the chromatographic system. The signal is evaluated as described in Section 1.5.

## 2.5.4. Optimization

Separating columns and operating parameters must be selected so suit the application. The quantities relevant to the problem can be assessed from a triangular scheme

$R_s$
△
$t$  cap.

$t$ = time for analysis, cap. = capacity of the column, $R_s$ = (see equ. 7)

In general, it is not necessary to optimize all three parameters. For example, in routine analyses, short analysis times, good resolution and small column capacity are required whereas preparative separations require large capacity and good resolution but not necessarily short analysis times.

Separation time and resolution are in functional correlation. The separation time of the process, can be optimized by increasing the number of theoretical plates for example by the use of packing material with small particle diameter. (Reducing the particle diameter from 100 μm to 10 μm improves separation efficiency by a factor of 50 ... 60). Another method uses gels with a larger internal volume. In this case, the separation efficiency is increased by an improvement in the phase ratio of $V_i$ to $V_0$. It should be noted that swollen gels with a large internal pore volume, e.g. polyvinyl acetate gels, can only be subjected to a low pressure because of their mechanical instability. The diffusion velocity increases with temperature so that, with an increase of the flow, an additional optimization of time is possible.

Increasing the length of the column improves the resolving power because of the proportionality between the square of the reciprocal peak width at base and the number of theoretical plates but this will result in longer separation times.

An increase in capacity can only be achieved by a considerable increase in total volume.

## 2.5.5. Examples

### 2.5.5.1. Determination of Molecular Weight Distribution of a Polystyrene Sample

For the determination of molecular weight distribution, a 0.5 per cent by weight solution in tetrahydrofuran is prepared. Before the analysis, the sample solution is pressed through a G-4 glass frit. By means of a sample injector valve, 0.5 ml of the purified solution are put into the separating columns.

For separation, three columns filled with styragel (exclusion limits: $1.5 \cdot 10^4$; $1.5 \cdot 10^5$; $10^6$ g · mol$^{-1}$) are connected in series. Each column has a length of 500 mm and an internal diameter of 9.5 mm. Tetrahydrofuran is used as eluent and the flow rate is 1 cm$^3$ · min$^{-1}$. The sample concentration in the eluate is measured by an RI detector. Fig. 62 shows the elution chromatogram obtained. For evaluating the chromatogram, the $\Delta n$-values, measured in arbitrary units, are taken at equidistant intervals on the $V_e$-axis (Fig. 62). They are called the chromatogram heights $h_i$. These values are normalized (equation 10):

$$h'_i = \frac{h_i}{\Sigma h_i} = \frac{dm}{dV_e} \tag{10}$$

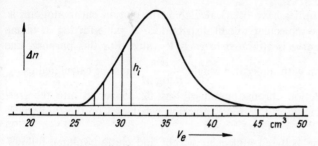

*Fig. 62*: Elution chromatogram of a polystyrene sample (experimental conditions correspond to those given in the caption to Fig. 57)

The normalized chromatogram heights $h'_i$ correspond to the differential change of the mass fraction dm with the elution volume $dV_e$. This must be converted into the change of mass fraction with molecular weight.

$$\frac{dm}{d \log M} = h'_i \cdot \frac{dV_e}{d \log M} \tag{11}$$

For this purpose, the correlation between molecular weight and elution volume must be determined.

If this relationship is linear (Fig. 57), the numerical value for the expression $\frac{dV_e}{d \log M}$ can easily be calculated as the reciprocal of the gradient of the calibration curve.

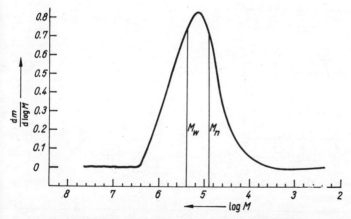

*Fig. 63*: True molecular weight distribution function of a polystyrene sample (converted elution chromatogram given in Fig. 62)

When all these quantities have been calculated, the elution chromatogram is converted into the true molecular weight distribution. A value for log $M$ taken from the calibration curve is allocated to each $V_e$-value. For this purpose, the change of mass fraction with molecular weight $\dfrac{dm}{d\log M}$ must be calculated using equation (11). The elution chromatogram of Fig. 62 converted into the true distribution curve is shown in Fig. 63. Evaluation was based on the calibration curve of Fig. 57.

When the correlation between molecular weight and elution volume follows equation (1) and is $S$-shaped, a graphical method for calculating the quantity $\dfrac{dm}{d\log M}$ is useful [12].

For this purpose, the normalized elution curve and the calibration function are represented together on one diagram, as is shown in Fig. 64. In order to

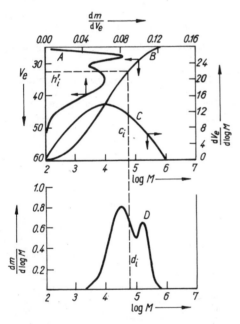

Fig. 64: Conversion of an elution chromatogram into the true molecular weight distribution for an $S$-shaped calibration curve

A: elution curve $h'_i = \dfrac{dm}{dV_e}$, B: calibration curve, C: differential calibration curve $c_i = \dfrac{dV_e}{d\log M}$, D: true molecular weight distribution $d_i = h'_i \cdot c_i$

# Gel Permeation Chromatography

determine the value of $\frac{dV_e}{d\log M}$ for each point on the calibration curve, the calibration function (curve $B$ in Fig. 64) is differentiated (Fig. 64, curve $C$). For each $h_i'$ the corresponding $\frac{dV_e}{d\log M}$-value $c_i$ is determined. The product of $h_i'$ and $c_i$ is the desired change of the mass fraction with the log of the molecular weight.

The averages $M_n$, $M_w$, $M_z$ and the molecular inhomogeneity $U$ as well as the polydispersity $g_w^2$ are obtained using the relationships listed in Table 33. $M_n$ and $M_w$ were calculated for the selected example and are shown in Fig. 63.

## 2.5.5.2. Determination of Molecular Weight Distribution of an Ethylene-Vinyl acetate Copolymer

The experimental method has already been described in 2.5.5.1. The chromatogram is represented in Table 34 in the form of pairs of measured values.

To evaluate the chromatogram, the $h_i'$ values corresponding to each $V_e$-value have to be calculated. In order to correlate elution volume and molecular weight, equation (3) is assumed to apply. When the limiting viscosity numbers in this relationship are replaced by the product of $K \cdot M^a$, equation (12) is obtained:

$$K_1 \cdot M_1^{(1+a)} = K_2 \cdot M_2^{(1+a)} \tag{12}$$

where index 1 or 2 are subscripts for the constants of the $[\eta]$-$M$ relationship of the calibration substance (1) polystyrene and the sample (2) in tetrahydrofuran, respectively.

Transforming equation (12) gives equation (13). Evaluation, requires only knowledge of equation (1).

$$\log M_2 = \frac{1+a_1}{1+a_2} \cdot \log M_1 + \frac{1}{1+a_2} \cdot \log \frac{K_1}{K_2} \tag{13}$$

The correlation between limiting viscosity number and molecular weight for the calibration substance and the sample is given by equations (14) and (15):

$$[\eta]_{PS}^{THF} = 6.8 \cdot 10^{-5} \, M_{PS}^{0.77} \tag{14}$$

$$[\eta]_{EVA}^{THF} = 4.16 \cdot 10^{-4} \cdot M_{EVA}^{0.67} \quad (42\% \text{ by weight of vinyl acetate}) \tag{15}$$

The value of $\log M_{PS}$ for each elution volume is obtained from the calibration curve shown in Fig. 57. A value of $\log M_{EVA}$ is thus obtained from Table 34 for each $V_e$-value. Plotting the $\log M_{EVA}$ values found in this way against the

*Table 33*: Equations for calculating the averages of the molecular weight, the molecular inhomogeneity, and the polydispersity

| Quantity to be determined | $\dfrac{dV_e}{d\log M}$ not constant | $\dfrac{dV_e}{d\log M}$ constant |
|---|---|---|
| $M_n$ | $\dfrac{\sum \dfrac{dm}{d\log M}}{\sum \dfrac{dm}{d\log M}\cdot M^{-1}}$ | $\dfrac{\sum \dfrac{dm}{dV_e}}{\sum \dfrac{dm}{dV_e}\cdot M^{-1}}$ |
| $M_w$ | $\dfrac{\sum \dfrac{dm}{d\log M}\cdot M}{\sum \dfrac{dm}{d\log M}}$ | $\dfrac{\sum \dfrac{dm}{dV_e}\cdot M}{\sum \dfrac{dm}{dV_e}}$ |
| $M_z$ | $\dfrac{\sum \dfrac{dm}{d\log M}\cdot M^2}{\sum \dfrac{dm}{d\log M}\cdot M}$ | $\dfrac{\sum \dfrac{dm}{dV_e}\cdot M^2}{\sum \dfrac{dm}{dV_e}\cdot M}$ |
| $U$ | $\dfrac{M_w}{M_n}-1$ | |
| $g_m^2$ | $\dfrac{M_z}{M_w}-1$ | |

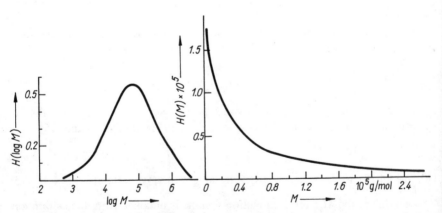

*Fig. 65*: Molecular weight distribution curve for an ethylene-vinyl acetate copolymer calculated from data given in Table 34

elution volume, gives the new calibration relationship. Further evaluation of the functional correlation between the log of the molecular weight and the elution volume is effected as described in 2.5.5.1.

Fig. 65 shows the molecular weight distribution curve for the sample, firstly in the usual form for *GPC* evaluation, $H(\log M)$ versus $\log M$, and secondly for comparison as the curve $H(M)$ versus $M$ deduced from equation (16).

$$H(M) = h' \cdot \frac{dV_e}{d \log M} \cdot \frac{\log e}{M} \tag{16}$$

Plotting $H(M)$ against $M$ gives a clearer indication of the actual conditions in the system and also permits a comparison with the distributions calculated using equations (20) or (30) in 2.1.5. (It should be noted that in the present form $H(M)$ is no longer normalized to unity.)

*Table 34*: Evaluation of the elution chromatogram of an ethylene-vinyl acetate copolymer with 42 per cent by weight of vinyl acetate according to [7]

| Elution volume (cm³) | $h_i$ (mm) | $h_i'$ (%) | $\log M_{PS}$ | $\log M_{EVA}$ | $M_{EVA} \cdot 10^{-3}$ (g · mol⁻¹) |
|---|---|---|---|---|---|
| 25 | 7.5 | 0.80 | 6.50 | 6.42 | 2620.3 |
| 26 | 17.0 | 1.82 | 6.30 | 6.21 | 1608.3 |
| 27 | 24.5 | 2.62 | 6.15 | 6.05 | 1115.3 |
| 28 | 33.0 | 3.53 | 5.95 | 5.84 | 684.6 |
| 29 | 42.5 | 4.55 | 5.80 | 5.68 | 474.7 |
| 30 | 53.5 | 5.72 | 5.65 | 5.52 | 329.2 |
| 31 | 65.5 | 7.01 | 5.50 | 5.36 | 228.2 |
| 32 | 78.0 | 8.35 | 5.35 | 5.20 | 158.3 |
| 33 | 88.0 | 9.42 | 5.15 | 4.99 | 97.2 |
| 34 | 91.0 | 9.47 | 5.00 | 4.83 | 67.4 |
| 35 | 91.5 | 9.79 | 4.85 | 4.67 | 46.7 |
| 36 | 79.0 | 8.45 | 4.65 | 4.46 | 28.7 |
| 37 | 69.0 | 7.38 | 4.50 | 4.30 | 19.9 |
| 38 | 57.0 | 6.10 | 4.35 | 4.14 | 13.8 |
| 39 | 46.0 | 4.92 | 4.15 | 3.93 | 8.5 |
| 40 | 32.5 | 3.48 | 4.00 | 3.77 | 5.9 |
| 41 | 22.0 | 2.35 | 3.80 | 3.56 | 3.6 |
| 42 | 15.5 | 1.76 | 3.65 | 3.40 | 2.5 |
| 43 | 10.0 | 1.07 | 3.45 | 3.19 | 1.5 |
| 44 | 6.5 | 0.70 | 3.25 | 2.97 | 0.9 |
| 45 | 3.5 | 0.37 | 3.10 | 2.82 | 0.7 |
| 46 | 1.5 | 0.15 | 2.90 | 2.60 | 0.4 |

In the examples, the zone broadening due to the equipment has not been taken into account. Where necessary, however, the elution curve must be corrected [4, 5] for the dispersion effect before conversion into the molecular weight distribution function.

## Bibliography

[1] BERGER, H. L. and A. R. SHULTZ, J. Polymer Sci. A **2** (1965), 3643
[2] ALTGELT, K. H., Advances in Chromatogr. **7** (1968), 3
[3] DETERMANN, H., *„Gelchromatographie, Gelfiltration, Gelpermeation, Molekülsiebe"*, Springer-Verlag Berlin—Heidelberg—New York 1967
[4] GLÖCKNER, G., *„Polymercharakterisierung durch Flüssigkeitschromatographie"*, VEB Deutscher Verlag der Wissenschaften Berlin 1980 und Dr. Alfred Hüthig-Verlag GmbH Heidelberg 1982
[5] TUNG, L. H., J. Appl. Polymer Sci. **10** (1966), 375
[6] BEREK, D., BAHRS, D., BLEHA, T. and L. SOLTES, Makromolekulare Chem. **176** (1975), 391
[7] BENOIT, H., REMPP, P. and Z. GRUBISIC, J. Polymer Sci., B, Polymer Letters **5** (1967), 753
[8] Group of authors, *„Grundlagen der Flüssigkeitschromatographie"*, Varian Aerograph USA 1971, Chapter 5.3.4.
[9] PERRY, S. G., AMOS, R. and P. I. BREWER, *„Praktische Flüssigkeitschromatographie"*, Plenum Press; New York—London (1972)
[10] MAJORS, R. E., J. chromatogr. Sci. **15** (1977), 333
[11] EVANS, J. M., Polymer Engng. and Sci. **13** (6) (1973), 401
[12] YAU, W. W. and S. W. FLEMMING, J. appl. Polymer Sci.**12** (1968), 2111
[13] Published Information by the firm of Waters Associates (1975)
[14] OUANO, A. C., Rubber Chemistry and Technology, Vol. 54 (3) (1981), 535
[15] KREMMER, T. and L. BOROSS, *„Gelchromatography. Theory, Methodology, Applications"*, Academic Verlag Budapest 1979

# 3. Microstructural Investigations

## 3.1. Introduction

The microstructure of a macromolecule is of particular importance to the properties of the polymeric material. Therefore, with a few exceptions, methods of microstructural investigation are among the standard methods of polymer characterization.

The term microstructure is defined as the structure of an individual molecule whereas the macromolecular or super-molecular structure describes the mutual arrangement of macromolecules or segments of a chain. The present Chapter deals only with selected methods of structure analysis applicable to determination of the microstructure.

In general, non-uniform chain structures are brought about during the polymerization steps. Even in the homopolymerization of a simple vinyl monomer a great number of structural units which differ from each other in their linkage and configurational isomerism can be expected. Chemical and stereochemical copolymers are insufficiently characterized by the use of average values of properties alone. The distribution of the components within a macromolecule must also be ascertained. Furthermore, information on branched and cross-linked structures is required because of their strong influence on the physical properties of the polymers.

The identification of the product has precedence over all special structural investigations. Once the type of material is known, specific methods of structural analysis can be selected.

For microstructural investigations, physical methods such as IR spectroscopy, high-resolution proton magnetic resonance, $^{13}$C-NMR spectroscopy and mechanical or dielectric relaxation measurements are generally preferred. For certain problems, however, chemical methods have been employed successfully.

The problems of the identification of polymer materials are presented in a generalized manner. With some knowledge of macromolecular chemistry, application to specific polymers will then be possible without any difficulty. This also applies to the characterization of network polymers and to the determination of branching because the same principles and modes of operation are applicable to other polymers. The n-ades principle for the representation of sequence and tacticity distribution is generally applicable.

The examples given below for the analysis of microstructure concentrate on the most important structural parameters of chain molecules and the methods required for the solution of the relevant problems. Other instrumental methods for the determination of microstructure, e.g. pyrolysis gas chromatography should be taken from the literature.

## 3.2. Identification of High-polymer Materials

### 3.2.1. Introduction

Chemical analysis of polymer materials is rendered more difficult because of the recent large increase in the number and types of such materials and because of a general trend towards modification and compounding of conventional polymers.

For exact identification of polymers it is particularly important for the samples to be available in the form of pure products without incorporated additives such as plasticizers, fillers or stabilizers. It is expedient to separate such additives by extraction or reprecipitation before identification. The solvents or mixtures of solvent and precipitant are substance-specific and should be chosen separately for each particular case (see, for example [1]).

In order to determine the quantitative composition of a polymeric material, the following sequence of operations is advisable:
— Comminution of the polymer sample,
— separation of additives,
— qualitative and quantitative investigation of the additives,
— identification and quantitative analysis of the isolated polymers.

### 3.2.2. Comminution of Polymer Samples

Mechanical comminution of the sample is necessary because the composition of polymer materials frequently shows inhomogeneities despite good processing techniques. Furthermore, some tests (e.g. monomer content and $H_2O$ content) are dependent on the particle size of the sample.

The products are comminuted with cutting tools such as shears, knives, or razor blades. Drilling, milling, etc. are also suitable. A smaller particle diameter can also be obtained by grinding precooled samples in mills. Depending on the elasticity characteristics of the sample, either dry ice or liquid nitrogen is used.

Identification of High-polymer Materials

After comminution, the samples must be conditioned over phosphorus pentoxide at room temperature for 24 h in a desiccator. In all mechanical loads and also in low temperature treatment, decomposition processes which may affect polymolecularity and sample composition must be taken into account. Since the analytical result may be affected, particular attention must be paid to the reproducibility and uniformity of the comminution processes.

### 3.2.3. Separation of Plasticizers and their Quantitative Determination

Normally, plasticizers are separated by extraction with diethyl ether and then identified. Stabilizers based on pure organic or organo-metallic compounds may only be partially separated. Extraction time depends on particle size and, above all, on the amount of plasticizer in the sample. 12 h are needed for particle size diameters up to 1 mm for products containing high levels of plasticizer and at least 24 h for products containing lesser amounts.

For the quantitative determination of the mass of plasticizer about 1 to 2 g of the comminuted sample are weighed exactly and extracted with anhydrous diethyl ether in a Soxhlet apparatus for the required time. After distilling off the ether and drying the extract at 105 °C to constant weight, the amount of ether soluble components is calculated from the difference in weight of the extraction flask before and after the extraction. Preparative separation of the plasticizer before identification of the polymer is performed in an analogous manner.

### 3.2.4. Identification of Plasticizers

Plasticizers include the esters of a few aliphatic and aromatic mono and dicarboxylic acids, aliphatic and aromatic phosphorus acid esters, ethers, alcohols, ketones, amines, amides and non-polar and chlorinated hydrocarbons.

Depending on their function, these additives are used in various mixtures. For their separation and qualitative detection, thin-layer chromatography ($TLC$) is preferred [2]. Usually Kieselgur plates, 0.25 mm thick, activated at 110 °C for 30 min, in the saturated vapour are used. Methylene chloride and mixtures of diisopropyl ether/petether at temperatures between 40° to 60 °C have been successfully used as the mobile phase.

For special separation problems, the necessary conditions should be taken from the relevant literature [1, 2]. Some common reagents for the detection of plasticizers are listed in Table 35. As evidence of the efficiency of $TLC$ separations, Table 36 shows the $R_f$-values for plasticizers based on o-phthalic acid or citric acid ester separated using methylene chloride or a mixture of diisopropyl ether/petether (70:30 volume %). Under the selected conditions, polymer plasticizers remain at the starting point. Although variation of the mobile and stationary phases and detection reactions allows selectivity of *TLC* to be increased, methods such as gas chromatography must be used for more complex plasticizer mixtures.

The gas-chromatographic separation of plasticizers can be effected directly or after conversion to low boiling point compounds [3]. This is usually achieved

*Table 35*: Reagents for the detection of plasticizers in TLC

| Type of plasticizer | Reagent |
|---|---|
| Phthalates, adipates | Ethanolic resorcinol solution (20%) with 1% zinc chloride (10 min at 100 °C); subsequently 4 N $H_2SO_4$ (20 min at 120 °C) |
| Phosphates | Ammonium molybdate solution made from 3 g of ammonium molybdate in 20 m*l* of perchloric acid (40%) and 5 m*l* of concentrated hydrochloric acid in 200 m*l* of water (10 min 100 °C); subsequently saturated hydrazine sulphate solution (20 min 110 °C); |
| Citrates | Ethanolic vanillin solution (20%) (10 min 80 °C); subsequently 4 N $H_2SO_4$ (10 to 20 min 110 °C) |

*Table 36*: $R_f$-values of phthalate and citrate plasticizers

| Plasticizer | $R_f$-value | |
|---|---|---|
| | Methylene chloride | Diisopropyl ether/petether 70:30 |
| dimethyl phthalate | 0.46 | 0.43 |
| diethyl phthalate | 0.49 | 0.58 |
| dibutyl phthalate | 0.60 | 0.76 |
| benzyl butyl phthalate | 0.67 | 0.65—0.75 |
| di(2-ethylhexyl) phthalate | 0.79 | 0.83 |
| | Methylene chloride | Diisopropylether/ petether 30:70 |
| triethyl citrate | 0.09 | 0.07 |
| tributyl citrate | 0.13 | 0.20 |
| tri(2-ethylhexyl) citrate | 0.30 | 0.57 |

Identification of High-polymer Materials 187

by a transesterification reaction with methanol or diazomethane. Relative retention volumes and working conditions for direct gas chromatographic analysis of some plasticizers are listed in Table 37. After separation of the plasticizers mixtures with liquid chromatography, identification by spectroscopic methods is not only still possible but advisable. For the common plasticizer combinations, the preceding separation can then often be omitted.

*Table 37*: Relative retention volumes of plasticizers in gas chromatographic separation

| Plasticizer | rel. retention volume | Working conditions |
|---|---|---|
| Dibutyl phthalate | 0.40 | Column filling: |
| Di(2-ethylhexyl)phthalate | 1.60 | 15% of Ultramoll III on Kieselgur*); |
| Benzyl butyl phthalate | 2.52 | 1 m glass column, diameter: 4 mm; |
| Dibenzyl phthalate | 14.40 | column temperature: 230 °C; carrier gas: He (170 m*l* per min) |
| Dibutyl adipate | 0.16 | Column filling: |
| Di(2-ethylhexyl)adipate | 1.00 | 15% of Ultramoll III on Kieselgur*); |
| Dibutyl acelate | 0.41 | 1 m glass column, diameter: |
| Di(2-ethylbutyl)acelate | 1.00 | 4 mm; column temperature: 240 °C; |
| Dihexyl acelate | 1.43 | carrier gas: He (196 m*l* per min) |
| Di(2-ethylhexyl)acelate | 2.56 | |

*) The composition of the stationary phase is given in [10].

## 3.2.5. Separation of Inorganic Fillers and Quantitative Determination

Since inorganic fillers are, in general, insoluble in organic solvents, they can be quantitatively separated from the soluble polymers by centrifugation using solutions of 5 per cent by weight in suitable solvents ([4] and Table 38) and subsequent pouring-off of the liquid. Dissolution has to be accelerated by shaking or stirring; nevertheless, it frequently takes 24 h. The conditions of centrifugation depend on the density and particle size of the fillers. In general, centrifugation at 4000 rpm for 2 hours will suffice. Carbon black cannot be separated completely, even under the most severe centrifugation and special methods are available for its quantitative separation [1]. The same applies to the fillers in insoluble polymers, for example, in thermosets and unsaturated polyesters. If required the fillers may be identified using the normal methods of qualitative inorganic analysis.

For the quantitative determination of fillers, about 2 g of the polymer material are weighed accurately and dissolved in 50 ml of solvent. After centrifuging at 4000 rpm for two hours, the solution is pipetted off from the centrifugate. The residue is suspended once more in the same solvent, centrifuged again and after isolation is washed once more with the solvent and then several times with methanol. The residue is dried at 105 °C to constant weight and then the solid matter is determined gravimetrically.

*Table 38*: Solubility of selected polymers

| Polymer | soluble | insoluble |
|---|---|---|
| Polyvinyl chloride | dimethyl formamide, tetrahydrofuran, cyclohexanone | alcohols, hydrocarbons, butyl acetate |
| Polyvinylidene chloride | ketones, tetrahydrofuran, dioxan, butyl acetate | alcohols, hydrocarbons |
| Polyamides | phenols, formic acid, trifluoroethanol | alcohols, hydrocarbons, esters |
| Polyethylene | tetralin, decalin, xylene, dichloroethylene at temperatures $>100$ °C | alcohols, petrol, esters |
| Polystyrene | ethyl acetate, benzene, acetone, chloroform, methylene dichloride | alcohols, water |
| Polyvinyl acetate | aromatic hydrocarbons, ketones, chlorinated hydrocarbons, alcohols | petrol |
| Polyvinyl alcohol | formamide, water | ether, alcohols, petrol, benzene, esters, ketones, hydrocarbons |
| Polyurethanes | dimethyl formamide | ether, alcohols, petrol, benzene, water |
| Polyacrylonitrile | dimethyl formamide, nitrophenols | esters, alcohols, ketones, hydrocarbons |
| Polyester | phenols, nitrated hydrocarbons, acetone, benzyl alcohol | esters, alcohols, hydrocarbons |
| Aminoplastics | benzylamine (160 °C), ammonia | organic solvents |
| Phenoplastics | benzylamine (200 °C) | |
| Cellulose, regenerated | SCHWEIZER's reagent | organic solvents |
| Cellulose ester | esters, ketones | aliphatic hydrocarbons |
| Polybutadiene | benzene | alcohols, petrol, esters |
| Polyisoprene | | ketones |
| Polyisobutylene | ether, petrol | alcohols, esters |
| Polymethacrylic acid | aromatic hydrocarbons, esters, ketones, chlorinated hydrocarbons | ether, alcohols, aliphatic hydrocarbons |

Apart from this method for quantitative determination of fillers and the special methods already indicated for thermosets, inorganic substances in polymers are frequently also determined by measurement of the „sulphate ash".

For this purpose, 1 to 2 g of the material are weighed exactly, put into a porcelain crucible and ashed at 500 °C for 30 minutes. Then the material is wetted with concentrated sulphuric acid and slowly heated with a Bunsen flame until white vapour is produced. This operation can be repeated several times as required. The residue is ignited to constant weight at a temperature between 500 and 700 °C in a muffle furnace.

$$\% \text{ of sulphate ash} = \frac{A}{E} \cdot 100 \tag{1}$$

$A$ = weight of the residue in g, $E$ = weight of the sample in g

Other inorganic additives such as buffering agents, prestabilizers and organometallic catalysts or stabilizers can also be detected by this method.

## 3.2.6. Identification of Stabilizers

Classification into heat stabilizers, light stabilizers (also known as ultraviolet absorbers) and antioxidants is indicative of the great number of compounds which have gained importance in this field.

Among the most important heat stabilizers are basic salts of heavy metals, metal salts of organic acids and nitrogenous organic compounds. Common antioxidants are phenols, aromatic amines, and benzimidazoles. Ultraviolet absorbers are substances which absorb strongly in the short-wavelength range but are transparent at wave numbers $< 25000 \text{ cm}^{-1}$ so that the stabilized material does not show any colouration. The hydroxybenzophenone derivatives, salicyl esters and benzotriazoles are particularly interesting examples.

The identification of stabilizers is a complicated problem because of their great number and the small amounts usually present. Added to this is the difficulty that stabilizers take part in transfer or rearrangement reactions during the moulding process so that only a portion of the stabilizer is found unchanged in the finished product [5]. Because of the intense toxicity of some of these decomposition products, their detection is of particular importance; however, this calls for special knowledge which goes beyond the scope of this book.

With few exceptions, stabilizers are identified after separation by solid-liquid extraction or after the removal of the polymer by precipitation from the diluted solution. Some extraction solvents for the most important stabilizers and polymers are given in Table 39; suitable solvent-precipitant mixtures are shown in Table 38.

*Table 39*: Extraction agents for stabilizers

| Polymer | Stabilizers to be extracted | Extracting agent |
|---|---|---|
| Polyvinyl chloride | organotin stabilizers | heptane:glacial acetic acid 1:1 |
| Polyvinyl chloride | N-containing organic stabilizers | methanol or diethyl ether |
| Polyethylene | antioxidants | chloroform |
| Polyoxymethylene | phenolic antioxidants | chloroform |
| Rubbers | stabilizers; accelerators | boiling acetone boiling water |

Of all the methods for the separation and detection of stabilizers, *TLC* has proved to be the best because of its high separation efficiency, rapidity and large variety of detection possibilities. Usually 0.5 mm thick silica-gel-G-plates are used, activated at 120 °C for 30 min, in a supersaturated atmosphere. Well-known techniques such as multiple separation in opposite or parallel direction allow the selectivities to be further increased. The selection of an appropriate mobile phase (see Table 40) determines the efficiency of separation. Advantage is taken of specific interactions and also of reactivity with the stabilizers under investigation. The best example so far of optimum separating conditions is that of the antioxidants which are first separated into 6 groups with different $R_f$-values using benzene and are then treated successively with 9 further solvent combinations. The combination of 4 different detection methods ensures reliable identification.

## 3.2.7. Identification of Polymers

The identification of polymers begins with a series of preliminary tests. In contrast to low molecular weight organic compounds, which are frequently satisfactorily identified simply by their melting or boiling point, molecular weight and elementary composition, precise identification of polymers is rendered difficult by the presence of copolymers, the statistical character of the composition, macromolecular properties and, in addition, by potential polymeric-analogous reactions. Exact classification of polymers is not usually possible simply from a few preliminary tests. Further physical data must be measured and specific reactions must be carried out in order to make a reliable classification. The efficiency of physical methods such as IR spectroscopy [4] and NMR spectroscopy [6] as well as pyrolysis gas chromatography [7] makes them particularly important for this purpose.

# Identification of High-polymer Materials

## 3.2.7.1. Preliminary Tests for Quantitative Identification

### 3.2.7.1.1. Exposure to Open Flames and Heating in a Glow Tube

Application of thermal energy causes pyrolysis. As a consequence, covalent bonds are broken. Fragments are produced whose chain length and structure are dependent upon the temperature, on the one hand, and on the type of bonds, on the other. If oxygen is present, oxidation occurs at the same time which may lead to the ignition of the sample. Since pyrolysis itself is an endothermic reaction, the required energy must be supplied, by an external heat source. The subsequent oxidation process is exothermic. If sufficient energy is released, the sample will burn spontaneously. If the temperature drops after removal of the flame, the polymer will be self-extinguishing. A survey of the burning behaviour of polymers and reactions during heating in a glow tube is given in Table 41.

Behaviour in an open flame can easily be observed by holding about 0.1 ... 0.2 g of sample with a suitable implement in the outer edge of a small Bunsen flame.

To test behaviour during dry heating, about 0.1 g of the sample is carefully heated in a 60 mm long glow tube with a diameter of 6 mm over a small flame. If heating is too vigorous, the characteristic phenomena can no longer be observed.

### 3.2.7.1.2. Depolymerization

Depolymerization, a special case of thermal degradation, can be observed particularly in polymers based on $\alpha, \alpha'$-disubstituted monomers. In these, degradation is a reversal of the synthesis process. It is a chain reaction during which the monomers are regenerated by an unzipping mechanism. This is due to the low polymerization enthalpy of these polymers. For the thermal fission of polymers with secondary and tertiary C-atoms, higher energies are required. In these cases elimination reactions predominate. This can be seen very clearly in PVC and PVAC. The polyethylene chain, however, is split statistically.

The depolymerization is in functional correlation with the molecular weight distribution and with the type of terminal groups which are formed in chain initiation and chain termination. The products of the thermal degradation of some well-known polymers are listed in Table 42.

Depolymerization, elimination and statistical chain-scission reactions can be used for polymer analysis. When the monomer is the main degradation product obtained, it can easily be identified by boiling point and refractive index.

Table 40: Experimental conditions for thin-layer chromatography

| Substances to be separated | Stationary phase | Mobile phase | Detection* |
|---|---|---|---|
| Organotin compounds | silica gel G | n-butanol/glacial acetic acid (98:2) | A. Reaction with $Br_2$ or UV irradiation<br>B. dithizone in chloroform (0.01%) |
| N-containing organic stabilizers | silica gel G | chloroform (free from ethanol) | p-dimethylaminobenzaldehyde; 1 g dissolved in 80 to 85 per cent phosphoric acid + 50 g of glacial acetic acid plus n-butanol to a total volume of 100 ml |
| Antioxidants | silica gel G | 1. benzene | 0.5% $Fe_2(SO_4)_3$ in 1 N $H_2SO_4$ + 0.2% $K_4[Fe(CN)_6]$ (1:1) |
| | | 2. petether/benzene/ glacial acetic acid (1:1:1) | |
| | | 3. glacial acetic acid/ acetone/methanol/benzene (10:5:20:70) | 2,6-dichloro-p-benzoquinone-4-chloride in ethanol (0.1%) |
| | | 4. glacial acetic acid/ water/ethanol (5:75:20) | diazotized 1-aminoanthraquinone (0.2%) + sodium acetate (0.01%) in water |
| | | 5. benzene/dioxan glacial acetic acid (125:25:4) | see item 1. |
| | | 6. cyclohexane/diethyl amine (4:1) | see item 1. |
| | | 7. benzene/ethyl acetate (100:5) | 0.4% p-methoxybenzaldehyde in methanol/conc. $H_2SO_4$ (95:5) |

| | | | |
|---|---|---|---|
| | | 8. hexane/acetone (80:20) | see item 1. |
| | | 9. hexane/methanol/diethyl amine (95:10:5) | see item 3. |
| | | 10. chloroform (1% ethanol) | see item 1. |
| Salicylates Resorcinol derivatives | aluminium oxide G impregnated with polyester according to data given in item [1] of bibliography | m-xylene/formic acid (98:2) | 0.02% diazotized 1-aminoanthraquinone + 0.01% sodium acetate in $H_2O$ |
| Benzophenones | silica gel G impregnated with 5% silicone | ethanol/water (7:3) | |
| Benzotriazoles | silica gel G | benzene/petether (7:3) | UV light or the same reagent as for salicylates |

*) The numbers given in parantheses refer to volume proportions

*Table 41*: Behaviour of selected polymers during heating in a glow tube and during combustion

| Polymer | Combustibility | Reaction of vapours | Appearance of flame | Smell of vapours |
|---|---|---|---|---|
| Polytetra-fluoroethylene | incombustible | strongly acid | | evaporates at red heat; pungent smell |
| Silicones | incombustible | | | |
| Aminoplastics (melamine and urea resins) | combustible with difficulty extinguish outside the flame | alkaline | slightly yellow with white edges | amines, ammonia and formaldehyde |
| Phenoplastics | | neutral to slightly acid | bright and sooting | phenol and formaldehyde |
| Polyvinyl chloride Polyvinylidene chloride | burns in the flame but difficult to ignite; goes out outside the flame | strongly acid | yellow-orange with green border | hydrochloric acid |
| Polyamides | combustible only in the flame | alkaline | yellow-orange with blue border | burnt horn |
| Polyvinyl alcohol | burns in the flame; extinguishes slowly outside the flame | neutral | bright | irritating |
| Polyvinyl formal | easy to difficult to ignite; | slightly acid | yellowish white | slightly sweet |
| Polyethylene | burns in the flame and continues to burn outside it | neutral | brilliant with blue core | paraffin (extinguished candle) |
| Polypropylene | | | | |
| Polyethylene-terephthalate | | | yellow-orange and sooting | sweetish aromatic |
| Polystyrene | easy to ignite; | neutral | brilliant, sooting | sweetish aromatic |
| Polymethyl methacrylate | burns in the flame and continues to burn outside of it | neutral | brilliant, crackling; yellow with blue core | sweetish fruit-like |
| Polyvinyl acetate | | acid | dark yellow brilliant; somewhat sooting | of acetic acid |

# Identification of High-polymer Materials

*Table 41* (continued)

| Polymer | Combustibility | Reaction of vapours | Appearance of flame | Smell of vapours |
|---|---|---|---|---|
| Butyl rubber; synthetic rubber vulcanized | combustible | neutral | yellow, sooting | of burnt rubber |
| Cellulose acetate | | acid | yellow-green with sparks | of acetic acid and burnt paper |
| Polyester (filled with glass fibres) | easy to difficult to ignite combustible | neutral | yellow, brilliant sooting | pungent smell — styrene |

Elimination and chain-scission reactions provide characteristic pyrograms which can often be identified by gas chromatography or IR spectroscopy.

For testing depolymerization behaviour, about 0.2 to 0.3 g of the polymeric substance is carefully and gently heated to a maximum of 500 °C in a small distillation flask. The distillate is collected in a receiver and its boiling point and refractive index are determined (see Table 42).

*Table 42*: Products of the thermal degradation of selected polymers

| Polymer | Degradation product | B. Pt. of the monomer at 1 atm. pressure (°C) | $n_D$ of the monomer |
|---|---|---|---|
| Polymethyl methacrylate | 90% monomer | 100.0 ... 101.0 | 1.4170 (20 °C) |
| Poly-α-methyl styrene | 90% monomer | 161.0 ... 162.0 | 1.5384 (17.4 °C) |
| Polystyrene | 65% monomer | 145.0 ... 145.8 | 1.5462 (20 °C) |
| Polyisobutylene | 20 ... 50% monomer | −6.9 | — |
| Polybutadiene | 50% monomer | −4.4 | — |
| Polyethylene | 1% monomer statistical fragments | — | — |
| Polymethyl acrylate | 1% monomer statistical fragments | — | — |
| Polyvinyl acetate | mainly acetic acid | — | — |
| Polyvinyl chloride | mainly hydrochloric acid | — | — |

## 3.2.7.2. Determination of Simple Physical Characteristic Values

### 3.2.7.2.1. Solubility

Solubility of high polymers in organic solvents is of great technological and analytical importance (see also 3.2.1.). Table 38 and [4] show the most widely used solvents for a variety of polymers. These solvent mixtures were usually devised empirically. In favourable cases, the chemical nature of the polymer can be suggested from a knowledge of the solubility.

For this purpose, about 0.1 g of the substance is added to 10 m$l$ of the solvent in a small stoppered vessel and then shaken at room temperature. Solution can take several hours and can usually be accelerated by slight heating.

### 3.2.7.2.2. Determination of Melting Point and/or Melting Range and of Density

Sharp melting points are exhibited only by substances having a crystalline structure. In polymers, however, there almost are always amorphous zones alongside the crystalline ones so that a definite and informative melting point can only be observed in a few cases. Amongst these are polyesters, polyamides, and polyethylenes and also vinyl polymers with a highly organized isotactic structure. Because of their amorphous character most polymers only have a more or less well defined softening range. The resulting softening or glass temperature can only rarely be

*Table 43*: Melting ranges and densities of crystalline polymers

| Polymer | Melting range (°C) | Density (kg/m$^3$) |
|---|---|---|
| Polyethylene-low density | 105 to 120 | 910 to 930 |
| Polyethylene-high density | 125 to 135 | 960 to 970 |
| Polypropylene, isotactic | 176 | 930 to 940 |
| Polybutene-1, isotactic | | |
| Modification I | 142 | 950 |
| Modification II | 126 | 960 |
| Modification III | 106 | — |
| Polyamide-6 | 215 to 220 | 1190 to 1200 |
| Polyamide-6.6 | 250 to 260 | 1240 |
| Polyester-2.10 | 72 to 79 | 1148 |
| Polyester-2.8 | 55 | 1281 |
| Polystyrene, isotactic | 235 to 250 | 1020 to 1060 |
| Polytetrafluoroethylene | 330 | 2120 to 2300 |

used for identification because it depends on the thermal history of the sample and the measuring techniques. A functional correlation between the microstructure and the molecular weight is also observed. Table 43 lists melting ranges and densities of crystalline polymers.

The melting point is determined in a capillary tube in a heating bath or electrically heated block. More sensitive measurements can be made using differential thermal analysis and differential scanning calorimetry.

For polymers, the density (Table 43) and all other properties are a statistical average. Furthermore, this parameter as well as the glass temperature are dependent upon the molecular weight and the microstructure. Nevertheless, it plays a part in the characterization of polymers and is frequently determined by the buoyancy method or in the density gradient column.

### 3.2.7.3. Determination of Chemical Characteristics

### 3.2.7.3.1. Saponification Value (S.V.)

Saponification value is defined as the amount of potassium hydroxide in mg which is required for the saponification of 1 g of ester. Determination of the $S.V.$ allows a classification of a comprehensive group of the C, H, O polymers. High saponification values are characteristic of ester groups in the product.

$S.V. < 200$
polyether, cellulose ether
polyvinyl acetals, phenoplasts,
polyaldehydes, epoxide
resins, polyvinyl alcohol
(completely saponified)

$S.V. > 200$
polyester, polycarbonates,
polymeric vinyl esters such as
polyvinyl acetate and polyacrylates
cellulose acetates

To determine the saponification value, duplicate experiments are carried out in which an exactly weighed amount of the sample (about 0.5 to 1 g) is placed in a 250 ml conical flask and mixed with 25.00 ml of 0.5 N alcoholic popassium hydroxide or sodium hydroxide solution. As a blank a third flask is prepared containing only 25.00 ml of the alcoholic potassium hydroxide solution. The mixtures are refluxed for 3 h. The hot samples are then titrated with 0.5 N hydrochloric acid using phenolphthalein as the indicator.

$$S.V. = \frac{(V_2 - V_1) \cdot 28.05}{W} \qquad (2)$$

$V_1$ = consumption in ml for the titration of the sample, $V_2$ = consumption in ml for the determination of the blank reading, $W$ = weight of sample in g

## 3.2.7.3.2. Hydroxyl Value (OH.V.) and Acid Value (A.V.)

In the analysis of polyesters, OH-value and $A.V.$ are additional important characteristics. OH-values are also of interest in the investigation of cellulose acetates, polyalkylene oxides and epoxide resins. The experimental determination is described in [1].

## 3.2.7.4. Chemical Determination

To illustrate the possibilities of chemical analysis, some group tests based on characteristic properties of various functional groups are discussed below.

Physical methods of identification are necessary for full characterization of the polymers of the C, H-group.

### 3.2.7.4.1. Chemical Determination of Heteroelements and Functional Groups

### 3.2.7.4.1.1. Detection of Nitrogen, Chlorine, Fluorine, Sulphur and Oxygen

Tests for elemental composition are among the most important reactions in polymer analysis. These tests have to be carried out with properly isolated polymers and with the exception of oxygen, the above mentioned hetero-elements must be determined in aqueous solutions of a sodium or potassium fusion.

In a dry semimicro test tube, about 0.1 g of the polymer is heated with an equal amount of sodium or potassium over a small flame and heated to red heat. After 30 to 60 s, the test tube is dropped into a beaker containing 5 to 10 ml of distilled water. The tube cracks and the melt is liberated. This has to be carried out with adequate safety measures because of the explosive reaction of the excess alkali metal with water. The solution thus obtained is filtered and tested as detailed below.

#### 3.2.7.4.1.1.1. Detection of Nitrogen

On a spot plate, 5 to 7 drops of the filtrate are mixed with one drop of an iron(II) salt solution and one drop of a strongly acidified (HCl) iron(III) salt solution. 10 per cent hydrochloric acid is then added until a slightly acid

# Identification of High-polymer Materials

response is indicated by litmus paper or universal test paper. If cyanide is present in the filtrate, a precipitate of Prussian blue will be formed indicating the presence of nitrogen in the original polymer.

### 3.2.7.4.1.1.2. Detection of Chlorine

On a spot plate, 5 drops of the filtrate are mixed with 5 drops of semi-concentrated nitric acid. Universal test paper is used to check that the solution is strongly acid. Three drops of 0.1 N silver nitrate solution are added. The formation of a white precipitate of silver chloride indicates the presence of chlorine in the polymer.

### 3.2.7.4.1.1.3. Detection of Fluorine

About 0.5 ml of the filtrate are mixed with the same amount of concentrated acetic acid. On a spot plate, a strip of zirconiumalizarin paper is wetted with 2 to 3 drops of 50 per cent acetic acid and 5 drops of the pre-treated filtrate are added. The formation of a yellow spot on the red paper or the yellow colouration of the solution not absorbed by the indicator paper is indicative of the presence of a fluorine-containing compound.

In addition to performing a blank experiment, a test with a fluoride-containing solution as a reference is advantageous.

### 3.2.7.4.1.1.4. Detection of Sulphur

1 to 2 ml of the filtrate are acidified with acetic acid and mixed with a few drops of an aqueous lead acetate solution. A black precipitate of lead sulphide is characteristic of the presence of sulphur in the polymer.

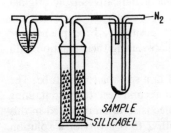

*Fig. 66*: Apparatus for the detection of oxygen

## 3.2.7.4.1.1.5. *Detection of Oxygen*

The sample is decomposed under inert gas and the carbon monoxide, produced from oxygen if present in the polymer, identified using a CO detection tube.

About 0.2 g of the polymer is placed in a decomposition apparatus (Fig. 66) which is then flushed with strong flow of nitrogen for 5 min.

After reducing the nitrogen supply until only 2 bubbles per second can be observed in the bubble counter, the bubble counter is replaced by a CO gas detection tube and the sample heated strongly. The presence of oxygen will be indicated by discolouration of the reaction layer in the detection tube.

## 3.2.7.4.2. *Group Test and Functional Groups*

### 3.2.7.4.2.1. *Detection of Ester Groups using Hydroxamic Acid*

Indications of the presence of a polyester or of ester groups in the polymer are indicated by high $S.V.$ (see 3.2.7.3.1.) and by the detection of the carboxylic acids released on saponification using the hydroxamic acid test.

Carboxylic acids react with hydroxylamine hydrochloride forming hydroxamic acids which in a slightly acid medium form intensely dark-red to violet complexes with $FeCl_3$.

About 50 mg of the comminuted polymer are mixed with 1 ml of a 0.5 N alcoholic hydroxylamine hydrochloride solution and 0.2 ml of 6 N sodium hydroxide solution. This mixture is heated for 30 s to boiling, then it is cooled and 2 ml of 1 N hydrochloric acid are added. After adding 1 to 2 drops of a 5 per cent iron(III) chloride solution, the test solution will become dark-red to violet in the presence of hydroxamic acids formed under these conditions.

### 3.2.7.4.2.2. *Detection of Phenols*

The detection of phenolic components in the sample is indicative of the original presence of condensation products of phenol and formaldehyde. Epoxides and polycarbonates based on bisphenol A also react positively. Free phenols or phenols produced by pyrolysis can be detected reliably using their reaction with diazo compounds to give azo dyes.

About 0.1 g of the sample material is pyrolyzed in a semimicro test tube. The decomposition products are distilled over into a second test tube containing 2 ml of a 2 N sodium hydroxide solution. The aqueous solution obtained in this way is mixed at a temperature of 5 to 10 °C with 2 to 3 drops of a solution of diazotized amine prepared in the following manner:

At a temperature below 5 °C, 1 ml of 2.5 M sodium nitrate solution is added dropwise to 2 ml of a 1 M solution of aniline in 3 N hydrochloric acid. Free nitrous acid may be checked for by means of starch iodide paper, spotting for blue colouration.

If phenol is present, the solution will turn red upon the addition of diazotized amine.

### 3.2.7.4.2.3. Detection of Formaldehyde

Formaldehyde is a constituent of the phenol and urea-formaldehyde condensates, of polyformaldehyde and polyvinyl formals. For detection, the aldehyde is releazed by acid splitting of the resin.

About 0.5 g of the comminuted sample, or a drop of a urea resin solution, are put into a test tube and heated to 60 °C in a water bath for 10 min with 2 ml of 81 per cent sulphuric acid and a few grains of pulverized chromatropic acid (1,8-dihydroxynaphthalene-3,6-disulphonic acid). The reagents must be added in the order specified. An intense violet colour indicates the presence of formaldehyde. A blank should be performed for comparison purposes.

### 3.2.7.4.2.4. Detection of the Epoxide Group

The qualitative test for the presence of epoxides in uncured products is effected using the ring opening reaction of the epoxide group with hydrochloric acid in ethers. This reaction is not applicable to cured epoxide resins because after curing, detectable epoxide groups will no longer be present. About 50 mg of the comminuted sample are dissolved in 2 ml of anhydrous neutralized dioxan and mixed with 3 drops of a 0.2 N solution of HCl in dioxan and 3 drops of cresol red solution as indicator. After 15 min, the colour of the sample is compared with a blank subjected to the same treatment. In the presence of expoxide resins, the colour will change to yellow due to the HCl consumption.

### 3.2.7.4.2.5. Detection of the Urethane Group

Pyrolysis of polyurethanes at temperatures above 150 °C, liberates, in addition to $CO_2$ and water, the diisocyanates, diamines and diols and their subsequent products. The composition of the gases is dependent on the pyrolysis conditions and temperature. The diisocyanates and diamines produced are used as indicators of the original presence of polyurethanes. Free isocyanate groups can be detected by their reaction with sodium nitrite in acetone.

About 0.1 g of the sample is comminuted and carefully heated in a test tube. The pyrolysis gases are conducted into anhydrous acetone. On adding 1 drop of a 10 per cent aqueous sodium nitrite solution, a yellow to red-brown colouration occurs if isocyanate groups are present.

The aromatic isocyanate components can be diazotized with sodium nitrite after acid decomposition of the products. When coupling with $\beta$-naphtholdisulphonic-acid sodium, characteristic colours develop. This reaction, however, is interfered with aromatic amines. Polyesters and polyether urethane can be distinguished from one another by the hydroxamic acid reaction (see 3.2.7.4.2.1.).

### 3.2.7.4.3. Specific Tests

### 3.2.7.4.3.1. Detection of Polystyrene

When subjected to heating, polystyrene easily depolymerizes into the monomer (see Table 42). At room temperature, the monomer can be converted into styrene dibromide with bromine in ethereal solution (melting point after recrystallization from petrol: 74 °C). On adding one grain of solid iodine to the solid polymer, polystyrene takes on a violet colour.

### 3.2.7.4.3.2. Detection of Polyvinyl Alcohol

Polyvinyl alcohol forms coloured inclusion compounds with iodine; however detection may be disturbed by dextrine and starch.

Two drops of a 0.1N solution of iodine in KI solution are added to 5 ml of neutral aqueous 1 per cent polyvinyl alcohol solution. This is diluted with distilled water until the colour developed (green, yellow-green, blue) can just be perceived. Subsequently, a spatula-tipful of borax is added to 5 ml of this diluted solution which is then shaken. After adding 5 drops of concentrated HCl, a green colouration occurs in the presence of polyvinyl alcohol which particularly visible in the undissolved borax.

### 3.2.7.4.3.3. Identification of Saturated Aliphatic Polymeric Chlorinated Hydrocarbons

The qualitative detection of this class of compounds is often achieved by means of the WECHSLER reaction.

According to WECHSLER, splitting off of HCl and addition of pyridine lead to the formation of compounds which continue to react when exposed to alkali and atmospheric oxygen. Eventually, cyanine dyes are produced whose colours depend on the chlorine-containing initial substance. The colourations are reproducible and characteristic and can therefore be used for identification (Table 44).

About 0.1 g of the sample is mixed with 1 ml of pyridine and allowed to stand for 3 min at room temperature; then, 2 to 3 drops of a 5 per cent methanolic sodium hydroxide solution are added. The colour occurring should be recorded immediately. A second sample is heated to boiling with 1 ml of pyridine for about 1 min and then 2 drops of a 5 per cent methanolic sodium hydroxide solution are added to the hot solution. The colour is also recorded.

The substance is identified according to Table 44.

*Table 44*: Colour reactions according to WECHSLER

| Polymer | Pyridine reaction in the cold | Pyridine reaction after heating and addition of alkali |
|---|---|---|
| Polyvinyl chloride | colourless | olive-green |
| Polyvinylidene chloride | black-brown | dark-brown |
| Post-chlorinated polyvinyl chloride | dark-red | dark-red |
| Post-chlorinated polyethylene | red-brown | brownish |
| Chlorinated rubber | olive-green | dark red-brown |
| Polychloroprene | colourless | colourless |
| Rubber hydrochloride | colourless | colourless |

## 3.2.7.4.3.4. *Identification of Cellulose and Its Derivatives*

The detection of cellulose compounds is based on the formation of furfural.

About 50 mg of the sample together with a few drops of syrupy phosphoric acid are carefully heated in a small porcelain crucible. The crucible is covered with filter paper which is wetted with one drop of aniline acetate (10 g of aniline in 100 ml of 10 per cent acetic acid) and weighted with a watch glass. In the presence of cellulose compounds, a pink-red spot appears on the filter paper. Lignin, sugar and other compounds, which also split off furfural or furfural derivatives when subjected to the this treatment, interfere.

Cellulose esters are identified by *TLC* after conversion to the respective hydroxamic acids [8]. A survey of the water-soluble cellulose esters is given in [9].

## 3.2.7.4.3.5. Detection of Urea with Urease

This identification is based on the ammonia produced by ureasesplitting of urea resins.

About 0.1 to 0.25 g of the pulverized sample material is heated in a 100-m$l$ conical flask together with 20 to 25 m$l$ of 5 per·cent sulphuric acid until there is no longer a smell of formaldehyde. Subsequently, it is neutralized with 6 N sodium hydroxide solution using phenolphthalein indicator and then 1 drop of 1 N sulphuric acid and 1 m$l$ of freshly prepared urease solution are added. A strip of red litmus paper is arranged so that it is exposed to the rising vapours in the flask. Then, the vessel is closed. A blue colouration of the litmus paper after a short time is indicative of ammonia, and, hence, of the presence of urea from urea resin.

## 3.2.7.4.3.6. Detection of Polyamides

Polyamides can be easily identified by the reaction of the amino terminal groups with dinitrofluorobenzene.

About 0.1 g of the comminuted sample is treated with 10 m$l$ of a 1 per cent solution of 1-fluoro-2,4-dinitrobenzene in ethanol to which about 5 m$l$ of a 2.5 per cent sodium hydrogencarbonate solution have been added. The solution may need to be warmed. Polyamides form bright yellow 2,4-dinitrophenyl derivatives while polyurethanes fail to react.

## 3.2.7.4.3.7. Reaction of Iodine with Polymers

By mixing polymers with elemental iodine, iodine inclusion compounds with the following colours are formed:

| | |
|---|---|
| Red to violet | polystyrene and styrene-copolymers |
| Pink | copolymers of vinyl chloride, chlorinated rubber, polychloroprene |
| Brownish | cellulose ether, cellulose ester, polyamides, polyisobutylene, polyvinylether |
| Dark-yellow | cellulose, polyvinyl chloride and its copolymers |
| Light yellow | polyvinyl alcohol |

To carry out the test, about 1 g of sample is mixed with 20 mg of solid iodine in a test tube and after one hour the colour is observed.

# Identification of High-polymer Materials

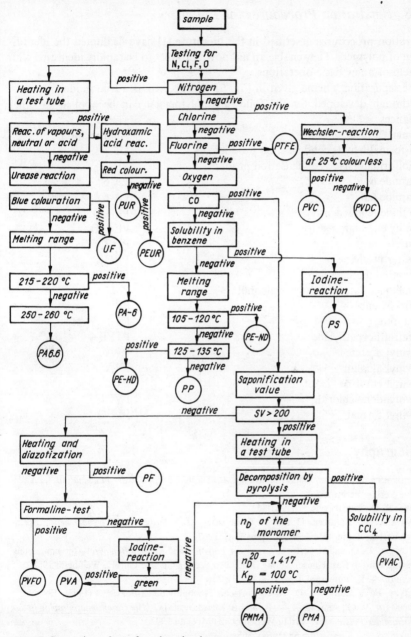

*Fig. 67*: Separation scheme for selected polymers

## 3.2.8. Separation Procedures for Polymers

Separation procedures described in the literature [1] have facilitated the identification of polymers. They enable an unknown product to be rapidly identified with the help of a few basic operations.

The separation scheme given in Fig. 67 is suggested as a trial model.

Although developed for the following polymers it can be applied to other substances.

| Polymer | Abbreviation |
|---|---|
| Urea-formaldehyde condensation product | UF |
| Phenolformaldehyde condensation product | PF |
| Polyamide 6 | PA-6 |
| Polyamide 6.6 | PA-6,6 |
| Polyethylene low density | PE-LD |
| Polyethylene high density | PE-HD |
| Polyurethane | PUR |
| Polyester urethane | PEUR |
| Polymethyl methacrylate | PMMA |
| Polymethacrylate with higher alcohols | PMA |
| Polypropylene | PP |
| Polystyrene | PS |
| Polytetrafluoroethylene | PTFE |
| Polyvinyl acetate | PVAC |
| Polyvinyl alcohol | PVA |
| Polyvinyl chloride | PVC |
| Polyvinylidene chloride | PVDC |
| Polyvinyl formal | PVFO |

## Bibliography

[1] SCHRÖDER, E., FRANZ, J. and E. HAGEN, „Ausgewählte Methoden der Plastanalytik", Akademie-Verlag, Berlin 1976
[2] BRAUN, D., Kunststoffe **52** (1962), 2
[3] WANDEL, M., TENGLER, H. and H. OSTROMOW, „Die Analyse von Weichmachern", Springer-Verlag, Berlin—Heidelberg—New York 1967
[4] HUMMEL, D. O. and F. SCHOLL, „Atlas der Polymer- und Kunststoffanalyse", 2nd completely revised edition Carl-Hanser-Verlag, München, Verlag Chemie GmbH, Weinheim 1981
[5] SCHRÖDER, E., Pure appl. Chem. **36** (1973), 233
[6] BOVEY, F. A. and G. V. D. TIERS, Fortschr. Hochpolymeren-Forsch. 3 (1963), 139
[7] BERESKIN, V. G., ALISCHOEV, V. R. and B. MEMIROVSKAJA, „Die Gaschromatographie in der Chemie der Polymeren", Verlag Wissenschaft, Moskau 1972
[8] HEKELER, W., Kunststoffe **58** (1968), 365

[9] GROSSE, L. and W. KLAUS, Z. analyt. Chem. **259** (1972), 195
[10] LEIBNITZ, E. and H. G. STRUPPE, „*Handbuch der Gaschromatographie*", Akademische Verlagsgesellschaft Geest & Portig KG, Leipzig 1966
[11] KRAUSE, A., LANGE, A. and M. EZRIN, „*Plastics Analysis Guide*", Carl-Hanser-Verlag, München 1983

## 3.3. Double Bonds in Polymers

### 3.3.1. Introduction

Double bonds in macromolecules are differentiated according to their position. Isolated unsaturated groups ($-CH=CH-$) in the chain are called vinylene, side-chain and terminal double bonds ($-CH=CH_2$) vinyl groups. If the preceeding unit is included ($-CH_2-CH=CH_2$), then this latter arrangement is termed the allyl group. If the hydrogen of the vinyl CH-group is replaced ($>C=CH_2$), the structural unit is termed the vinylidene group.

Numerous reactions lead to the incorporation of double bonds into macromolecules. They are not always desirable.

When the number of double bonds in a polymerizable monomer is greater than unity, the resulting polymer contains unsaturated structural units. Also the catalytic ring fission polymerization of unsaturated cyclic monomers leads to double bonds in the polymer chain.

$$CH=CH \atop (CH_2)_n \quad \overset{k_1}{\underset{k_2}{\nearrow \searrow}} \quad \begin{array}{l} -CH=CH-(CH_2)_n- \\ \\ -CH-CH- \\ \phantom{-CH-}(CH_2)_n \end{array}$$

$k_1$ = ring fission polymerization, $k_2$ = vinyl polymerization

Furthermore, due to termination and chain transfer to the monomer in the free radical vinyl polymerization, unsaturated structures in the terminal groups of the macromolecules [1] can be expected. Thermal decomposition reactions can also lead to double bonds in the polymer chains. In some vinyl polymers, vinylene groups appear (e.g. in irradiated high-pressure polyethylene) whose existence is traced back to a splitting-off of side-groups (hydrogen) during the process.

In high-pressure polyethylene, 68% of the unsaturated structures are in the form of vinylidene groups whereas the terminal vinyl group in Ziegler-catalyzed polyethylene comprises 43%, in Phillips polyethylene 94% and in high-pressure polyethylene 15% of all double bonds. These differences in the various polyethylene types clearly arise from the different polymerization methods.

Of particular industrial importance are the unsaturated polyester resins formed by polycondensation of maleic or fumaric acid and maleic anhydride with polyhydroxy compounds. The double bonds in the chain enable the polymer to

Table 45: Unsaturated structures in polymers

| Type of the double bond | Basic unit | Origin |
|---|---|---|
| isolated in the chain | $-CH=CH-$ | polymerization of diolefins and cyclo-olefins; polycondensation of unsaturated dicarboxylic acids (acid anhydrides); process-conditioned structural irregularities due to the splitting-off of a side group |
| located in the side-group; terminal | $-CH=CH_2$ | polymerization of diolefins; transfer reaction to the monomer and disproportionation termination in free radical polymerization |
| conjugated | $-CH=CH-CH=CH-$ | decomposition by reaction of the side groups; |
| | (bicyclic diene structure) | polyvinyl methyl ketone after treatment by bases |
| | (naphthyridine structure) | thermally aftertreated polyacrylonitrile under anaerobic conditions |
| $C=N$ | (triazine structure) | polycondensation of melamine, cyanuric acid |
| aromatic | (benzene ring) | polymerization of styrene and its derivatives; polycondensation of aromatic dicarboxylic and tetracarboxylic acids and phenols; polyaddition of aromatic diisocyanates and aromatic amines |

# Double Bonds in Polymers

copolymerize with suitable vinyl or allyl compounds forming hard cross-linked products. This property of double bonds in macromolecules to form networks under certain conditions is also a precondition for the industrial and commercial use of elastomers.

Aromatic double bonds will not be dealt with in this section (see Table 45) because they behave differently from aliphatic double bonds. $C=N$-groups in melamine and cyanuric acid derivatives are included because they exhibit a reactivity comparable with that of the vinyl, vinylene and vinylidene groups or with that of sequences of conjugated double bonds.

The position of double bonds in polymer chains is dependent on the polymerization and degradation mechanism. In polymers whose structural elements consist of diolefins, different steric configurations with respect to the double bond are to be expected. Like ring opening polymerization of unsaturated cyclic monomers, 1,4-addition leads to double bonds in the chain (vinylene groups). In this case, there is the possibility of cis-trans-isomerism, that is to say, all carbon atoms are in cis or trans position with respect to the double bond:

$$-CH_2\diagdown_{H}C=C\diagup^{CH_2-CH_2}_{H} \quad \diagdown_{H}C=C\diagup^{CH_2-}_{H} \quad \text{1,4 cis}$$

$$-CH_2\diagdown_{H}C=C\diagup^{H}_{CH_2-CH_2} \quad \diagdown_{H}C=C\diagup^{CH_2-}_{H} \quad \text{1,4 trans}$$

The 1,2 or vinyl propagation reaction leads to double bonds in the sidegroup (vinyl groups). In this case a distinction is made between isotactic and syndiotactic structures (see Section 3.5.). Stereoisomeric centres of adjacent basic structural elements which show the same steric arrangements give rise to isotactic subchains while inverse steric arrangements give rise to syndiotactic subchains.

$$-(CH_2-\underset{\underset{CH_2}{\overset{\|}{CH}}}{CH}-)_n \quad \text{1,2 iso(syndio) tactic}$$

Longer sequences of conjugated double bonds are formed by the thermal or alkaline aftertreatment of some special vinyl polymers (see Table 45). For example, by slowly heating polyacrylonitrile to 200 °C under exclusion of air carbon fibres are formed which are used as reinforcing material for unsaturated polyesters because of their high elasticity modulus. Thermal degradation reactions in which side groups are split off lead to conjugated double bonds in the case of longer polyvinyl chloride, polyvinyl acetate and polyvinyl alcohol sequences. The degradation reaction is indicated by a decolouration from white to orange and brown through to black [4], because even 5 to 7 double bonds in conjugation absorb in the visible region of the spectrum. In the presence of oxygen an oxidation reaction occurs simultaneously to the elimination leading, via hydroperoxides to carbonyl groups in the chain thereby accelerating degradation. The carbonyl groups incorporated in this way act as chromophores and influence the colour intensity of the degraded products.

Double bond determinations in polymers are of particular importance because of the information they provide about the reaction processes that have taken place. An excellent example of this is the analysis of the isomer distribution in synthetic rubbers. The IR spectrum of these products provides quantitative data on position and concentration of the unsaturated groups and, hence, knowledge of their microstructure. These analytical measurements allow conclusions to be drawn about the influence of polymerization conditions and about the state of incorporation of the monomer in chemical and stereochemical copolymers [5].

In general, the chemical composition of diolefin copolymers (and also of EPT rubbers) can be determined quantitatively via the double-bond content. These characteristics are of particular importance for the user because the technological properties of the synthetic rubbers depend largely on chemical and stereochemical composition.

In the analytical chemistry of unsaturated polyesters, double bonds play a particular role for three reasons:

Firstly, the copolymerization component is accessible quantitatively. Secondly, the unsaturated group content of the polyester can be determined. This is important because the network density in the final product can be regulated by co-condensing unsaturated with saturated compounds. Thirdly, the degree of isomerization is accessible to direct determination. This indicates the content of fumaric acid double bonds arising from isomerization of the maleic acid.

Methods are available [6] for the determination of these parameters.

Double-bond determination is also used to monitor ageing because unsaturated structures, developed as a reaction product diminish performance.

## 3.3.2. Methods for Determination of Double Bonds

### 3.3.2.1. Chemical Methods

Electrophilic addition reactions which lead to trans-substituted products [2] are characteristic of the high reactivity of aliphatic double bonds in organic polymers. An exception to this situation are the double bonds in $\alpha, \beta$-unsaturated carbonyl groups which can only be detected by a catalyzed bromination ($HgSO_4-H_2SO_4$).

Sufficient concentrations of vinylene and vinyl groups as well as C=N-double bonds can be detected qualitatively by discolouration of bromine solution. For this purpose, chlorinated aliphatic and aromatic hydrocarbons should be used as a solvent provided the polymer is soluble or capable of swelling in it. Possible further reaction of halogen must be taken into account in quantitative measurements. The addition of bromine and iodine-monochloride is of interest analytically.

The bromine-vapour method (ROSSMANN [3]) provides data about ageing in polyvinyl chloride [4] via the gravimetric determination of the amount of bromine taken up and via the double bond content in melamine and cyanuric acid condensates. For determination of the total proportion of unsaturated structural units in various types of rubber (Buna-S, butyl rubber, diene-polymerisates), the volumetric addition of iodine-monochloride is useful.

In general, in volumetric determination a surplus of the reagent (ICl-, Br/BrO$_3$-, Br$_2$-solution) is added to the dissolved sample and the consumption of reagent is determined by back titration with sodium thiosulphate solution after addition of potassium iodide. Alternatively, mercury(II) acetate can also be used as an electrophilic reagent (e.g. for the determination of styrene content in unsaturated polyesters).

As measure of the number of double bonds, the iodine value, defined as the amount of iodine in g which is chemically bonded by 100 g of substance, is calculated.

Other electrophilic reagents such as oxygen and chemically bonded oxygen in peroxy acids and peroxides are unsuitable for quantitative double-bond analyses in macromolecules. These substances initiate degradative reactions preferentially. Such selective oxidation reactions of unsaturated basic structural units in the polymer chain are used in structure analysis especially when determining the sequence length distribution in butadiene copolymers.

The double-bond content in unsaturated polyester resins is determined polarographically after hydrolytic degradation. This method permits the simultaneous determination of fumaric and maleic acids. The half-wave potential of maleic acid is $-1.36$ V and that of fumaric acid is $-1.60$ relative to a saturated calomel electrode. After calibration, the mass fractions of unsaturated acids and, hence, the degree of isomerization have to be determined from the wave heights.

### 3.3.2.2. Physical Methods

IR spectroscopy is a good method for double-bond analysis. The occurrence of the C=C bond is indicated in three places in the spectrum:

1. A stretching vibration of the band assigned to the =CH or =CH$_2$ group. These absorptions are not used for analytical purposes because they are obscured by CH stretching vibrations.
2. The stretching vibration of the C=C group. The frequency and intensity of this band are influenced by conjugation and substituents.
3. By the wagging vibration of the =CH group whose frequency and intensity is largely dependent on the environment and steric arrangement of the substituents.

Using the assignments in Table 46, unsaturated groups can be detected by IR spectroscopy and used for structure elucidation.

Table 46: Characteristic vibrations of the C=C and C=N groups in the IR spectrum

| Type of vibration | Wave number cm$^{-1}$ | Intensity |
|---|---|---|
| $v$ (=CH) | 3020 | m |
| $v_a$ (=CH$_2$) | 3080 | m |
| $v_s$ (=CH$_2$) | 3000 | m |
| $v$ (C=C)$_{olef.}$ | 1700 ... 1600 | v |
| $v$ (C=C)$_{arom.}$ | 1600 ... 1500 | v |
| $v$ (C=N) | 1690 ... 1600 | m |
| $\gamma_w$ (CH$_2$=CH—) | 998 | m |
| $\gamma_w$ (—CH=CH—)$_{trans}$ | 967 | s |
| $\gamma_w$ (CH$_2$=CH—) | 910 | s |
| $\gamma_w$ (—CH=CH—)$_{cis}$ | about 740 | s |

$m$ = medium; $s$ = strong; $v$ = variable

The electronic spectra of macromolecules with unsaturated structural units are not so well resolved as the vibrational spectra and their interpretation is thus more complicated. For isolated double bonds, the energy difference between the ground state ($\pi$) and the excited state ($\pi^*$) is so large that one can detect only the long-wave end of the absorption peak. For polymers with longer polyene sequences or with aromatic $\pi$ systems, additional information about the arrangement of the unsaturated structures in the chain are obtained by UV spectroscopy. Since in mesomeric resonance structures, the $\pi$ electrons no longer belong to individual bonds, it is impossible to allocate the absorption bands to the individual chromophores. The measured spectrum is always typical of the entire electron system.

Position and intensity of the absorptions of the phenyl ring can be influenced by substituents and neighbouring units. In order to simplify evaluation, specific absorptions (see Table 47) of the total complex are only used for analytical purposes.

*Table 47*: Characteristic wave numbers for double bond determination in the ultraviolet spectral region

| Polyene | Calibration substances [10] | | Degraded PVC [11] | |
|---|---|---|---|---|
| | $\tilde{\nu}_{max}$ (cm$^{-1}$) | $\lambda$ (nm) | $\tilde{\nu}_{max}$ (cm$^{-1}$) | $\lambda$ (nm) |
| $R_1-(C=C-)_2R_2$ | 46 080 | 217 | — | |
| $R_1-(C=C-)_3R_2$ | 38 760 | 258 | — | |
| $R_1-(C=C-)_4R_2$ | 33 780 | 296 | 32 470 | 308 |
| $R_1-(C=C-)_5R_2$ | 29 850 | 335 | 31 250 | 320 |
| $R_1-(C=C-)_6R_2$ | 27 780 | 360 | 29 410 | 340 |
| $R_1-(C=C-)_8R_2$ | 24 100 | 414 | 25 770 | 388 |
| $R_1-(C=C-)_{11}R_2$ | 21 280 | 470 | 22 030 | 454 |
| polymers with aromatic structure elements | $\tilde{\nu}$ (cm$^{-1}$) | | $\lambda$ (nm) | |
| polystyrene | 37 180; 38 310; 39 060; | | 269; 261; 256 | |
| polyalkylenterephthalate | about 41 400 | | 242 | |

## 3.3.3. Examples of Double Bond Determination

### 3.3.3.1. Volumetric Determination of Double Bond Content and Composition of Buna-S with Iodine Monochloride

The double bonds in a polymer due to butadiene incorporation are determined by electrophilic addition of iodine monochloride. The iodine value (*I.V.*) is calculated according to equation (1):

$$I.V. = 12.69 \cdot \frac{(V_2 - V_1)}{W} \tag{1}$$

$V_1$ = m*l* 0.1 N thiosulphate solution for the sample, $V_2$ = m*l* 0.1 N thiosulphate solution for the blank, $W$ = sample weight in g

0.20 g of the extracted sample are dissolved in chloroform in a 300 m*l* conical flask. After complete dissolution, 20.00 m*l* of a 0.2 N iodine monochloride solution is added. The iodine monochloride solution is prepared by dissolving the calculated amounts of iodine trichloride and iodine in a mixture of carbon tetrachloride and glacial acetic acid (1:2.3). The conical flask is stoppered and allowed to stand in darkness for 1 h. 100 m*l* of freshly prepared 10 per cent potassium iodide solution (free from iodine) is then added.

The mixture is titrated with 0.1 N thiosulphate solution with continuous shaking until decolouration of the aqueous phase. After adding 5 ml of starch solution, the flask is shaken vigorously in order to extract the iodine from the organic solvent layer. Titration is carried out with 0.1 N thiosulphate until both layers are decolourized. A blank is prepared and evaluated.

This method can also be used for products which are initially cross-linked. In order that the analytical value should be reproducible the sample must swell strongly in the solvent.

The iodine values for different concentrations of comonomers in styrene butadiene rubbers are given in Table 48. The great variation in the analytical values for pure polybutadiene and for copolymers with a low percentage of styrene is due to reactions occuring during latex processing.

Table 48: Iodine values of Buna-S types of different composition

| Composition % styrene | Iodine value |
|---|---|
| 0 | 442 ... 385 |
| 21 | 340 ... 360 |
| 36 | 291 |
| 66 | 142 |

## 3.3.3.2. Gravimetric Determination of Double Bond in Aged PVC (ROSSMANN's method [4])

The amount of added bromine is determined gravimetrically by the bromine-vapour method described by ROSSMANN. The iodine value ($I.V.$) is calculated from equation (2):

$$I.V. = \frac{100(A - W) \cdot f}{W} \qquad (2)$$

$A$ = weighed-out quantity in mg, $W$ = originally weighed-in quantity in mg,

$f = \dfrac{\text{molecular weight of iodine}}{\text{molecular weight of bromine}} = 1.588$

About 40 to 60 mg of the finely comminuted sample are distributed over an accurately preweighed glass plate and the quantity of the sample found by re-weighing. The plate is placed on a glass stand which is put into black desiccator. A dish containing 1 to 2 ml of liquid bromine and another dish containing the drying agent $P_2O_5$ are accommodated in the desiccator. After one hour, the sample is put under a blackened beaker for 5 min to remove excess bromine and then placed in a vacuum desiccator over calcium chloride and potassium carbonate for one hour. The sample is heated in an oven (65 ... 70 °C) to constant weight.

Due to the occurrence of consecutive reactions in the polymer, the number of measurable double bonds is not always comparable with the degree of degradation measured by other methods. This method for following the thermal degradation of a polymer, is therefore limited to the initial stage of the reaction.

Assuming that each macromolecule possesses the same number of double-bond sequences with comparable length distribution, the average number of double bonds per macromolecule can be calculated from the iodine value.

$$\text{Average number of double bonds} = \frac{I.V. \cdot M_{n,\text{PVC}}}{253.82 \cdot 100} \tag{3}$$

### 3.3.3.3. IR-Spectrometric Double Bond Determination: Isomer Analysis of Polybutadiene

In the IR spectrum, polybutadienes show intense absorptions which are due to the wagging vibration of the $=CH$-group (see Table 46). These bands are characteristic of the content of the 1,4 or 1,2 linked monomer units. As analytical

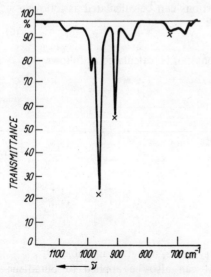

*Fig. 68*: A part of the IR-spectrum for isomer analysis of a low temperature polybutadiene

bands, the vibrations at 740 and 967 cm$^{-1}$ are selected for the 1,4 cis- and 1,4 trans-structure and at 910 cm$^{-1}$ for the 1,2 structure (Fig. 68). The overlapping of the absorptions at 967 cm$^{-1}$ is insignificant and is neglected.

Quantitative evaluation is effected by the swelling or simultaneous method by KIMMER [7]. This method provides relative values, the sum of the spectroscopically determined isomeric proportions corresponds to the sum of the double bonds.

Assuming 100% unsaturation the sum of the isomeric proportions is equal to unity:

$$\frac{c_{1,4\,cis} + c_{1,4\,trans} + c_{1,2}}{c_g} = 1 \tag{4}$$

$c_g$ = total concentration (primary mol $\cdot$ $l^{-1}$), $c_{1,4\,cis}$, $c_{1,4\,trans}$, $c_{1,2}$ = concentration of the various isomers (primary mol $\cdot$ $l^{-1}$)

If the BEER-LAMBERT law is assumed, the isomeric proportions are proportional to the absorbance:

$$A_{\tilde{v}(1,2,3)} = \log\left(\frac{I_0}{I}\right)_{\tilde{v}(1,2,3)} = \varepsilon_{(1,2,3)} \cdot c_{(1,2,3)} \cdot d \tag{5}$$

$\tilde{v}_1 = 740$ cm$^{-1}$ = 1,4 cis-structure, $\tilde{v}_2 = 967$ cm$^{-1}$ = 1,4 trans-structure, $\tilde{v}_3 = 910$ cm$^{-1}$ = 1,2 structure, $\varepsilon_{(1,2,3)}$ = absorbance coefficient of the absorption at $\tilde{v}_{(1,2,3)}$ in $l \cdot$ primary mol$^{-1}$ $\cdot$ cm$^{-1}$, $c_{(1,2,3)}$ = concentration of the structures corresponding to the respective wave numbers in units of primary mol $\cdot$ $l^{-1}$, $d$ = layer thickness in cm

With equation (4) the structural proportions can be calculated as follows:

$$c_{(1,2,3)} = \text{isomeric proportion}_{(1,2,3)} \cdot c_g \tag{6}$$

Therefore, the structural unit concentrations can be calculated as follows:

(a) $\quad 1,4$ cis % $= \dfrac{A_1}{\varepsilon_1} \cdot \dfrac{100}{K}$

(b) $\quad 1,4$ trans % $= \dfrac{A_2}{\varepsilon_2} \cdot \dfrac{100}{K}$

(c) $\quad 1,2$ % $= \dfrac{A_3}{\varepsilon_3} \cdot \dfrac{100}{K}$ $\tag{7}$

(d) $\quad K \quad = \dfrac{A_1}{\varepsilon_1} + \dfrac{A_2}{\varepsilon_2} + \dfrac{A_3}{\varepsilon_3}$

The relative absorbance coefficients $\varepsilon_1$, $\varepsilon_2$ and $\varepsilon_3$ are given in Table 49.

With slight modification, this method can also be applied to butadiene copolymers [8].

*Table 49*: Relative absorbence coefficients for rubber analysis

| Wave number (cm$^{-1}$) | Structural unit | Relative absorbance coefficient |
| --- | --- | --- |
| 910 | 1,2 linked | $\varepsilon_3 = 1.000$ |
| 967 | 1,4 trans | $\varepsilon_2 = 0.647 \pm 2\%$ |
| 740 | 1,4 cis | $\varepsilon_1 = 0.175 \pm 4.6\%$ |

About 0.1 g of the sample is properly swollen in carbon disulphide in a stoppered weighing bottle. The glassy gelatinous product is pressed between two KBr discs in a variable path length cell.

The maximum layer thickness is adjusted by means of the knurled nut of the cell so that the transmission ratio of the most intense peak appearing most distinctly is to be between 20 and 80%. The partial spectrum is recorded within the range from 600 to 1100 cm$^{-1}$ (Fig. 68). It is evaluated according to the base line method.

Table 50 shows the isomeric contents calculated from the spectrum of Fig. 68 using equation (7a—d).

## 3.3.3.4. UV-Spectrometric Determination of Polyene Sequences in Degraded PVC

While the method of bromine addition (3.3.3.2.) only provides data about the double bond content in degraded PVC, data about length and distribution of the polyene sequences formed during the reaction $R'-(C=C-)_n R''$ are obtained from UV measurements.

The characteristic wave numbers in the electronic spectrum (see Fig. 69) are classified according to Table 47, i.e., the electronic spectrum of the polymer

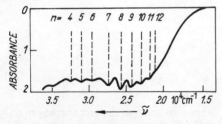

*Fig. 69*: Electronic spectrum of a PVC sample in THF solution aged thermally at 170 °C under an inert atmosphere ($X = 0.35\%$; $c_p = 0.144$ primary mol $l^{-1}$; $d = 2$ cm)

*Table 50*: Isomer distribution in a low temperature rubber

| Structural unit | Absorbance | Concentration (%) |
|---|---|---|
| 1,2 linked | 0.234 | 18.2 |
| 1,4 trans | 0.607 | 73.0 |
| 1,4 cis | 0.020 | 8.8 |

is considered as a superimposed absorption spectrum of polyenes with sequence lengths $n_1$, $n_2$ etc. [9]. For the absorbance of the absorption maximum assigned to the polyene sequence with $n$ double bonds:

$$A_n = \log\left(\frac{I_0}{I}\right)_n = \varepsilon_n \cdot c_n \cdot d \tag{8}$$

$\varepsilon_n$ = absorbance coefficient for structures with $n$ double bonds in $l \cdot \text{mol}^{-1} \cdot \text{cm}^{-1}$, $c_n$ = concentration of the polyene with $n$ double bonds in $\text{mol} \cdot l^{-1}$, $d$ = layer thickness in cm

If the intensity of the absorption maximum of a polyene is assumed proportional to the number $n$ of conjugated double bonds, then:

$$\varepsilon_n = \varepsilon_g \cdot n \tag{9}$$

With $\varepsilon_g = 3.4 \cdot 10^4 \, l \cdot \text{mol}^{-1} \cdot \text{cm}^{-1}$ (absorbance coefficient per double bond)

The following relationship exists between the primary molar concentration of the polymer $c_P$ and the molar concentration of the polyene sequence with $n$ double bonds $c_n$:

$$c_n = c_P \cdot \frac{X_n}{n} \tag{10}$$

$X_n$ = amount of HCl eliminated due to formation of polyene sequences with $n$ double bonds

We thus obtain for the quantity $X_n$:

$$X_n = \frac{A_n}{c_P \cdot \varepsilon_g \cdot d} \tag{11}$$

The numerical value of the constant $\varepsilon_g$ takes into account the fact that the maximum of the respective polyene is influenced by the tails of adjacent absorption bands.

Relating $X_n$ to the total conversion $X$ gives the percentage of the polyene sequence with $n$ double bonds:

$$\frac{X_n}{X} = \frac{A_n}{c_P \cdot \varepsilon_g \cdot d \cdot X \cdot 10^{-2}} \cdot 100 \tag{12}$$

$$\sum_{i=1}^{n} \frac{X_i}{X} = 100\%$$

About 100 mg of the PVC-sample are reduced to powder, weighed precisely, and thermally degraded in a nitrogen stream at 170 °C. The hydrogen chloride eliminated is transferred by means of the inert gas into a conductivity cell. The HCl concentration is measured conductometrically. The degradation conversion is calculated as the fraction $X$ of HCl eliminated relative to the total amount that can be eliminated. When conversion has reached 0.2 to 0.4%, the reaction is interrupted and the degraded product immediately dissolved in 10 ml of freshly distilled tetrahydrofuran.

The electronic spectra are measured in tetrahydrofuran solution by means of an UV spectrometer. The path length of the cell is 2 cm. Evaluation is by the base line method.

Table 51 lists the percentages of polyene sequences with $n$ double bonds relative to total degradation conversion calculated from the spectrum represented in Fig. 69. Polyene sequences with $4 < n < 12$ comprise about 50 per cent of total degradation. Short sequences $n < 4$ present in the product cannot be evaluated from the spectrum (see Fig. 69). The absence of absorption at wave numbers $< 15000$ cm$^{-1}$ shows that the maximum length of the polyene sequences is 25 to 30 conjugated double bonds.

*Table 51*: Content of the polyene sequences relative to the total degradation conversion calculated from the spectrum shown in Fig. 69

| Number of conjugated double bonds $n$ | Absorbance | Content of the total degradation conversion (%) |
|---|---|---|
| 4 | 1.77 | 5.2 |
| 5 | 1.81 | 5.3 |
| 6 | 1.76 | 5.2 |
| 7 | 1.88 | 5.5 |
| 8 | 1.98 | 5.8 |
| 9 | 1.93 | 5.6 |
| 10 | 1.85 | 5.4 |
| 11 | 1.72 | 5.0 |
| 12 | 1.55 | 4.5 |

## Bibliography

[1] ULBRICHT, J., „Grundlagen der Synthese von Polymeren", Akademie-Verlag Berlin 1978
[2] Group of authors, „Organikum" (5th revised edition), VEB Deutscher Verlag der Wissenschaften Berlin 1965
[3] ROSSMANN, E., Ber. Dtsch. chem. Ges. **65** (1932), 1847
[4] THINIUS, K. and E. HAGEN, Trans. J. Plastics Inst. (1966), 11
[5] DECHANT, J., „Ultraspektroskopische Untersuchungen an Polymeren", Akademie-Verlag, Berlin 1972
[6] SCHRÖDER, E., FRANZ, J. and E. HAGEN, „Ausgewählte Methoden der Plastanalytik", Akademie-Verlag, Berlin 1976
[7] KIMMER, W. and E. O. SCHMALZ, Z. analyt. Chem. **181** (1961), 229
[8] KIMMER, W., Plaste und Kautschuk **9** (1962), 135
[9] THALLMAIER, M. and D. BRAUN, Makromolekulare Chem. **108** (1967), 241
[10] BORSDORF, R. and M. SCHOLZ, „Spektroskopische Methoden in der organischen Chemie" (4th edition), Akademie-Verlag, Berlin 1982
[11] MINSKER, K. S., KRAI, O. E. and I. K. PACHOMOVA, Vysokomol, Soed. A **12** (1970), 483

## 3.4. Sequence Length Distribution in Copolymers

### 3.4.1. Introduction

Binary copolymers, consisting of the primary structural elements $A$ and $B$, may have different microstructures despite having the same chemical composition. For interpretation of physical properties and for characterizing the copolymerization process, data on the average composition of the copolymer alone will not therefore suffice. For this purpose, the intramolecular distribution of the two monomer units should be known.

The chemical microstructure of copolymers can be related to copolymerization statistics [1]. The probabilities $P_{i,j}$ ($i, j = A$ or $B$) for the addition of the monomers $A$ and $B$ to an active chain ending in $A$ or $B$ can be calculated making the following assumptions:

a) The probability of the addition of a monomer unit to the growing chain end is dependent only on the last monomer unit (terminal model).
b) The growth steps are irreversible.
c) The chain growth takes place according to a head-to-tail mechanism
d) The Steady State Approximation holds, that is to say, $\dfrac{dR\cdot}{dt} = r_i - r_t = 0$

   $r_i$ = rate of initiation, $r_t$ = rate of termination, $R\cdot$ = radical concentration
e) The monomer ratio at the reaction site remains constant.

# Sequence Length Distribution in Copolymers

Table 52 lists equations for the correlations between the probabilities $P(A)$, $P(B)$[1]), the addition probabilities $P_{i,j}$ and the kinetic parameters.

When the copolymerization mechanism does not follow the terminal model, i.e., when the penultimate unit influences the addition probabilities (penultimate

*Table 52*: Equations for the relationships between the $P_{i,j}$, $P(A)$, $P(B)$ and the kinetic parameters

| Addition probability | as a function of $r_A$, $r_B$ and $F_0$ | as a function of $x$ and $y$* | as a function of the run number $R$ |
|---|---|---|---|
| $P_{AA}$ | $\dfrac{r_A \cdot F_0}{r_A \cdot F_0 + 1}$ | $\dfrac{x}{x+1}$ | $\dfrac{[A] - \dfrac{R}{2}}{[A]}$ |
| $P_{AB}$ | $\dfrac{1}{r_A \cdot F_0 + 1}$ | $\dfrac{1}{x+1}$ | $\dfrac{\dfrac{R}{2}}{[A]}$ |
| $P_{BA}$ | $\dfrac{1}{\dfrac{1}{F_0} \cdot r_B + 1}$ | $\dfrac{1}{y+1}$ | $\dfrac{\dfrac{R}{2}}{[B]}$ |
| $P_{BB}$ | $\dfrac{r_B \dfrac{1}{F_0}}{r_B \cdot \dfrac{1}{F_0} + 1}$ | $\dfrac{y}{y+1}$ | $\dfrac{[B] - \dfrac{R}{2}}{[B]}$ |
| $P(A)$ | $\dfrac{r_A \cdot F_0 + 1}{r_A \cdot F_0 + 2 + r_B \dfrac{1}{F_0}}$ | $\dfrac{x+1}{2+x+y}$ | $\dfrac{[A]}{[A]+[B]}$ |
| $P(B)$ | $\dfrac{r_B \cdot \dfrac{1}{F_0} + 1}{r_A \cdot F_0 + 2 + r_B \cdot \dfrac{1}{F_0}}$ | $\dfrac{y+1}{2+x+y}$ | $\dfrac{[B]}{[A]+[B]}$ |

\* $x = r_A \cdot F_0 = \dfrac{F - 1 + \sqrt{(F-1)^2 + 4 r_A r_B F}}{2}$

\* $y = \dfrac{r_B}{F_0} = \dfrac{2 r_A \cdot r_B}{(F-1) + \sqrt{(F-1)^2 + 4 r_A r_B F}}$

$F_0 = \dfrac{[A]_0}{[B]_0}$ = molar composition of the initial mixture

$F = \dfrac{[A]}{[B]}$ = molar composition in the copolymer

$r_A$; $r_B$ = copolymerization parameters

[1]) $P(A)$, $P(B)$ is the probability that an arbitrarily selected unit in the polymer chain is $A$ or $B$.

effect), quantities characteristic of the sequence distribution can also be combined with the respective copolymerization parameters [1]. Characteristic values of the sequence distribution are the average sequence length $l$, the run number $R$, and the persistence ratio $\varrho$. The boundary conditions mentioned above, however, are not completely applicable to many industrial processes. In these cases, deviations from the distributions calculated on a statistical basis will occur because the simplified copolymerization model no longer applys.

## 3.4.2. Characteristic Values of Sequence Distribution

The representation of sequences of identical primary structural elements can be approached in two ways: the $n$-ades distribution and the sequence distribution.

The principle of the $n$-ades distribution consists of dividing the macromolecules into equally long subchains (diades, triades, tetrades, ..., $n$-ades = sequences of $n$ monomer units) which are still distinguishable experimentally. The percentages of all possible combinations of monomer units in such an $n$-ade are then calculated (Fig. 70). When the monomer unit consists of two structural elements, as is the case with the vinyl monomers for example, then each $n$-ade contains two partial structures with $(2n - 1)$ structural elements. The monomer $n$-ades must be distinguished from the linkage $n$-ades; each monomer $n$-ade exhibits $(n - 1)$ configurational linkages (see Chapter 3.5.).

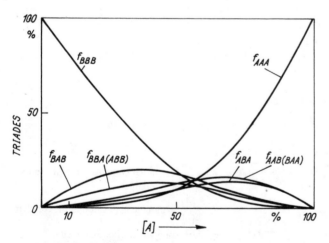

*Fig. 70*: Triade distribution as a function of the composition of a copolymer ($r_A \cdot r_B = 0.5$); for A-centred triades the following holds: $f_{AAA} = P_{AA}^2 \cdot P(A)$, $f_{AAB(BAA)} = P_{AA} \cdot P_{AB} \cdot P(A)$, $f_{BAB} = P_{AB}^2 \cdot P(A)$

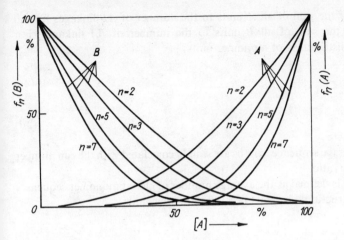

Fig. 71: Length distribution of the closed $A(B)$ sequences ($n = 2, 3, 5, 7$) as a function of the composition of a copolymer ($r_A \cdot r_B = 0.5$); $f_n(A) = P_{AA}^{n-1} \cdot P(A)$

Alternatively, the length distribution of sequences which consist only of one monomer type (A or B) is considered (Fig. 71). The sequence length distribution function of component $A$ or $B$ is given by equations (1a) or (1b) respectively:

$$W(A)_n = P_{AA}^{n-1} \cdot (1 - P_{AA}) \quad \text{(a)}$$
$$W(B)_n = P_{BB}^{n-1} \cdot (1 - P_{BB}) \quad \text{(b)} \tag{1}$$

With these relationships, the frequency of occurrence of the $A(B)$ sequences from $n$ and only $n$ structural element is defined. The sequence is limited by $B(A)$ units ($BA^nB$; $AB^nA$). The number-average sequence lengths $\bar{l}_A$ and $\bar{l}_B$ are calculated from these distribution functions according to equations (2a) and (2b):

$$\bar{l}_A = \frac{\sum_{n=1}^{\infty} nW(A)_n}{\sum_{n=1}^{\infty} W(A)_n} = \frac{1}{P_{AB}} \tag{2a}$$

$$\bar{l}_B = \frac{\sum_{n=1}^{\infty} nW(B)_n}{\sum_{n=1}^{\infty} W(B)_n} = \frac{1}{P_{BA}} \tag{2b}$$

$\bar{l}_A$ is the ratio of the sum of all $A$ units to the number of $AB$ linkages [2] and $\bar{l}_B$ is the ratio of the sum of all $B$ units to the number of $BA$ linkages. For statistically ideal distribution of monomer units:

$$\bar{l}_A = \frac{1}{P(B)} \qquad (2\text{c})$$

$$\bar{l}_B = \frac{1}{P(A)} \qquad (2\text{d})$$

The number-average sequence lengths are simply correlated with the run number and the persistence ratio.

The run number is defined as the average number of closed monomer sequences of the two basic structural elements per 100 monomer units.

$$R = \frac{200}{2 + r_A F_0 + r_B \frac{1}{F_0}} = \frac{200}{\bar{l}_A + \bar{l}_B} \qquad (3)$$

This is equal to the average number of $A - B$ and $B - A$ linkages between the two basic structural elements. For statistically ideal distribution of $A$ and $B$ units:

$$R_{i,s} = \frac{[A] \cdot [B]}{50} \qquad (4)$$

For alternating copolymers $R = 100$, for statistically distributed monomer units $R < 100$; when there is a tendency to block formation, the value of $R$ becomes very small ($R \to 0$). When a statistically ideal distribution is related to the run number, then for alternating structural units $R > R_{i,s}$ and for homogeneous blocks, $R < R_{i,s}$.

The persistence ratio is defined as the ratio of the real mean length of a closed $A$ or $B$ sequence to the average sequence length calculated for a statistically ideal distribution of monomer units for equal values of $P(A)$ or $P(B)$ [3].

$$\varrho = \bar{l}_A \cdot P(B) = \bar{l}_B \cdot P(A) \qquad (5\text{a})$$

or

$$\varrho = \frac{\bar{l}_A \cdot \bar{l}_B}{(\bar{l}_A + \bar{l}_B)} \qquad (5\text{b})$$

Experimentally determined persistence ratios can be compared with those from a BERNOUILLI distribution. If the copolymerization follows ideal statistics, then $\varrho = 1.0$ but if an alternating tendency is predominant in the polymer chain then $\varrho = 0.5$. When block structures exist in copolymers then $\varrho$ becomes considerably greater than $1$ ($\varrho \to \infty$).

Comparison with statistically ideal quantities can be avoided by calculating the product of the reactivity relations from the number-average sequence lengths using equation (6):

$$(\bar{l}_A - 1) \cdot (\bar{l}_B - 1) = r_A \cdot r_B \tag{6}$$

For statistically ideal copolymers $r_A \cdot r_B = 1$, with tendency to an alternating structure $r_A \cdot r_B < 1$ and for block formation $r_A \cdot r_B > 1$. Copolymers with the same product of reactivity relations $r_A \cdot r_B$ and identical composition exhibit the same sequence distribution.

### 3.4.3. Experimental Determination of Sequence Distribution

The sequence distribution of a polymer chain consisting of $A$ and $B$ is experimentally accessible if the frequency occurrence of any structural feature of $A$ or $B$ is measurable. Chemical or physical methods can be used.

The proportion of $A - A(B - B)$ or $A - B(B - A)$ bonds can be directly ascertained by chemical methods. They can be used wherever reproducible, analytically measurable structures of the same type result from the chemical reaction. Suitable methods involve oxidation, hydrolysis or pyrolysis [4].

In contrast, the proportion of $A$ and $B$ units in the centre of an $n$-ade is in general measured by physical methods [4]. This means that the physical property under consideration, which can be traced back to an $A(B)$ unit in a polymer chain containing $A$- and $B$-units, is dependent upon the neighbours of the $A(B)$ unit. The analytical methods most frequently used are IR spectroscopy and high resolution nuclear magnetic resonance ($^1$H and $^{13}$C NMR). The results are obtained in the form of $n$-ades and thus the $n$-ades distribution is chosen for the representation of the sequence distribution.

In IR spectroscopic evaluation of copolymer spectra for sequence analysis, frequency and intensity dependences are usually followed. Changes observed in the band profile and the width at half height should also be associated with sequence shortenings. The sequence-dependent frequencies (wave numbers) and their assignment in a few selected copolymers are listed in Table 53.

The influence of sequence distribution on absorption intensities can be detected using IR spectroscopy as follows. In a copolymer which is built up of $A$- and $B$-units, equations (7) and (8) are applicable to two bands which do not mutually superimpose:

$$A_A = \varepsilon_A \cdot c_A \cdot d \qquad A_B = \varepsilon_B \cdot c_B \cdot d \tag{7}$$

$$c_A + c_B = c_M \tag{8}$$

Substituting $c_A$ and $c_B$ in equation (8) by equation (7), leads to:

$$\frac{A_A}{c_M \cdot d} = -\frac{\varepsilon_A A_B}{\varepsilon_B \cdot c_M \cdot d} + \varepsilon_A \tag{9}$$

If $A_A/c_M \cdot d$ is plotted against $A_B/c_M \cdot d$ a linear dependence is obtained if the two measured absorptions are not sequence-sensitive. A non-linear dependence is obtained for sequence-sensitive absorptions.

If the intensity of the analytical band is influenced by the nearest neighbour, then the measured absorbance $A$ is the sum of the products of absorbance coefficients and concentrations multiplied by the layer thickness $d$.

$$A = (\varepsilon_1 \cdot c_{AAA} + \varepsilon_2 \cdot c_{AAB} + \varepsilon_3 \cdot c_{BAB}) \cdot d \tag{10}$$

$\varepsilon_1; \varepsilon_2; \varepsilon_3$ = absorbance coefficient of the triades $AAA$; $AAB$; $BAB$; in cm$^2 \cdot$ g$^{-1}$, $c$ = concentration of the respective triades in g $\cdot$ cm$^{-2}$ ($n$-ades of a higher order cannot be distinguished using IR spectroscopy), $d$ = layer thickness in cm

Frequently, it is desirable to make measurements relative to the absorbance of a band that is not sequence-dependent, because the percentage of triades can be immediately determined from the spectrum (equation (11)):

$$\frac{A}{A_u} = \frac{\varepsilon_1}{\varepsilon_u} \cdot \frac{f_{AAA}}{P(A)} + \frac{\varepsilon_2}{\varepsilon_u} \cdot \frac{f_{AAB}}{P(A)} + \frac{\varepsilon_3}{\varepsilon_u} \cdot \frac{f_{BAB}}{P(A)} \tag{11}$$

$\varepsilon_u$ = absorbance coefficient of the sequence-independent band in cm$^2 \cdot$ g$^{-1}$, $f$ = proportions of the respective triades

Equation (11) links the relative intensities of the absorptions in the spectrum with the chain structure. The ratios $\frac{\varepsilon_i}{\varepsilon_u}$ ($i = 1, 2, 3$) are determined using model substances.

The simultaneous presence of a large number of inseparable structures in the polymer chain also shows up as a superposition of partial spectra in the NMR spectrum. The number of these partial spectra superimposed, (i.e. the length of the subchains ($n$-ades), whose central protons or $^{13}$C-atoms just have different partial spectra) is dependent on measuring frequency. In general, the higher the measuring frequency, the longer the subchains which are distinguishable. The information relevant to sequence analysis given by an NMR spectrum is dependent on recording conditions and the correct choice of solvent. The parameters which allow evaluation of the NMR spectrum relevant to the microstructure of the polymer chain are shown in Table 54 (KELLER [6]).

The signals in the spectrum can be classified (Table 55):
— By comparison with the NMR spectra of similar copolymer systems [6],
— by comparison with the NMR spectra of copolymers of different compositions,
— by comparison with the NMR spectra of homopolymers.

*Table 53*: Sequence-dependent vibrations of selected copolymers and their assignment

| Copolymer | Assignment | $n$-ade | Band position ($cm^{-1}$) |
|---|---|---|---|
| Vinyl chloride (VC) vinilydene chloride (VDC) | $CH_2$-deformation vibration | VC-VC | 1434 |
| | | VC-VDC | 1420 |
| | | VDC-VDC | 1405 |
| | CH-deformation vibration | VC—VC—VC | 1247 |
| | | VC—VC—VDC | 1235 |
| | | VDC—VC—VDC | 1197 |
| Vinyl chloride isobutylene (IB) | $CH_2$-deformation vibration | VC—VC | 1425 ... 1435 |
| | | VC—IB | 1451 |
| | | IB—IB | 1470 |
| Trifluorochloroethylene (TFCF) — tetrafluoroethylene (TFE) | C—Cl stretching vibration | TFCE—TFCE—TFCE | 971 |
| | | TFCE—TFCE—TFE | 967 |
| | | TFE—TFCE—TFE | 957 |
| Ethylene (E) — propylene (P) | $CH_2$-rocking vibration | E—E | 723 |
| | | E—P | 735 |
| | | P—E | 752 |
| | | P—P | 815 |

The scheme prepared by PRIMAS [7] also furnishes information about the classification. It can also be used to calculating chemical displacements for high polymers [6].

Table 55 shows the line classification of the $CH_2$-proton groups for the respective diades in ethylene-vinyl acetate copolymers. Depending on whether the influence of the substituents extends over 3, 5 or 7 bonds from the protons of the $CH_2$-group under consideration, the spectrum must be interpreted in diades, triades or tetrades. The relative intensities of the separated partial spectra are then identical with the concentrations of the partial structures.

*Table 54*: NMR data used for characterization and quantitative evaluation [6]

a) position of the lines in the spectrum (chemical shift measured in ppm relative to a reference substance)
b) intensities (areas under the peaks whose sizes are proportional to the number of the protons $^{13}C$-atoms attaining resonance);
c) coupling constants $J$ (Hz) (only in the case of $^1H$-NMR)
d) line widths (Hz or ppm)
e) multiplicities of the signals in the case of equivalent nuclei

*Table 55*: Possible arrangements of the methylene protons in ethylene-vinyl acetate copolymers in diades and their classification

| Diade* | Probability | Symbols | Triple chain | $\delta$, chemical shift in ppm (relative to TMS) |
|---|---|---|---|---|
| AA | $P_{AA} \cdot P(A)$ | 00 00 | (0)0 0 | 1.27 |
| AA | $P_{AA} \cdot P(A)$ | 00 00 | 0 0 0 | |
| AB | $P_{AB} \cdot P(A)$ | 00 10 | (0)0 0 | |
| AB | $P_{AB} \cdot P(A)$ | 00 10 | 0 0 1 | 1.50 |
| BA | $P_{BA} \cdot P(B)$ | 10 00 | 1 0 0 | |
| BB | $P_{BB} \cdot P(B)$ | 10 10 | 1 0 1 | 1.88 |

Monomer units under considerations.
Symbols: A — unit                       B — unit
$-CH_2-CH_2-$                        $-CH_2-\underset{\underset{OCOCH_3}{|}}{CH}-$
  0     0                                 0     1

*) $\delta = \dfrac{\nu_{substance} - \delta_{standard}}{\nu_0}$      $\nu$ = resonance frequency
                                                $\nu_0$ = operating frequency of the spectrometer

Criteria helpful in taking the correct decisions for effecting separation are given in [8, 9]. In these publications, the correlation between experimentally determinable quantities and quantities characterizing the sequence distribution is described in detail.

### 3.4.4. Examples

### 3.4.4.1. NMR Determination of Sequence Distribution using Ethylene-Vinyl Acetate Copolymers with Low Vinyl Acetate Content [9] as an example

Before the analysis, the antioxidants contained in the sample are removed by precipitating the polymer from a one per cent solution in chloroform using methanol. The sample purified in this way is prepared as a solution of 7 per cent by weight by shaking with chloroform. The spectrum is recorded on a high resolution NMR spectrometer equipped with a variable temperature sample head. A measuring frequency of 100 MHz and a temperature between 100 and 120 °C are used. An important factor in recording the spectra is the homogeneity of the magnetic field since the symmetry of the resonance lines depends on it. Tetramethylsilane or hexamethyl disiloxane is used as a reference.

Fig. 72 shows the spectrum obtained for an ethylene-vinyl acetate copolymer with a composition of 12.2 mol % of vinyl acetate and 87.8 mol % of ethylene. The assignment of the signals is given in Table 55. The evaluation method is explained with reference to the spectrum shown in Fig. 72.

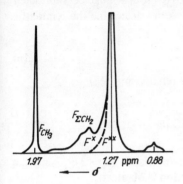

*Fig. 72*: Part of a NMR spectrum of an ethylene-vinyl acetate copolymer containing 12.2 mol % of vinyl acetate

The run number $R$ is determined by separating the partial spectra of the homogeneous $A$-sequences (1.27 ppm) from the remaining $CH_2$-spectrum. This separation is effected as illustrated in Fig. 72.

$$F^* = F_{-CH_2-\text{total}} - F^{**} \tag{12}$$

$F^*$ = area of the non-symmetric separated $CH_2$-spectrum, $F^{**}$ = area of the symmetric separated $CH_2$-spectrum

The areas of the parts of the spectrum isolated in this way are functions of the parameters characterizing the sequence distribution. $D_{\text{exp}}$ is defind as:

$$D_{\text{exp}} = \frac{F_{CH_3}}{F^*} = \frac{I_{CH_3}}{I^*} \tag{13}$$

$F_{CH_3}$ = peak area of the acetate group

This quantity $D$ can also be expressed as a relationship between the intensities of the protons of the respective $n$-ades, and as a relationship between the addition probabilities $P_{i,j}$ and the mole fractions $P(A)$ and $P(B)$. The number of protons are introduced as intensity factors. The intensity relative to the total number of all monomer units of the acetate group in the copolymer is obtained from equation (14)

$$I_{CH_2} = 3 \cdot P(B) \tag{14}$$

while the intensity [10] relative to the total number of $CH_2$ groups in all monomer units is given by equation (15).

$$I_{CH_2} = 2\{2 \cdot P(A) + P(B)\} \tag{15}$$

In order to obtain $I^*$, the intensities of the $CH_2$-groups must be subtracted from the $I_{CH_2}$-values because the $CH_2$-group intensities determine the symmetric part of the methylene proton spectrum.

$$I^{**} = 2 \cdot P(A) \cdot P_{AA}^{n-2} \cdot (1 + P_{AA}) \tag{16}$$

In this way, using equation (13) we obtain the relationship (17):

$$D^{n\text{-ade}} = \frac{3 \cdot P(B)}{2\{2 \cdot P(A) + P(B) - P(A) \cdot P_{AA}^{n-2} \cdot (1 + P_{AA})\}} \tag{17}$$

which applies to any $n$-ades with $n > 2$.

The spectrum shown in Fig. 72 is evaluated in triades ($n = 3$). Using the equations of Table 52 the following $D$-value function is obtained:

$$D^{\text{Triade}} = \frac{3 \cdot [A] \cdot [B]}{2 \cdot [A] \cdot [B] + 3 \cdot [A] \cdot R - \frac{1}{2} R^2} \tag{18}$$

Since the calculated intensity relationships are equivalent to the measured area relationships, equation (18) reflects the correlation between $R_{\text{exp}}$ and $D_{\text{exp}}$ if the value of $D_{\text{exp}}$ established by planimetry is used.

$$R_{\text{exp}}^{\text{Triade}} = 3 \cdot [A] - \sqrt{9 \cdot [A]^2 - 2 \cdot [A] \cdot [B] \cdot \left(\frac{3}{D_{\text{exp}}} - 2\right)} \tag{19a}$$

For the spectrum shown in Fig. 72 the following values characterizing the sequence distribution are obtained:

$D_{\text{exp}} = 0.431 \quad \bar{l}_A = 8.36 \quad \varrho = 1.02$

$R_{\text{exp}} = 21 \quad \bar{l}_B = 1.16 \quad r_A \cdot r_B = 1.18$

This shows the presence of a statistical distribution of the $A$- and $B$-units.

This example demonstrates clearly the linking of microstructural quantities with spectroscopic data.

The availability of $^{13}$C-NMR spectroscopy has also led to progress in the field of sequence analysis. While, in the above example, the effects of the sequence distribution must be estimated from the poorly resolved spectrum of the methylene protons, in the $^{13}$C-NMR spectrum of the EVA copolymers, completely separated resonances in both the methine atom (CH) region and the methylene carbon atom

($CH_2$) region can be distinguished. The chemical shift $\delta$ permits the nature of the $n$-ade to be stated (in Fig. 73 interpreted as triade). The composition of the copolymers and the $n$-ades distribution can be evaluated from the intensities (peak areas) using easily derived equations [11].

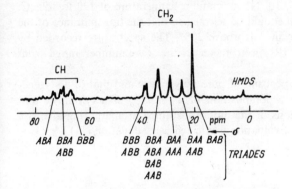

*Fig. 73*: Proton-noise decoupled 22.63-MHz-$^{13}$C-PFT-NMR spectrum of an ethylene-vinyl acetate copolymer (43.0 mol % of ethylene) showing the line assignments for the determination of microstructure

When the chemical composition of the copolymers is known, the run number $R_{exp}$ is determined using the peak areas of the methine carbon atom resonances according to equations (19b, c):

$$R_{exp}(CH) = \frac{2F_{ABA} + F_{ABB}}{F_{ABA} + F_{ABB} + F_{BBB}} \cdot B_{Vac} \qquad (19b)$$

and

$$R_{exp}(CH) = \frac{2F_{ABB}}{F_{ABB} + 2F_{BBB}} \cdot B_{Vac} \qquad (19c)$$

$B_{Vac}$ = mol % of vinylacetate in the copolymer

Further equations for calculating $R_{exp}$ are given in [11]. A quantitative evaluation of the spectrum shown in Fig. 73 results in the following values characterizing the sequence distribution:

$R_{exp} = 49 \quad \bar{l}_A = 1.8 \quad \bar{l}_B = 2.3 \quad \varrho = 1.01 \quad r_A \cdot r_B = 1.04$

## 3.4.4.2. IR Determination of Sequence Distribution using Styrene-Acrylonitrile Copolymers [12] as an Example

The sample is prepared in the form of a pressed foil by means of a hydraulic press using a pressure of $2.5 \cdot 10^7$ N $\cdot$ m$^{-2}$ and a temperature of 120 to 150 °C. The layer thickness of the foil should be selected so that the transmittance of the absorption recorded at 1603 cm$^{-1}$ is above 20%. The spectra are recorded by means of a high resolution IR spectrometer in the wave-number range from 1550 to 1650 cm$^{-1}$.

Fig. 74 shows a part of a spectrum of pure polystyrene and that of a styrene-acrylonitrile copolymer with a composition of 44 mol % of acrylonitrile and 56 mol % of styrene. Vibrations of the monosubstituted aromatic ring give rise to the measured absorptions. Evaluation is effected using equation (11).

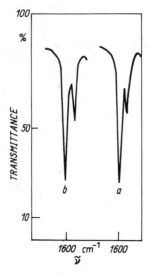

*Fig. 74*: Intensity changes in the polystyrene doublet at 1585 and 1603 cm$^{-1}$
a — polystyrene, b — styrene-acrylonitrile copolymer with 44 mol % of acrylonitrile

The absorbance of the band at 1585 cm$^{-1}$ is corrected by $0.05 \cdot A_{1603}$ in order to take into account the error caused by band superimposition. The absorbance coefficients of the absorption at 1603 cm$^{-1}$ relative to the absorbance coefficients of the sequence-insensitive band at 1585 cm$^{-1}$ can be obtained from Table 56.

# Sequence Length Distribution in Copolymers

They are dependent on the microstructure of the polymer chain. The normalization

$$\frac{f_{SSS}}{P(S)} + \frac{f_{SSA(ASS)}}{P(S)} + \frac{f_{ASA}}{P(S)} = 1 \tag{20}$$

results in the following equation for the proportion of triads consisting of styrene units only:

$$f_{SSS} = \frac{\dfrac{A_{1603}}{A_{1585}} - \varepsilon_{ASA}}{\varepsilon_{SSS} - \varepsilon_{ASA}} \cdot P(S) \tag{21}$$

*Table 56*: Reduced absorbance coefficients of the SSS-, SSA(ASS)- and ASA-triads in styrene-acrylonitrile copolymers

| Triade | reduced absorbance coefficients $\dfrac{\varepsilon_{1603}}{\varepsilon_{1585}}$ |
|---|---|
| SSS | 3.62 |
| SSA (ASS) | 2.75 |
| ASA | 2.75 |

Using the equations given in Table 52, the following relationship is obtained for the correlation between the measured ratio of the absorbances and the run number $R$:

$$R = 2\left\{[S] - [S]\left(\frac{\dfrac{A_{1603}}{A_{1585}} - \varepsilon_{ASA}}{\varepsilon_{SSS} - \varepsilon_{ASA}}\right)^{1/2}\right\} \tag{22}$$

From the partial spectrum of the copolymer shown in Fig. 74 the following characteristic values are obtained for the sequence distribution:

$\dfrac{A_{1603}}{A_{1585}} = 2.8 \qquad \bar{l}_S = 1.32 \qquad \varrho = 0.58$

$R = 85 \qquad \bar{l}_A = 1.035 \qquad r_A \cdot r_S = 0.011$

This shows the presence of an alternating structure. Whether an evaluation on the basis of the product of the reactivity relationships from the average sequence length is meaningful depends on the accuracy of the experimentally determined run numbers. It is advisable to estimate the magnitude of the relative error $\left|\dfrac{\Delta(r_A \cdot r_B)}{r_A \cdot r_B}\right|$ as a function of the composition of the sample [9].

## Bibliography

[1] CHONG WHA PYUN, J. Polymer Sci. A-2 **8**, (1970), 1111
[2] HARWOOD, H. J. and W. M. RITCHEY, J. Polymer Sci. B Polymer Letters **2** (1964), 601
[3] COLEMAN, B. D. and T. G. FOX, J. Polymer Sci., A **1** (1963), 3183
[4] HARWOOD, H. J., Angew. Chem. **77** (1965), 405, 1124
[5] DECHANT, J., *„Ultrarotspektroskopische Untersuchungen an Polymeren"*, Akademie-Verlag Berlin 1972
[6] KELLER, F. and H. ROTH, Plaste und Kautschuk **15** (1968), 800
[7] PRIMAS, H., ERNST, R., ARND, R. and P. BOMMER, paper on „International meeting of molecular spectroscopy", Bologna, Sept. 1959
[8] KELLER, F. and M. ARNOLD, Plaste und Kautschuk **19** (1972), 269
[9] KELLER, F., Plaste und Kautschuk **18** (1971), 657
[10] MAYO, F. R., Ber. Bunsenges. physik. Chem. **70** (1966), 233
[11] ROTH, H.-K., KELLER, F. and H. SCHNEIDER, *„Hochfrequenzspektroskopie in der Polymerforschung"*, Akademie-Verlag Berlin 1984
[12] KIMMER, W. and R. SCHMOLKE, Plaste und Kautschuk **19** (1972), 260

## 3.5. Tacticity Determination in Polymers

### 3.5.1. Introduction

According to NATTA [1], tactic (stereoregular) polymers consist of macromolecules whose basic structural elements follow a nonstatistical law in their steric configuration. Depending on the definition, these polymers contain either at least one centre for stereoisomerism or one double-bond in each basic structural element of the polymer chain (see 3.3.).

The asymmetric C-atoms in the primary molecules may already be present in polymerizable monomers or, they may develop during polymerization as, for example, in the head-to-tail structure of vinyl monomers. Due to the transformation of the planar substituent arrangement at double bonded C-atoms in the vinyl monomer into a tedrahedral one in the polymer chain, so-called pseudoasymmetric C-atoms are brought about [2]. Although they do not show optical activity because of the chemical environment, two different steric enantiomorphous arrangements of the substituents are possible as in the case of asymmetric C-atoms of the same constitution.

When we consider the relative position of the stereo centres immediately following each other in the chain, then the (pseudo) asymmetric C-atoms may be of the same or opposite configuration. In the first case, the structure is termed isotactic, in the second case syndiotactic. Polymers with short, frequently alternating isotactic or syndiotactic sequences are called heterotactic or atactic, respectively. The NATTA or FISCHER projection [2] or occasionally the NEWMAN projection are used for representation.

In general, the basic structural element of a molecular chain contains only one stereoisomeric centre. The resulting monotactic polymers as opposed to polytactic (polymers where the primary structural element contains several stereoisomeric centres) have aroused great commercial interest because the technological properties of these products are in essence determined by the stereoregular structure of the macromolecules. In general, the stronger influence will be exerted by the tactic part, the distributions involved being of minor importance with the exception of stereo-block copolymers. Density, crystallinity, freezing and melting characteristics, solubility, and chemical reactivity of the macromolecule depend distinctly on the stereoregularity [3]. For crystallization, however, a minimum sequence length must additionally be present. In this case, the tactic structure is certainly a necessary but not sufficient condition for crystallizability of the macromolecules.

## 3.5.2. *Formation of Tactic Polymers; Characteristic Bond Distribution Values in the Polymer Chain*

The tactic configuration of the macromolecules is established by the propagation steps during polymerization. In ionic-coordination polymerization, the initiator system controls stereoregular chain propagation whereas in ionic polymerization, the polarity of the solvent is important. In free radical polymerization, syndiotactic monomer addition is favoured over isotactic addition only at low polymerization temperatures ($<0$ °C) because the differences in activation enthalpies [4] are insignificant. Table 57 shows the reaction conditions leading to highly ordered structures for certain commercially interesting polymers. On careful observation of the microstructure of the substances synthesized in this way, it is found that none of the specified initiator systems produce sterically pure macromolecules without configuration defects.

Characterization of tactic polymers only in terms of isotactically and syndiotactically linked monomer units related to the sum of all basic structural elements (isotactic $(I)$ + syndiotactic $(S)$ = 1), is therefore frequently inadequate. In order to establish the distribution of the bonds over the whole chain, a further independent quantity is required. It is to be found in the relative frequency of heterotactic units $H^*$ which only occur at the ends of the tactic sequences. These indicate the mean sequence lengths $\bar{l}_i$ and $\bar{l}_s$. The greater $H^*$ becomes, the shorter are the tactic sequences.

*Table 57*: Reaction conditions for the synthesis of some tactic polymers

| Polymer | Reaction mechanism | Initiator | Tactic Structure |
|---|---|---|---|
| Polypropylene | ionic-coordinative | AlEt$_3$(AlEt$_2$Cl) + α(γ) TiCl$_3$ | iso |
| Polybutene-1 |  |  | iso |
| Polypropylene | ionic-coordinative | AlEt$_2$Cl + VCl$_4$  −80 °C | syndio |
|  |  | AlEt$_2$Cl + VCl$_4$  +50 °C | hetero |
| Polybutadiene | ionic-coordinative | AlEt$_2$Cl + VCl$_4$(VCl$_3$) | 1,4 trans |
|  |  | AlEt$_3$ + Cr(acac)$_3$ | 1,2 iso |
|  |  | AlEt$_3$ + Cr(CO)$_6$ |  |
|  |  | Al/Cr < 1 | 1,2 syndio |
|  |  | Al/Cr > 1 | 1,2 iso |
|  |  | π-allyl-nickel-complexes | 1,4 cis |
| Polymethyl methacrylate | ionic | n-butyl-lithium, n-amyl-sodium in non-polar hydrocarbons at 0 °C | iso |
|  |  | n-butyl-lithium; n-amyl-sodium in polar hydrocarbons at −70 °C | syndio |
| Polyvinyl chloride | radical | radical former +50 °C | hetero |
|  |  | −78 °C | syndio |

The aforesaid characteristic values are related to the rate constants of the propagation reactions (see 3.4.) by the probabilities of forming isotactic or syndiotactic units $P_{i,j}$ ($i, j$ = isotactic or syndiotactic). The relationships between the parameters characterizing the tactic structure and the $P_{i,j}$ are shown in Table 58. Using the stated equations, the weight fraction $w_{i,j}(n)$ of the tactic sequences of length $n$ in the entire macromolecule can be calculated as a function of $n$ (Fig. 75) [5].

## 3.5.3. Experimental Determination of Tacticity

### 3.5.3.1. NMR Spectroscopy

For practical purposes the NMR spectrum of a tactic polymer should be considered as that of a copolymer (see 3.4.), that is to say, as a superposition of isotactic, heterotactic and syndiotactic subchain spectra. The magnitude of the chemical shift between the distinguishable structures is dependent on the screening

Tacticity Determination in Polymers

*Table 58*: Equations for the relationships between quantities characterizing the tactic structures and the probabilities of forming isotactic or syndiotactic units $P_{i,j}$

| Parameter | Probability of forming isotactic or syndiotactic units |
|---|---|
| $I$ (diade) | $P_i$ |
| $S$ (diade) | $P_s$ |
| $H^*$ (triade) | $2 \cdot P_i \cdot P_{is}$; $2 \cdot P_s \cdot P_{si}$ |
| $I^*$ (triade) | $P_i \cdot P_{ii}$ |
| $S^*$ (triade) | $P_s \cdot P_{ss}$ |
| $\bar{l}_i$ | $P_{is}^{-1} + 1$ |
| $\bar{l}_s$ | $P_{si}^{-1} + 1$ |
| $w_i(n)$ | $n \cdot P_i \cdot \dfrac{P_{is}^2 \cdot P_{ii}^{n-1}}{1 - P_{is} \cdot P_{si}}$ |
| $w_s(n)$ | $n \cdot P_s \cdot \dfrac{P_{si}^2 \cdot P_{ss}^{n-1}}{1 - P_{si} \cdot P_{is}}$ |
| $w_{ii}(n)$ | $P_{si} \cdot w_i(n)$ |
| $w_{is}(n)$ | $P_{ii} \cdot w_s(n)$ |
| $w_{si}(n)$ | $P_{ss} \cdot w_i(n)$ |
| $w_{ss}(n)$ | $P_{is} \cdot w_s(n)$ |

of the nuclei by the electron cloud and, thus, on the chemical environment of the nucleus. The resolution of the signals is a function of the measuring frequency.

For spectrum interpretation and quantitative evaluation of the tactic parts in the molecule chain, the same information is therefore gathered and evaluated from the NMR spectrum as if a sequence analysis were being performed. The most important quantities to be determined are the position of the resonances (chemical

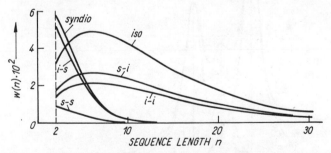

*Fig. 75*: Weight fractions of the sequences with n basic structural elements as a function of the sequence length $n$ in polymethyl methacrylate ($n \geq 2$)
$P_i = 0.76$;   $P_s = 0.24$;   $P_{ii} = 0.85$;   $P_{is} = 0.15$;
$P_{ss} = 0.55$;   $P_{si} = 0.45$

shift) and the areas under the peaks as a measure of the concentration of the structural units. It should be noted that, in contrast to proton resonance, the signals in $^{13}$C-NMR are in general not proportional to concentration because of the nuclear Overhauser effect ($NOE$). For the relative chemical shift of the same nuclei, which are distinguished only by the configurational arrangement, this effect is so small that it can be neglected. Furthermore, the coupling constants and the multiplicities of the signals are also used in the interpretation. According to the multiplet rule:

$$\text{multiplicity} = 2 \cdot N \cdot I + 1 \tag{1}$$

$I$ = nuclear spin quantum number, $N$ = number of equivalent magnetic nuclei

The signals split into several lines. For example, the two protons of the $CH_2$-group in vinyl polymers with syndiotactic configuration are magnetically equivalent and, therefore, also show the same chemical shift. The isotactic linkage cancels the symmetry of the methylene protons [6]. Each proton shows a resonance signal of its own which is resolved into a multiplet due to the spin coupling. If such differentiation is possible, the methylene protons are called heterosteric. In a homosteric $CH_2$-group, the two protons are equivalent. If NMR resolution is insufficient, then interpretation of the steric arrangement in the chain is frequently possible using the spectra of deuterated polymers.

In addition to proton resonance, $^{13}$C-NMR furnishes valuable information about the configurational microstructure of the macromolecules. The range of

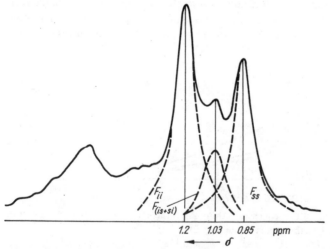

Fig. 76: NMR spectrum of the α-methyl protons in a mainly isotactic polymethyl methacrylate

the chemical shift of $^{13}$C-nuclei is about 30 times that of protons. Thus very small differences in chemical and steric environment of a $^{13}$C-nucleus are visible in the spectrum.

Figs. 76 and 77 show, as an example, the proton spectrum and the $^{13}$C spectrum of the α-methyl group of polymethyl methacrylate. While the resonances of the isotactic, syndiotactic and heterotactic structures are superimposed in the proton spectrum, complete resolution is achieved in the $^{13}$C spectrum. In Fig. 78, the chemical shifts in the spectrum of the $^{13}$C-nuclei are represented as a function of the substituent in the respective vinyl polymers. Table 59 shows the effect of configuration on the relative chemical shift [7].

NMR spectra are usually discussed in terms of $n$-ades. An increase in measuring frequency sometimes permits an interpretation of the spectra in terms of higher $n$-ades (tetrades, pentades). For example, the α-proton spectrum of polyvinyl chloride becomes sensitive to a pentade structure when the measuring frequency is 100 MHz [8]. The same effect occurs in $^{13}$C spectroscopy as has been demonstrated in polypropylene.

The resolution of the signals is limited by the fact that intermolecular interactions hinder segment mobility. As a consequence of this, the band width of dissolved macromolecules is considerably larger than that of low molecular weight substances. Band broadening can be inhibited increasing the temperature.

In general, the signals are classified with reference to polymer standards of known tactic structure or with reference to polymers with quite different steric arrangements of the primary, structural elements. Frequently, low molecular

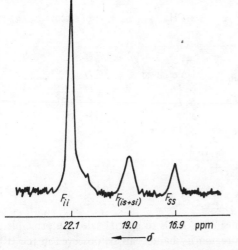

*Fig. 77*: $^{13}$C spectrum of the α-methyl group in mainly isotactic polymethyl methacrylate

weight model substances are used, for example, polyvinyl chloride and polypropylene oxide to aid spectral interpretation.

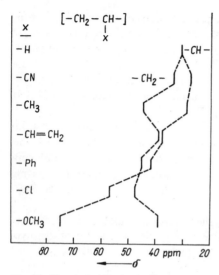

Fig. 78: Chemical shifts in the $^{13}$C NMR spectrum for some vinyl polymers [7]

Table 59: Effect of steric arrangement on the relative chemical shift, arranged in order of increasing magnetic field

| Substituent of the vinyl monomer $X$ | $-\overset{\bullet}{C}H-$ | $-CH_2-$ | α-C-atom |
|---|---|---|---|
| $-H$ | — | — | — |
| $-CN$ | $S^*; I^*;$ | not resolved | $I^*; S^*$ |
| $-CH_3$ | not resolved | $S; I$ | $I^*; S^*$ |
| $-CH=CH_2$ | not resolved | $S; I$ | $I^*; S^*$ |
| $-Ph$ | not resolved | $S; I$ | $I^*; S^*$ |
| $-Cl$ | $S^*; I^*;$ | $S; I$ | — |
| $-OCH_3$ | not resolved | $S; I$ | — |

## 3.5.3.2. IR Spectroscopy

The ordered arrangement of monomer units leads to special selection principles for infrared vibrations so that stereoregularity bands can be observed in the IR spectra of the polymers both in the solid state and in solution. According to

DECHANT [10], characteristic group frequencies are dependent on local configuration whereas symmetry, selection principles, intensities, polarization, and intramolecular interactions are determined by conformation. Correlations existing between the conformation of the molecule chains and the steric arrangement of the substituents have been demonstrated for PVC by GERMAR [11]. The connection between stereoregularity and the existence of regular chains, leads to the appearance of characteristic regularity bands in the IR spectrum. KÖNIG [12] has classified IR absorptions of tactic polymers into three characteristic types according to their behaviour in the transition from syndiotactic to isotactic materials. This classification represented graphically in Fig. 79.

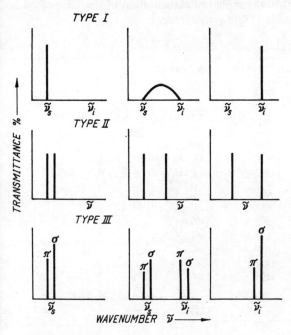

Fig. 79: Classification of the bands depending on stereoregularity [12]

The frequency of absorption of a type-I band in a given stereoregular polymer is dependent upon the preferred configuration. Syndiotactic and isotactic polymers absorb at different frequencies $\tilde{v}_s$ and $\tilde{v}_i$ whereas heterotactic structures only absorb weakly at an intermediate frequency. Separation and intensity are functions of the isosteric sequence lengths. Type-II bands are expected where there is a high degree of intramolecular interactions as with helix structures. The band splits into two components with the same or opposite polarization $\sigma$. The separation increases with growing isotactic sequence length and the intensity depends on the number of stereosequences of the particular length. Type III bands have, irrespective of the steric purity of the polymer, components that are polarized perpendicularly to each other. The selection rules for this band splitting are determined by the chain conformation resulting in a characteristic frequency splitting. In the case of syndiotactic or isotactic configuration, two series of split bands are observed. The stronger series appears with the frequencies of the prevailing type of configuration. Since heterotactic polymers exhibit only short stereosequences, which preclude helix formation, no absorption is observed for these structures. The majority of IR absorptions used in the evaluation of stereoregularity belong to type III. In publication [12], the explained classification of the IR method is applied to polypropylene, and in [13] to polystyrene.

*Table 60*: Tacticity dependent IR absorptions of selected polymers

| Polymer | Wave number ($cm^{-1}$) | | Assignment |
|---|---|---|---|
| | iso- | syndio- | |
| Polypropylene | 998 | | $\gamma_r(CH_3) + v(C-CH_3) + \delta(CH) + \gamma_t(CH_2)$ |
| | 1168 | | $\gamma(C-C) + \gamma_r(CH_3)$ |
| | | 866 | helix $\gamma_r(CH_2)$ |
| | | 830 | zig zag conform. $\gamma_r(CH_2)$ |
| | | 1233 | zig zag conform. $\gamma_w(CH_2) + \delta_0(CH)$ |
| Polyvinyl chloride | 624 | 603 | $v(C-Cl)$ |
| | 633 | 613 | $v(C-Cl)$ |
| | 695 | 639 | $v(C-Cl)$ |
| | | 647 | $v(C-Cl)$ |
| | | 677 | $v(C-Cl)$ |
| | 1434 | 1428 | $\delta(CH_2)$ |
| | | 1434 | $\delta(CH_2)$ |
| Polymethyl methacrylate | 759 | 749 | $\gamma_r(CH_2)$ coupled with skeletal valence vibrations |
| | 951 | 967 | $\gamma_r(\alpha\text{-}CH_3)$ |
| | | 1060 | $v(C-C)$ |
| Polystyrene | 566 | | ring vibrations |
| | 1053 | | $v(C-C)_{helix}$ |
| | 1084 | | ring vibrations |
| | 1187 | | $v(C-C)_{helix}$ |
| | 1297 | | $\gamma_t(CH_2)$ |
| | 1312 | | $\gamma_t(CH_2)$ |
| | 1364 | | $\delta(CH)$ |
| Polybutene-1 | 1220 | | $\gamma_t(CH_2) + \delta(CH) + \gamma_t(ethyl)$ |

The prevailing type of stereoregularity is identified by IR spectroscopy, especially by comparison with spectra of standard polymers of known tacticity and with spectra of low molecular weight model substances; symmetry considerations and normal coordinate analysis are also used. The assignments of some IR absorptions to the appropriate tacticity type is given for selected polymers in Table 60.

IR spectra of partially deuterated polymers are particularly useful for spectral interpretation. If regularity bands are chosen for the analysis, then only relative tacticity may be determined. The helical content determined from these absorptions is rarely identical with the absolute degree of tacticity since short regular sequences and tactically linked primary structural units, which are not in the helix, are not included.

### 3.5.3.3. Further Experimental Methods of Ascertaining Tactic Structure

X-ray structural analysis of partially crystalline polymers, enables a qualitative identification of the type of stereo-regularity since crystal structure and tacticity are related [14]. This method, however, furnishes only limited quantitative information because the average molecular weight, the distribution of configurative defects and the thermal history of the sample exert a noticeable influence on the crystallinity. The connection between crystallinity and stereo-sequence length has already been pointed out (see 3.5.1.).

Solvent extraction furnishes a relative stereo-regularity index provided it is carried out under comparable conditions on chemically similar polymers with a comparable average molecular weight and distribution function. This technique will not give an absolute degree of tacticity but separates fractions occurring during synthesis which are stereoregularly uniform.

The solution properties of several tactic polymers have been examined carefully by light scattering, viscosity and osmotic pressure measurements, by phase separation, ultracentrifugation and dipole-moment measurements. While the limiting viscosity numbers in thermodynamically good solvents differ insignificantly, definite tendencies are found in the theta temperatures (see 4.1.),

isotactic < heterotactic < syndiotactic,

and in FLORY's entropy parameters $\psi$ (see 4.1.) (Table 61),

isotactic > heterotactic > syndiotactic.

The relationships between tacticity and optical activity (when true asymmetric C-atoms are present in the backbone) are not unequivocal in most cases. Even when

Table 61: Theta temperatures and $\psi$-values of some stereoregular polymers

| Polymer | Tacticity | Theta-temperature (K) | $\psi$ | Solvent |
|---|---|---|---|---|
| Polystyrene | iso | 356.7 | — | cyclohexanol |
|  | hetero | 358.7 | — |  |
| Polypropylene | iso | 418.4 | 1.414 | diphenylether |
|  | hetero | 426.5 | 0.986 |  |
| Polymethyl methacrylate | iso | 349.1 | 2.320 | n-propanol |
|  | hetero | 357.6 | 1.940 |  |
|  | syndio | 358.4 | 1.850 |  |
|  | iso | 299.7 | 0.590 | n-butyl chloride |
|  | hetero | 308.2 | — |  |
|  | syndio | 308.2 | 0.510 |  |
| Polybutene-1 | iso | 362.3 | 0.956 | anisol |
|  | hetero | 359.4 | 0.740 |  |

stereoselective catalysts are used, the highly stereoregular polymers obtained fail to show a significant rotation of polarized light due to the poor optical purity of the initial monomers [15]. An equimolar mixture of the homopolymers of the two antipodes is formed. This is optically inactive because of intermolecular compensation.

## 3.5.4. Examples of Determination of Tactic Constituents

### 3.5.4.1. NMR Determination of Isotactic, Syndiotactic and Heterotactic Triades using Polymethyl methacrylate as an Example

NMR determination of the tactic constituents is carrried out using solutions of 5 per cent by weight of the samples in deuterated chloroform. The spectrum is recorded on a high resolution NMR spectrometer with variable temperature sample head at a frequency of 90 MHz and at a temperature of 60 °C. Suitable equipment allows other solvents to be used and measuring frequencies to be varied.

$^{13}$C-NMR is performed on saturated solutions of the sample in deuterated chloroform. In the present example a concentration of about 30 per cent by weight is used. The spectrum is obtained by means of a $^{13}$C-NMR spectrometer operating

# Tacticity Determination in Polymers

at 22.635 MHz using Pulse-FOURIER-Transform (*PFT*) technique with proton noise decoupling.

The reference substance used in both cases is hexamethylene disiloxane. Typical NMR spectra of the $\alpha$-$CH_3$-group of the polymethyl methacrylates are shown in Figs. 76 and 77.

The resonances are assigned as follows:

In the proton spectrum:      $\delta = 1.20$ ppm isotactic triades
(reference substance: HMDS)    $\delta = 1.03$ ppm heterotactic triades
                                        $\delta = 0.85$ ppm syndiotactic triades
in the $^{13}C$-spectrum:          $\delta = 22.1$ ppm isotactic triades
(reference substance: HMDS)    $\delta = 19.0$ ppm heterotactic triades
                                        $\delta = 16.9$ ppm syndiotactic triades

*Table 62*: Relationships between the probabilities of forming isotactic or syndiotactic units ($P_{i,j}$) and the peak areas

| Probability of forming iso- or syndiotactic units | Corresponding area expression |
|---|---|
| $P_{ii}$ | $\dfrac{F_{ii}}{F_{ii} + \dfrac{1}{2} F_{(is+si)}}$ |
| $P_{ss}$ | $\dfrac{F_{ss}}{F_{ss} + \dfrac{1}{2} F_{(is+si)}}$ |
| $P_{is}$ | $\dfrac{\dfrac{1}{2} F_{(is+si)}}{F_{ii} + \dfrac{1}{2} F_{(is+si)}}$ |
| $P_{si}$ | $\dfrac{\dfrac{1}{2} F_{(is+si)}}{F_{ss} + \dfrac{1}{2} F_{(is+si)}}$ |
| $P_i = \dfrac{P_{si}}{P_{si} + P_{is}}$ | $\dfrac{F_{ii} + \dfrac{1}{2} F_{(is+si)}}{F}$ |
| $P_s = \dfrac{P_{is}}{P_{is} + P_{si}}$ | $\dfrac{F_{ss} + \dfrac{1}{2} F_{(is+si)}}{F}$ |
| | $F = F_{ss} + F_{(is+si)} + F_{ii}$ |

The integrated peak intensities, which are proportional to the number of tactically linked primary structural units, are termed $F_{ii}$, $F_{ss}$ and $F_{(is+si)}$. The areas of the three resonances, relative to the total intensity of the α-methyl group, indicate directly the relative occurrence the isotactic, syndiotactic and heterotactic triades.

$$I^* = \frac{F_{ii}}{F} \qquad (2)$$

$$S^* = \frac{F_{ss}}{F} \qquad (3)$$

$$H^* = \frac{F_{is} + F_{si}}{F} \qquad (4)$$

In order to calculate the tactic content, the areas are measured planimetrically. Evaluation of the proton spectra requires that the individual partial spectra of the triades, as shown in Fig. 76, be separated. Relating the areas to the probabilities of forming isotactic or syndiotactic units, the relationships given in Table 62 are obtained.

From the $^{13}$C-spectrum shown in Fig. 77, 65% isotactic, 13% syndiotactic and 22% heterotactic triades were found using equations (2) to (4). This corresponds to 76% isotactic and 24% syndiotactic monomer linkages according to the relationships given in Table 62.

The equations given in Table 58 are used for calculating the sequence distribution. Using the relationships given there, the distributions of the tactic sequences shown in Fig. 75 are obtained.

## 3.5.4.2. IR Determination of the Proportion of Isotactic and Syndiotactic Monomer Linkages using Polyvinyl Chloride as an Example

For the determination of isotactic and syndiotactic diade content in the sample, a 2.5 per cent by weight solution in 1,1,2,2 tetrachloroethane is prepared at 80 °C. The spectra are recorded with a high resolution double-beam spectrometer over the range 1400 to 1470 cm$^{-1}$ at a temperature of 318 K. The thickness of the cell is 0.5 mm. The solvent bands are compensated by a cell filled with pure solvent placed in the reference beam of the spectrometer. In Fig. 80, the partial spectrum of a commercial PVC-S and that of a preparation polymerized at $-78$ °C are shown. The base line method is used for evaluation.

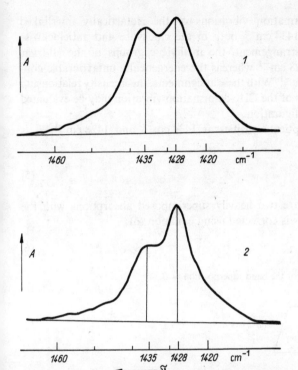

*Fig. 80*: Absorption spectrum of the CH$_2$-scissor vibrations in PVC
1 — PVC-S, 2 — PVC synthesized at −78 °C, Temperature: 318 K

Calculation of the degree of tacticity is based on the different degree of occupation of the two possible conformations for syndiotactic monomer linkage. For the syndiotactic chain, the energetically favoured conformation is the planar carbon skeleton, the all-trans conformation. At higher energies, a second conformation can be reached. In this structure, the planar chain is twisted. If this twisting process is continued the chain would return to itself (gauche-gauche conformation). These structures can thus appear only in isolation. The numerical ratio of the planar to the twisted structures present at any moment is fixed at a particular temperature, irrespective of the tacticity, by the BOLTZMANN factor alone. The isotactically linked monomer units form a helix coiled to the right or to the left with a threefold helical axis. Some conformations of a syndiotactic vinyl polymer diade are represented in Fig. 81 in the NATTA-projection.

Counting of the isotactic and syndiotactic structural units is effected from the CH$_2$-deformation vibration spectrum which is assigned as follows. The wave numbers of the $CH_2$-deformation vibrations of the isotactically interlinked monomer units are about 1434 cm$^{-1}$ both in the clockwise and anticlockwise helix. In the syndiotactic arrangement, the methylene groups of the all-trans conformation vibrate at 1428 cm$^{-1}$ whereas the energetically unfavourable conformation absorbs at 1434 cm$^{-1}$. With these assignments, the intensity relationship between the two absorptions of the $CH_2$-deformation vibration may be evaluated [11] as a function of the configuration.

First, the ratio of the optical densities at 1434 cm$^{-1}$ and 1428 cm$^{-1}$, $\lambda$, is determined:

$$\lambda = \frac{A_{1434}}{A_{1428}} \tag{5}$$

As the analytical bands are two heavily superimposed absorptions with the same line shape, the $\lambda$-values is corrected using equation (6):

$$\lambda_{corr.} = \frac{\lambda - \gamma_{corr.}}{1 - \lambda \beta_{corr.}} \tag{6}$$

$\gamma_{corr.} = \beta_{corr.} =$ correction terms for the band superposition $= 0.34$

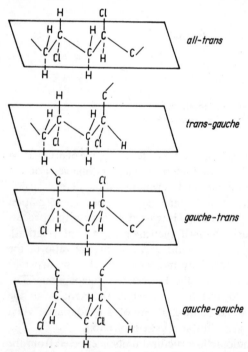

Fig. 81: Some syndiotactic conformations of the $CH_2$ group in PVC

# Tacticity Determination in Polymers

The correction factors are obtained by graphical decomposition of the recorded complex band. The relationship between the measured value $\lambda_{corr.}$ and the syndiotactic content in the sample is given by equation (7):

$$P_s = \frac{1 + \left(\exp - \frac{\Delta E}{RT}\right)}{1 + \lambda_{corr.}} \tag{7}$$

$\Delta E$ = energy difference between the two possible rotation conformers of the syndiotactically linked primary structural elements = 8400 J · mol$^{-1}$, $R$ = gas constant (8.314 J · mol$^{-1}$ · degree$^{-1}$), $T$ = temperature

From the spectra shown in Fig. 80 and applying equation (7), a syndiotactic content ($P_s$) of 0.57 ($\lambda_{corr.}$ = 0.82) is found for the commercial PVC-S, and of 0.75 ($\lambda_{corr.}$ = 0.39) for the PVC polymerized at $-78$ °C.

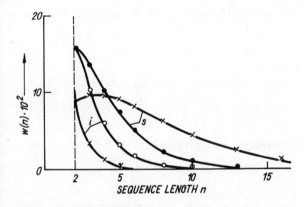

*Fig. 82*: Weight fractions of sequences with $n$ basic elements in PVC as a function of the sequence length $n$ ($n \geq 2$)
x: $P_i = 0.25$     x: $P_s = 0.75$
o: $P_i = 0.43$     o: $P_s = 0.57$

The sequence distribution is computed from the calculated $P_s$-values assuming that BERNOULLI statistics ($P_s = P_{ss} = P_{is}$) are applicable. This assumption is justified by NMR investigations of polymethyl methacrylate (BOVEY [4]). The distributions of the syndiotactic sequences for the determined $P_s$-values are shown in Fig. 82 as a function of the length $n$. The determination of the degree of tacticity of PVC by NMR spectroscopy is described in detail in [16, 17].

## Bibliography

[1] NATTA, G. and F. DANUSSO, J. Polymer Sci. **34** (1959), 3
[2] ULBRICHT, J., „*Grundlagen der Synthese von Polymeren*", Akademie-Verlag, Berlin 1978
[3] JOHNSON, J. F. and R. S. PORTER, in „*The Stereochemistry of Macromolecules*" by KETLEY, A. D., M. Dekker Inc., New York 1968, Vol. 3, 213ff.
[4] BOVEY, F. A. and G. V. D. TIERS, Fortschr. Hochpolymeren-Forsch. **3** (1963), 139
[5] JOHNSEN, U., Kolloid-Z. — Z. Polymere **178** (1961), 161
[6] BOVEY, F. A., „*Polymerconformation and -configuration*", Academic Press New York 1969
[7] SCHAEFER, J., Topics in $^{13}$C-NMR **1** (1974), 149
[8] JOHNSEN, U. and K. KOLBE, Kolloid-Z. — Z. Polymere **221** (1967), 64
[9] JOHNSON, L. F., Analyt. Chem. **43** (2) (1971), 28 A
[10] DECHANT, J., „*Ultrarotspektroskopische Untersuchungen an Polymeren*", Akademie-Verlag, Berlin 1972
[11] GERMAR, H., HELLWEGE, K.-H. and U. JOHNSEN, Makromolekulare Chem. **60** (1963), 106
[12] KÖNIG, J. L., WOLFRAM, L. E. and J. G. GRASSELLI, Spectrochim. Acta **22** (1966), 1223, 1233
[13] KRIMM, S., Fortschr. Hochpolymeren-Forsch. **2** (1960), 51
[14] CORRADINI, P., in „*The Stereochemistry of Macromolecules*" by KETLEY, A. D., M. Dekker Inc., New York 1968, Vol. 3, 1ff.
[15] TSURUTA, T., INOUE, S. and I. TSUKUMA, Makromolekulare Chem. **84** (1965), 198
[16] ROTH, H.-K., KELLER, F. and H. SCHNEIDER, „*Hochfrequenzspektroskopie in der Polymerforschung*", Akademie-Verlag, Berlin 1984
[17] BOVEY, F. A., in „*High Resolution NMR of Macromolecules*", Academic Press, New York—London 1972

## 3.6. Determination of Branching in Polymers

### 3.6.1. Introduction

Besides linkage-, sequence- and stereo-isomerism, chain branching gives rise to another isomeric form of macromolecules. Depending on the length of the side chains, a distinction is made between short-chain branching ($SCB$) and long-chain branching ($LCB$). Short-chain branchings in general consist of one or two primary structural units, whereas long-chain branchings may have a length up to one quarter of the backbone chain. Subbranchings are formed if the side chains are further branched. Depending on the arrangement of the side chains with respect to the main chain, various structural models are defined.

To be able to characterize the structure of a branched polymer exactly, a large number of parameters (Table 63) are required, not all of which are accessible to the normal methods of analysis. For a few polymers, the experimental determination of $n_e$ and $n$ as well as a distinction between short-chain or long-chain

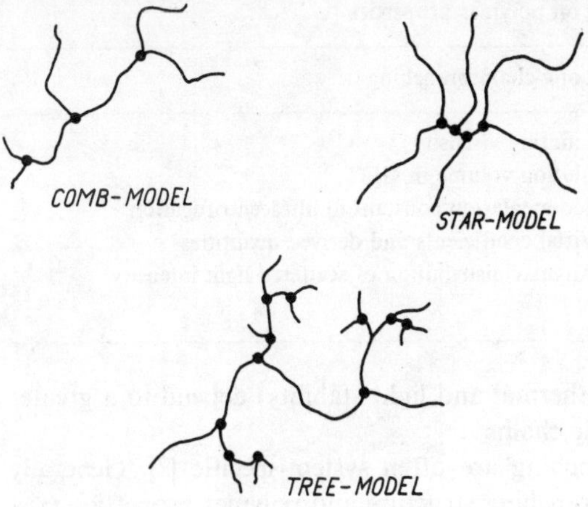

Determination of Branching in Polymers

*Fig. 83*: Structural models for branched polymers

branching can be established with satisfactory accuracy. In the case of short-chain branching, data on branch lengths and distribution and on the functionality $f$ may be obtained in selected instances. A prerequisite for any method is a measurable change of a property appropriate to the structural change in question. This leads to a differentiation into summative and selective methods of branching determination.

Alongside average molecular weight and its distribution, chain branching is one of the most important quantities which relate to technological and physical properties of polymers (Table 64).

While long-chain branchings particularly affect the rheology and thermodynamic behaviour of solutions and melts [1], the degree of order (and hence the

*Table 63*: Structural parameters for describing branching structures

| Parameter | Symbol |
| --- | --- |
| Number of branching-points/molecule | $n$ |
| Functionality of the branching points | $f$ |
| Number of branching end points | $n_e$ |
| Average length of the side branch | $\bar{b}_l$ |
| Distribution of the branch lengths | $h(b_l)$ |
| Length distribution of the non-branched chain segment | $h(d_b)$ |
| Average molecular weight of the non-branched chain segment | $\bar{M}_c$ |
| Distribution of the average molecular weight of the non-branched chain segment | $h(\bar{M}_c)$ |

*Table 64*: Influence of chain branching on polymer properties

| Short-chain branching | Long-chain branching |
| --- | --- |
| Crystallinity | Limiting viscosity |
| Density | Elution volume in GPC |
| Melting point | Sedimentation constant in ultracentrifugation |
| Dielectric constant | Virial coefficients and derived quantities |
| Creep and fracture | Angular distribution of scattered light intensity |
| Gas permeability | |

mechanical behaviour as well as thermal and light stability) depend to a greater extent on the number of short side chains.

Data on the influence of branching are often system-specific [2]. Generally accepted relationships between branching structure and polymer properties exist only in special cases, e.g. change of density and melting point of polyolefins as a function of number and length of short branchings [1].

The determination of the number of branchings of a macromolecule is not one of the standard methods of polymer characterization, requiring as it does a broad theoretical knowledge and comprehensive laboratory facilities.

## 3.6.2. Formation of Branched Structures in Polymers

Branched macromolecules can be formed by deliberate synthesis or by side reactions. In contrast to the products formed in a deliberate manner, neither the number nor chain length of branchings formed by side reactions (Fig. 84) can be predicted from knowledge of the synthesis process. They must therefore be determined experimentally.

The most important side reactions in free radical and ionic polymerization which lead to chain branchings are inter- and intramolecular transfer reactions to the polymer. The number of branchings formed by intermolecular transfer depends essentially on temperature, the concentration and structure of the polymer, on the monomer, the initiator and, in ionic polymerization, on the electron density at the double bond of the monomer. Intermolecular transfer reactions lead to a measurable broadening of the molecular weight distribution because the mass rather than the number of molecules changes considerably. Intramolecular transfer (self-transfer) leads to short-chain branchings (Fig. 84 reactions 2 and 6). The branching number is dependent on temperature, on polymer structure and on reaction pressure especially in the high-pressure polymerization of ethylene but (in contrast to intermolecular transfer) is independent of the amount of conversion.

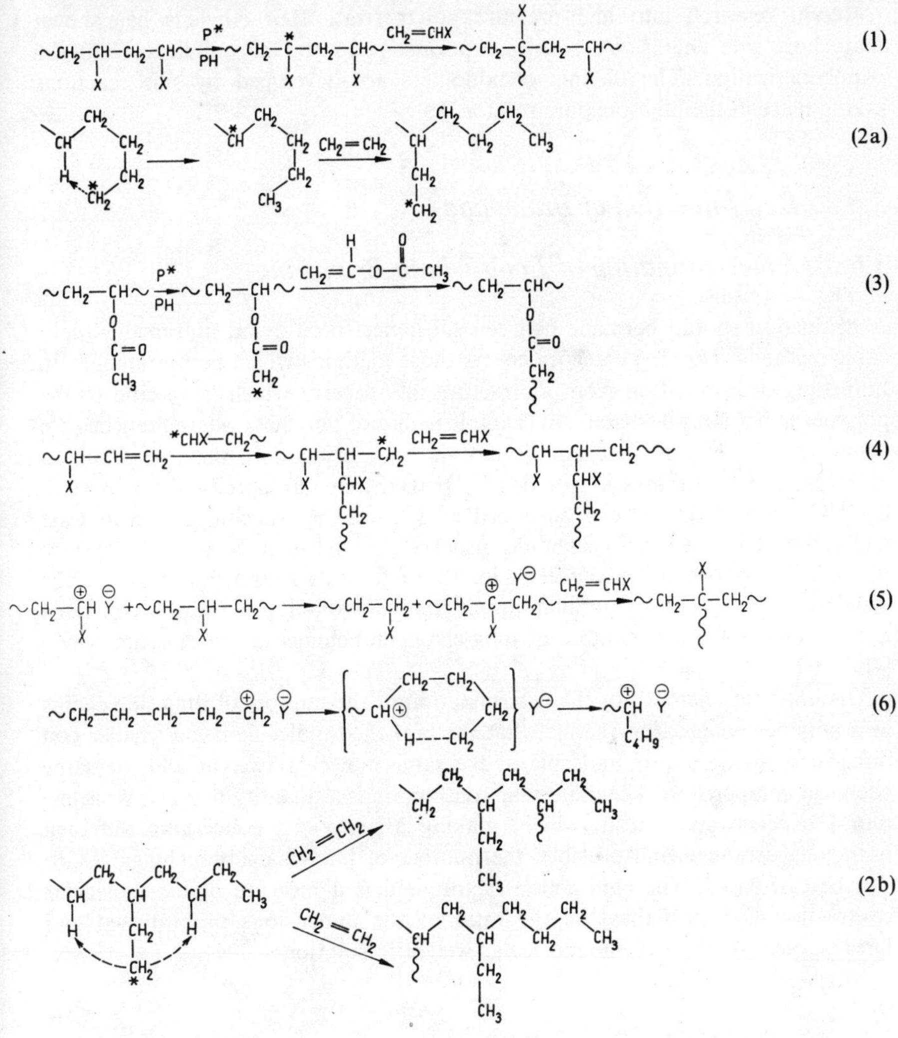

Fig. 84: Formation of branched structures in polymers
(1): Intermolecular transfer to polymer; (2a, b): intramolecular transfer to polymer; (3): transfer to estergroups; (4): adelition of radicals to double-bonds; (5): intermolecular transfer in cationic polymerization; (6): intramolecular transfer in cationic polymerization.

Recent research into high-pressure polymerization of ethylene has shown that short side chains are also incorporated into the polyethylene chain by copolymerization. The olefinic comonomers are developed by side reactions taking place in the high-pressure reactor.

### 3.6.3. Determination of Branching

### 3.6.3.1. Determination of Long-Chain Branching

A distinction should be made between substance-specific and universally-applicable methods. The first category covers those methods which permit long-chain branching determination from a structure or property which is specific to the polymer under consideration. An example is that of polymers where branches are connected with the main chain (PVac) via ester bridges. In the determination of long-chain branchings in low density polyethylene [3], specific resonances in the $^{13}$C-NMR spectrum can be traced back to chain branchings with at least 6 C-atoms. Indirect methods should also be included in this category. Because of the known correlation between kinetics of formation and molecular weight distribution, they allow determination of the relative polymer transfer constants and hence the average number of long-chain branchings in the macromolecule [4].

The universal method for the determination of the number of long side chains in a polymer is based on the fact that the branched molecule has a smaller coil dimension than a linear molecule of the same molecular weight and the same chemical composition. The molecular contraction is a quantity that can be determined experimentally from which, making assumptions concerning the long branching arrangements possible, the number of long-chain branchings ($LCB$) can be calculated. The contraction factor used is a measure of the molecular contraction and is defined as the ratio of the dimensions of branched and linear molecules of the same molecular weight in solution:

$$g = \frac{\langle s^2 \rangle_b}{\langle s^2 \rangle_l} \tag{1a}$$

where $s$ is the radius of gyration

and $h = \dfrac{R_{h,b}}{R_{h,l}}$

where $R_h$ is the hydrodynamic radius or (since $[\eta] \sim R_h^3$)

$$g' = h^3 = \frac{[\eta]_b}{[\eta]_l} \tag{2a}$$

# Determination of Branching in Polymers

Since the dimensions of dissolved macromolecules are a function of thermodynamic properties, related to the forces of interaction of the polymer solution, the contraction factors defined by equations (1a) and (2a) must be dependent on the state of the solution. Therefore, contraction factors $g_0$ and $g'_0$ are introduced for an interaction-free state (quasi force-free, undisturbed state, theta state, see Chapter 1.7.2. and 4.1.).

$$g_0 = \frac{\langle s^2 \rangle_{0,b}}{\langle s^2 \rangle_{0,l}} \tag{1b}$$

$$g'_0 = \frac{[\eta]_{0,b}}{[\eta]_{0,l}} \tag{2b}$$

In the undisturbed state, equation (9) of Chapter 1.7. is applicable to the correlation between $\langle s^2 \rangle_0$ and $[\eta]_0$. If this is inserted into equation (2b) and FLORY's universal constants for branched and linear molecules are assumed to be equal ($\Phi_{0,b} = \Phi_{0,l}$), equation (3a) is obtained:

$$g'_0 = \frac{\Phi_{0,b}}{\Phi_{0,l}} \left\{ \frac{\langle s^2 \rangle_{0,b}}{\langle s^2 \rangle_{0,l}} \right\}^{3/2} = g_0^{3/2} \tag{3a}$$

For the general case, equation (3a) can be formulated as a power law where the exponent $b$ assumes values between 0.5 and 1.5 and need not be constant.

$$g' = g^b \tag{3b}$$

$b = 0.5$ star-like branched, $b = 1.5$ comb-like branched

In order to calculate the number of $LCB$ from the measured value of the contraction factor, the correlation between contraction factor and number of $LCB$ must be known. Some of the relationships $g_0 = f(n)$ which result from model

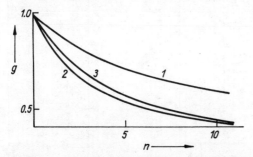

*Fig. 85*: $g$-factors for trifunctional (1) and tetrafunctional (2) branching points and for monodisperse (1) and polydisperse trifunctional (3) systems

Table 65: Equations for determination of branching by comparison of the undisturbed dimensions

| Model | Relationships between $g$ and the number of branching points per molecule | No. | Conditions of validity |
|---|---|---|---|
| Comb | $g(n) = \left\{\left(1 + \dfrac{n}{7}\right)^{1/2} + \dfrac{4n}{9\pi}\right\}^{-(1/2)}$ | (5) | $f = 3$; $U = 0$; statistical distribution of $\bar{M}_c$ and $n$ |
| Comb | $g(n) = \left\{\left(1 + \dfrac{n}{6}\right)^{1/2} + \dfrac{4n}{3\pi}\right\}^{-(1/2)}$ | (6) | $f = 4$; the same boundary conditions as with equation 5 |
| Comb | $g(n_w) = \dfrac{6}{n_w}\left\{\dfrac{1}{2}\left(\dfrac{2+n_w}{n_w}\right)^{1/2} \ln\left[\dfrac{(2+n_w)^{1/2} + n_w^{1/2}}{(2+n_w)^{1/2} - n_w^{1/2}}\right] - 1\right\}$ | (7) | $n_w$ = weight-average number of branchings per mole; $f = 3$; statistical distribution of the molecular weight, of the $\bar{M}_c$ and $n$-values |
| Comb LCB SCB | $g(n) = 1 - \left(\dfrac{ny}{N^3}\right)\{(n-1)(n-2)y^2 + 2(n-1)xy + x^2\}$ | (8) | $N$ – degree of polymerization $x(y)$ – number of monomer units in the backbone chain (side branch) statistical distribution of $\bar{M}_c$ and $n$ branchings of equal length |
| Star | $g(p) = \dfrac{3}{p} - \dfrac{2}{p^2}$ | (9) | $p$ = number of star-like branchings emanating from a centre; branchings of equal length; $y$ = segment number of branchings = constant |

| | | | |
|---|---|---|---|
| Star | $g(p) = \dfrac{3y_w}{py_n} - \dfrac{2y_w y_z}{p^2 y_n^2}$ | (10) | statistical distribution of $y$ |
| Star | $g(p_n) = \dfrac{3p_n - 1}{p_n^2 + 3p_n + 1}$ | (11) | statistical distribution of $p$; $y$ = constant |
| Comb (SCB) | $g = \dfrac{1}{s+1}\{1 + s(1 - 2a + 2a^2 - 2a^3) + s^2(-a + 4a^2 - a^3)\}$ | (12) | $s$ = number of branches per maromolecule; $a$ = number of C-atome in a branch relative to number of C-atoms in the backbone chain |

calculations of dimensions of branched molecules in the undisturbed state are given in Table 65 [5]. The number of *LCB* can only be stated if $g_0$ can be determined. This means that the measured value $g$ must be converted into $g_0$ or $g'$ into $g'_0$. Any method capable of determining molecular size may be used to obtain $g$ or $g'$. The most important are listed in Table 66 where the essential pre-conditions for their application are indicated.

Although coil contraction is mainly by long-chain branching, the influence of *SCB* cannot be ignored if their number is high. In this case, the $g$-value (equation (4)) is equal to the product of $g_{LCB}$ with $g_{SCB}$:

$$g = g_{LCB} \cdot g_{SCB} \tag{4}$$

In first deriving the relationship between contraction factor and number of branching points per mole, $n$, it is assumed that the side chains can be arbitrarily replaced by the branch distances. All isomers resulting from this are assumed to have the same probability, the excluded volume is neglected, and free chain mobility assumed. In more recent model calculations (Table 65), the branching model, the branch length distribution, and the distribution and functionality of the branching points are taken into consideration. Above all, the functionality $f$ of the branching points and molecular inhomogeneity $U$ of the polymer [1] influence the $g$-value significantly. When comparing the contraction factors calculated using equations (5) and (6) or (5) and (7) (Table 65), it becomes evident that $g$ for tetrafunctional branching points or polydisperse products (Fig. 85) decreases with increasing branching number more rapidly than it does for trifunctional branching points or for uniform products.

Besides long-chain branching, solution effects or macromolecular structural factors (e.g. microgels and associates [6]) may lead to a diminution of molecular dimension, that is to say, to a molecular contraction in the sense of equation (1a). Separation of these effects from that of contraction caused entirely by *LCB* is not normally possible. Therefore, at present, calculation of branching numbers from contraction factors is only partially justified and this should certainly not be attempted if the branching structure is not known. If boundary conditions are applicable to the selected model, however, then the branching frequency $\lambda$ is accessible via the branching number $n$

$$\lambda = \frac{n}{M} \tag{13}$$

and the average molecular weight of the non-branched chain segments $\bar{M}_c$:

$$\bar{M}_c = \frac{M}{(f-1) \cdot n + 1} \tag{14}$$

## 3.6.3.2. Determination of Ester Linked Side Chains

Determination of the mean number of branching points in polymers is relatively simple when the side branches are connected with the main chain by an ester bridge. The mean degree of branching $\bar{\varphi}$ of this type of branching can be measured by saponification [7]. The number of ester bridges broken by saponification, which is equal to number of branches, is obtained from the difference between the number of molecules before and after reaction. $\bar{\varphi}$ is related to the number-average degree of polymerization $P_n$ by equation (15):

$$\bar{\varphi} = \frac{P_{n,0}}{P_n} - 1 \tag{15}$$

$P_{n,0}$ = number-average degree of polymerization before saponification, $P_n$ = number-average degree of polymerization after saponification

The mean number of branching points per monomer unit $\bar{n}_0$ is then given by:

$$\bar{n}_0 = \frac{1}{P_n} - \frac{1}{P_{n,0}} \tag{16}$$

## 3.6.3.3. Determination of Total Branching

All methods for the determination of total branching involve following changes of physical and chemical properties due to end groups or tertiary C-atoms compared with those due to C-atoms in the backbone. This, however, requires knowledge of the chemical structure of the macromolecule. Reliable results are currently available for polyethylenes, a few ethylene copolymers and for special vinyl polymers.

### 3.6.3.3.1. IR Method

Quantitative IR determination of the degree of branching involves determining the number of end groups relative to 100 or 1000 C-atoms in the polymer chain. It is assumed that the terminations of all branches consist of the same, IR-spectrometrically measurable structural unit. This functional group must not be a constituent of the main chain. The two end groups of the macromolecule may be neglected for degrees of branching greater than 10 and for high molecular weights.

This method is generally used for the determination of the branching numbers in polyethylenes. A suitable absorption band is the $\delta_s(CH_3)$-vibration at 1378 cm$^{-1}$ but the proportion of individual alkyl branches ($SCB$) (see Table 67) can also be determined. In the latter case, the method belongs to the selective methods described in 3.6.3.4. In the quantitative evaluation of the $\delta_s(CH_3)$-vibration, corrections must be made for overlapping by the neighbouring vibrations of the methylene group at 1368 and 1352 cm$^{-1}$.

*Table 66*: Determination of the number of long-chain branchings by methods based on contraction

| Method | Quantity to be measured | Equation | | Prerequisites |
|---|---|---|---|---|
| Light scattering | $\langle s^2 \rangle_b, \langle s^2 \rangle_l$ $\bar{M}_b = \bar{M}_l$ | $g = \dfrac{\langle s^2 \rangle_b}{\langle s^2 \rangle_l}$ | $g \to g_0$ | $g = g_0$ or $g$ is to be converted with relationships of the 2-parameter theory; |
| | | | $g_0 \to n$ | branching models applicable |
| Viscometry | $[\eta]_b, [\eta]_l$ $\bar{M}_b = \bar{M}_l$ | $g' = \dfrac{[\eta]_b}{[\eta]_l}$ | $g' \to g$ $(g' \to g_0)$ $g \to g_0$ $g_0 \to n$ | $g' = g'_0$ and $g'_0 = g_0^b$ or: $g' = g^b$ and $g \to g_0$ |
| Sedimentation | $s_b^*, s_l^*$ $\bar{M}_b = \bar{M}_l$ | $h = (s_{c=0}^*)_l/(s_{c=0}^*)_b$ $h^3 = g'$ | | like viscometry |
| GPC — $[\eta]$ [14] | $[\eta]_b = f(V_e)$ or: $[\eta]_b$ of the total sample | $[\eta]_b \cdot M_b = [\eta]_l \cdot M_l \to M_b$ (each $V_e$) | | universal calibration correct (Chapter 2.5., equation (3)) |
| | | $[\eta]_l$ at $M_b$ calculated $\to g'$ | | like viscometry |
| | | $[\eta]_b = \sum_i H_i [\eta]_{b,i}$ $= K \sum_i H_i M_i^a g(M_i, \lambda)$ $\to \lambda$ (iteration method) | | correlation between $n$ and $M$ known (e.g. $\lambda = n/M =$ constant); like viscometry |
| GPC-light-scattering | $\bar{M}_b = f(V_e)$ | $[\eta]_b \cdot M_b = [\eta]_l \cdot M_l$ $= K \cdot M_l^{a+1}$ from $M_b \to [\eta]_{l, M_b} = K \cdot M_b^a$ $([\eta]_b/[\eta]_{l, M_b}) \cdot K \cdot M_b^{1+a}$ $= K \cdot M_l^{1+a}$ $g' = (M_l/M_b)^{1+a}$ ($V_e =$ const.) | | universal calibration correct like viscometry |

$K, a$ = constants of the $[\eta] - \bar{M}$ equation
$s^*$ = sedimentation coefficient
$H_i$ = normalized elution chromatogram heights
Index $b$ = branched, Index $l$ = linear

Table 67: Characteristic absorption bands of statistically distributed alkyl branches in LDPE and their assignment in the IR-spectrum

| Side chain | Absorption band | Assignment |
|---|---|---|
| methyl | 935 | $\gamma_r$-$CH_3$ |
|  | 1150 | $\gamma_w$-$CH_3$ |
|  | 1378 | $\delta_s$-$CH_3$ |
| ethyl | 762 (770, 780) | $\gamma_r$-$CH_2$–$CH$–$(CH_2)_1$–$CH_3$ |
|  | 895 | $\gamma_r$-$CH_3$ |
|  | 1380 | $\delta_s$-$CH_3$ |
| butyl | 745 | $\gamma_r$-$CH_2$–$CH$–$(CH_2)_3$–$CH_3$ |
|  | 890 | $\gamma_r$-$CH_3$ |
|  | 1378 | $\delta_s$-$CH_3$ |
| hexyl | 890 | $\gamma_r$-$CH_3$ |
|  | 1378 | $\delta_s$-$CH_3$ |

## 3.6.3.3.2. NMR Method, $^1$H-NMR-Spectroscopy

NMR measurements can be used to determine directly the number of branching points provided certain assumptions are made. The basic assumption is that each branching point in a vinyl polymer is surrounded by three methylene groups. The signal of these methylene protons is displaced with respect to the other $CH_2$-protons because of the different chemical environment. As a consequence, a doublet will occur. However evaluation is made more difficult by the fact that this assumption is not proven and that the methylene protons are intensely coupled with the methine protons. $^{13}$C-NMR is more informative when it comes to determining tertiary C-atoms. Details of experimental procedures and data evaluation can be found in [8]. In general, $CH_3$-protons are currently used in total branching determination in PE. This involves measurement of the number of branching end points as described in 3.6.3.3.1.

## 3.6.3.4. Specific Methods for the Determination of Short-Chain Branching

### 3.6.3.4.1. $^{13}$C-Nuclear Magnetic Resonance Spectroscopy

$^{13}$C-NMR provides considerably more information about short-chain branching structures than IR or PMR spectra. $^{13}$C-spectroscopy is more sensitive to chemical shifts (see 3.5.) than high resolution proton resonance. The influence of the chemical environment over 5 carbon-carbon bonds on the resonances of the structural unit under consideration (Fig. 86) is critical in the detection of the various branching lengths. Fig. 86 shows the characteristic resonances [9] for different branch lengths in high-pressure polyethylene and their assignment in the $^{13}$C-spectrum.

In order to obtain a signal-noise ratio which enables the type and amount of *SCB* to be determined Pulse-FOURIER-Transform-(*PFT*)-technique must be used. The number of accumulations is between $1 \cdot 10^4$ and $2 \cdot 10^4$. Difficulties in the

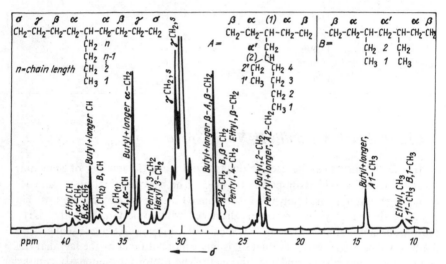

*Fig. 86*: $^{13}$C-spectrum of a high pressure polyethylene (20 per cent by weight solution in trichlorobenzene; reference substance: TMS)

Table 68: Length distribution of short-chain branching in a high-pressure polyethylene determined from the $^{13}$C-NMR Spectrum

| Branch length | Proportion in % |
|---|---|
| n-butyl | 67 |
| n-pentyl | 12 |
| n-ethyl | 8 |
| n-hexyl and longer | 5 |
| 2-ethylhexyl | 4 |
| pairs of ethyl | 4 |

quantitative evaluation occur because of differences in the relaxation times of $^{13}$C-nuclei which are in different chemical environments. Instrumental recording conditions are therefore particularly important.

The samples are prepared in the form of 20 to 25 per cent by weight solutions usually in ortho-dichlorobenzene. In general, measurements are carried out at temperatures between 110 and 140 °C. Quantitative evaluation of the spectrum shown in Fig. 86 results in the branch length distribution represented in Table 68. This indicates that, in high-pressure polyethylene, 95 per cent of all branches are short and only 5 per cent are long.

### 3.6.3.4.2. Radiolysis Gas Chromatography

During the γ-radiation of branched polyolefins, hydrogen is produced. In addition, low-molecular weight hydrocarbons are always found in the gaseous scission products [3, 9]; the chain lengths of these hydrocarbons correlate with the original branching structure. The fact that γ-irradiation chain splitting occurs at the tertiary carbon atom is due to the lower binding energy of these structural units. Decomposition gases of the irradiated high-pressure polyethylenes, for example, contain 75 to 85 per cent of hydrogen and 15 to 25 per cent of hydrocarbons whereas those of linear polymethylene contain up to 99.9 per cent of hydrogen.

Gas-chromatographic analysis of the decomposition gases showed mainly saturated $C_1$ to $C_7$ compounds. The exclusive occurrence of hydrocarbons with less than 8 C-atoms in the decomposition gases can be explained by the FRANK-RABINOWITSCH (cage) effect. According to this, only alkyl radicals with less than 8 C-atoms can diffuse away from the site of bond fission and recombine with the excess hydrogen radicals. For longer chain fragments, recombination of the macroradicals is the main process. The selectivity of the method is not found at the point of radiolytic degradation but in the diffusion-controlled separation of bond fission products. Fig. 87 shows the radiolysis gas chromatograms of ethyl-branched and linear polyethylenes and of a low density polyethylene.

The percentage of hydrocarbons in the decomposition gas is independent of radiation source, sample preparation and, below 34 °C, of temperature. It is dependent on the dose of the absorbed radiation, which is selected as a function of the amount of the sample and the detection sensitivity of the gas chromatograph.

Gas chromatograph, combined with a mass-spectrometer can be used for separation and identification of decomposition gases. For continuous (on stream) investigations, coupling of mass spectroscopy with gas chromatography is advisable for economic reasons.

Quantitative evaluation of gas chromatograms provides data on the proportion of different short side chains in the polymer. Evaluation requires knowledge of the

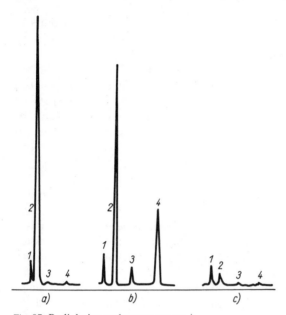

*Fig. 87*: Radiolysis gas chromatograms
a) ethyl-branched polyethylene, b) low density polyethylene, c) high density polyethylene
(1: methane; 2: ethane; 3: propane; 4: butane)

quantum yield $G$ of the respective alkyl branching. The $G$-value is defined as the number of molecules changed (converted, formed or consumed) per 100 eV. This quantity is only obtainable by calibration.

For this purpose, model polymers branched in the defined manner and whose branching number has been measured by the methods of 3.6.3.3.1. or 3.6.3.3.2., are irradiated. The gaseous irradiation products are determined as a function of the number and type of branching and the $G$-value is calculated from these measured values. An average scission coefficient $X_{100}$ can also be calculated in this way. This value is the ratio of short-chain branches split off as hydrocarbons to the total number of these branches per 1000 C-atoms when subjected to a radiation dose of 1 MGy.

As a guide we may assume that at doses of 0.2 MGy and 1.6 MGy about 0.25% and 1% respectively of alkyl branches present in the polymer are split off. Differences in the scission capacity are due to differing lengths of the side chains.

In contrast to radiation degradation, thermal degradation (pyrolysis) does not lead to splitting-off of side branches but to characteristic fragmentation of the polymer chain. By comparison with pyrograms of calibration samples, quantitative conclusions can be drawn about the original branching structures.

## 3.6.4. Examples

### 3.6.4.1. Determination of the Contraction Factor $g$ from Viscosity Measurements

For analytical determination of molecular contraction arising from long-chain branching, a linear high density polyethylene and branched low density polyethylene are fractionated (see 2.3.3.).

The fractions are characterized by means of light scattering and viscometry. The following relationships apply to the high density polyethylene in the solvent o-dichlorobenzene at 410 K:

$$[\eta]_{0-DCB}^{410\,K} = 1{,}67 \cdot 10^{-4} \cdot M^{0{,}76} \,; \quad [\eta] \text{ in } dl \cdot g^{-1} \tag{17}$$

$$\langle s^2 \rangle_z = 4{,}60 \cdot 10^{-20} \cdot M_w \,; \quad \langle s^2 \rangle \text{ in } m^2 \,; \quad M_w \text{ in } g \cdot mol^{-1} \tag{18}$$

Since squares of the radii of gyration or limiting viscosity numbers of fractions of the same molecular weight are related to each other, the $[\eta]_l$ and/or $\langle s^2 \rangle_l$ values for the linear average molecular weight of polyethylene fractions must be calculated using equations (17) and (18). From the experimentally determined $\langle s^2 \rangle_b$ and $[\eta]_b$, the contraction factors $g$ and/or $g'$ (Table 69) are then obtained along with $\langle s^2 \rangle_l$ and $[\eta]_l$ calculated for $\bar{M}_b$. From these values, the exponent $b$ in equation (3b) can be calculated. It lies between 1.0 and 1.1 and usually values of $1.0 \pm 0.3$ are quoted [10].

Comparison of $g$ and $g'$ provides information about the shape of the macromolecular coil. For this purpose, equation (3b) is written in a form analogous to equation (3a) [11]:

$$\frac{[\eta]_b}{[\eta]_l} = g^b = \frac{\Phi_b}{\Phi_l} \cdot g^{3/2} \tag{19}$$

The ratios $\Phi_b/\Phi_l$ calculated from the contraction factors are also included in Table 69.

It can be clearly seen that an increase of the value of $\dfrac{\Phi_b}{\Phi_l}$ is associated with a decrease of the ratio $\dfrac{[\eta]_b}{[\eta]_l}$. This means that the FLORY universal factor changes considerably with increasing branching number indicating a change in the shape of the coils. The branched product is more nearly spherical than linear. For a spherical molecule, the FLORY universal constant is $\Phi_0 = 13{,}56 \cdot 10^{24}\,mol^{-1}$, and for a linear molecule it is $\Phi_0 = 4{,}18 \cdot 10^{24}\,mol^{-1}$ (relative to the square of the radius of gyration).

Table 69: Measured values and calculated quantities for long-chain branched polyethylene

| $\bar{M}_b \cdot 10^{-5}$ g·mol$^{-1}$ | $[\eta]_b$ dl·g$^{-1}$ | $[\eta]_l$ dl·g$^{-1}$ | $\langle s^2 \rangle_b$ $10^{-14}$ m$^2$ | $\langle s^2 \rangle_l$ $10^{-14}$ m$^2$ | $g'$ | $g$ | $b$ | $\Phi_b/\Phi_l$ |
|---|---|---|---|---|---|---|---|---|
| 1.6 | 0.60 | 1.51 | 0.324 | 0.736 | 0.40 | 0.44 | 1.1 | 1.37 |
| 2.4 | 0.76 | 2.05 | 0.440 | 1.10 | 0.37 | 0.40 | 1.1 | 1.46 |
| 3.75 | 0.94 | 2.88 | 0.630 | 1.72 | 0.33 | 0.37 | 1.1 | 1.47 |
| 5.2 | 1.10 | 3.67 | 0.741 | 2.39 | 0.30 | 0.31 | 1.0 | 1.74 |
| 7.4 | 1.31 | 4.82 | 0.955 | 3.40 | 0.27 | 0.28 | 1.0 | 1.82 |

## 3.6.4.2. Determination of the Branching Number of a Low Density Polyethylene [12]

Before taking any measurements the antioxidants contained in the high-pressure polyethylene are quantitatively extracted by means of chloroform using a Soxhlet apparatus. The purified sample is prepared as a film by means of a hydraulic press, operating at a pressure of $2 \cdot 10^7$ N·m$^{-2}$ and a temperature of 120 to 150 °C. The layer thickness of the film should lie between $5 \cdot 10^{-2}$ and $12 \cdot 10^{-3}$ cm. The average layer thickness is measured with a micrometer or obtained from the weight per unit area. Alternatively $\beta$-radiation absorption measurement may be used. The spectra are recorded by means of a high-resolution doublebeam spectrometer over the wave number range from 1330 to 1400 cm$^{-1}$.

The methylene doublet at 1368 and 1352 cm$^{-1}$ is compensated using a polymethylene compensation wedge [12] prepared as follows.

On a rectangular polymethylene film ($15 \times 50 \times 0.04$ mm), synthesized from diazomethane, another film $15 \times 45 \times 0.04$ mm is placed and pressed together at a pressure of about $2 \cdot 10^7$ Nm$^{-2}$ and a temperature of 180 °C. This procedure is repeated several times using progressively smaller polymethylene films so that a wegde is produced with layer thickness which increases linearly from $4 \times 10^{-3}$ to $20 \times 10^{-3}$ cm. The position of the wedge film is adjusted in the beam of the spectrometer until the peaks at wave numbers 1368 cm$^{-1}$ and 1400 cm$^{-1}$ have the same intensity. The spectrum is then scanned over the range 1330 to 1400 cm$^{-1}$. The recorded methyl absorption (1378 cm$^{-1}$) is evaluated quantitatively by the base line method. The absorbance is divided by the layer thickness and the number of $CH_3$-groups per 1000 C-atoms is obtained from a calibration curve or calculated from the equation of a straight-line fit to the latter.

Mixtures of polypropylene or polybutene-1 and linear polyethylene are used for calibration. The amounts of polypropylene per 100 g of polyethylene are calculated to give the required number of $CH_3/1000$ C-atoms and weighed accurately. The substances are mixed and pressed to produce films. To homogenize the polypropylene in the polyethylene, the films are repressed several times at a temperature of 150 to 160 °C and a pressure of $2.5 \cdot 10^7$ N m$^{-2}$. The spectra of the calibration films and their layer thicknesses are obtained in the manner described above. The measured absorbances divided by the layer thickness are plotted versus the known number of $CH_3/1000$ C-atoms. In Table 70, some calibration values are listed from which the following calibration relationships was derived:

$$CH_3/1000 \text{ C-atoms} = 0.8 \cdot \frac{A_{1378}}{d} \qquad (20)$$

$A_{1378}$ = absorbance at 1378 cm$^{-1}$, $d$ = layer thickness in cm

*Table 70*: Calibration values for IR determination of the number of methyl groups per 1000 C-atoms

| $CH_3$ per 1000 C-Atome | g PP per 100 g PE | $A_{1378}$ | $d/10^{-3}$ cm | $\frac{A}{d}$/cm$^{-1}$ |
|---|---|---|---|---|
| 2.5 | 0.78 | 0.029 | 9.8 | 2.96 |
| 5.0 | 1.56 | 0.048 | 7.8 | 6.15 |
| 10 | 3.13 | 0.103 | 8.2 | 12.56 |
| 15 | 4.68 | 0.120 | 6.3 | 19.06 |
| 20 | 6.26 | 0.190 | 7.4 | 25.70 |
| 25 | 7.80 | 0.180 | 5.9 | 30.55 |
| 30 | 9.38 | 0.260 | 6.8 | 38.20 |

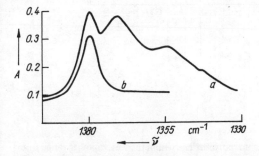

*Fig. 88*: Partial IR spectrum of a high pressure polyethylene
a) not compensated, b) compensated

The spectrum of a high-pressure polyethylene shown in Fig. 88, was obtained using a pressed film of thickness $7.15 \cdot 10^{-3}$ cm. The measured absorbance of 0.21 indicates (from equation 20) that the number of branching end points is 23 $CH_3$/1000 C-atoms. The error is estimated at $\pm 1$ $CH_3$ per 1000 C-atoms.

This method of evaluation ignores the fact that the absorbtion coefficient of the $\delta_s(CH_3)$-vibration is influenced by the lengths of the short side branches. Since in high-pressure polyethylene different short branches are found side by side, a correction method is necessary for the exact determination of the number of $CH_3$-groups per 1000 C-atoms; this method calls for the knowledge of the various proportions of the short side branches.

When evaluating the band ratio of the methyl group (1378 cm$^{-1}$) to the methylene group (1368 cm$^{-1}$), the lower limit of detection is about 3 $CH_3$/1000 C-atoms. Since the intensity of the reference band is dependent on the crystallinity [13], the samples must be measured at a temperature above 140 °C. The sensitivity can be increased so that less than 0.5 $CH_3$/1000 C-atoms can be determined if the second derivative of the band complex (Fig. 89) is recorded between 1330 and 1400 cm$^{-1}$.

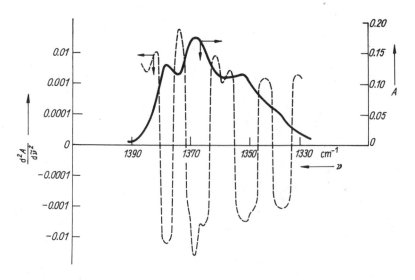

*Fig. 89*: Partial IR spectrum of a high pressure polyethylene within the wave number range from 1330 to 1400 cm$^{-1}$

——————— original spectrum, ————— second differential according to wave number

### 3.6.4.3. NMR Determination of the Number of Methyl Groups in an Ethylene-butene-1 Copolymer

For the determination of the number of $CH_3$/1000 C-atoms, a 5 per cent by weight solution of the sample in ortho-dichlorobenzene is prepared by shaking at a temperature between 100 and 120 °C. The spectrum is recorded using a high resolution NMR-spectrometer with a variable temperature sample head. The *PFT*-technique is used with 300 accumulations at 100 °C. Tetramethyl silane or hexamethylene disiloxane is used as a reference. In the PMR spectrum, the methyl protons appear at lower fields than the methylene protons (Fig. 90). The individual peaks are separated as shown in Fig. 90 and the areas isolated in this way are measured by a planimeter. Low $CH_3$-concentrations call for high amplification of the signal. In this case, the rotational side bands on either side of the $CH_3$-resonances as well as the $^{13}C$-satellites may no longer be neglected and must also be separated.

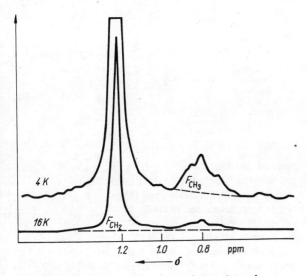

*Fig. 90*: PMR spectrum of an ethylene-butene-1 copolymer

| Group | chemical shift |
|---|---|
| $CH_2$ | 1.23 |
| $CH_3$ main peak | 0.81 |
| $CH_3$ secondary peak | 0.85 |
| $CH_3$ secondary peak | 0.73 |

The number of $CH_3/1000$-atoms is calculated according to equation (21):

$$CH_3/1\,000\,\text{C-atoms} = \frac{\frac{1}{3}F_{CH_3}}{\frac{1}{2}\frac{2^b}{2^a}\cdot F_{CH_2} + \frac{1}{3}F_{CH_3}} \cdot 1000 \qquad (21)$$

$F_{CH_3}$ = area of the methylene proton peak, $F_{CH_2}$ = area of the methyl proton peak

The factor $\frac{2^b}{2^a}$ results from the different amplification of the various signals (the computer used for the accumulation calculates in powers of two). The following areas are obtained for the spectrum shown in Fig. 90:

$F_{CH_3} = 0.51$ Au   $F_{CH_2} = 1.02$ Au

Au = area unit

For the amplification factors:
$a = 2$ and $b = 4$. This gives 77 $CH_3/1000$ C-atoms in the chosen example corresponding to a composition of 23 mol % of butene-1 in the copolymer.

# Bibliography

[1] SCHRÖDER, E., Plaste und Kautschuk **20** (1973), 241
[2] DEXHEIMER, H., FUCHS, O. and H. SEEHR, Dtsch. Farben-Z. **11** (1968), 481
[3] CROMPTON, T. R., „The Analysis of Plastics", Pergamon Press 1984
[4] FRIIS, N. and HAMIELEC, A. E., J. Appl. Polym. Sci. **19** (1975), 97
[5] SMALL, P. A., Advances Polym. Sci. **18** (1975), 11
[6] SCHRÖDER, E. and B. ANDREJS, Plaste und Kautschuk **19** (1972), 747
[7] STEIN, D. J. and G. V. SCHULZ, Makromolekulare Chem. **52** (1962), 249
[8] AXELSON, D. E., MANDELKERN, L. and G. C. LEVY, Macromolecules **10** (1977), 1032
[9] BOWMER, T. N. and J. H. O'DONNELL, Polymer **18** (1977), 1032
[10] CASPER, R., BISKUP, U., LANGE, H. and U. POHL, Makromolekulare Chem. **177** (1976), 1111
[11] HOFFMANN, M. and R. KUHN, Makromolekulare Chem. **174** (1973), 149
[12] WILLBOURN, A. H., J. Polymer Sci. **34** (1959), 569
[13] DECHANT, J., „Ultrarotspektroskopische Untersuchungen an Polymeren", Akademie-Verlag, Berlin 1972
[14] DROTT, E. E. and R. A. MENDELSON, J. Polymer Sci. A 2 **8** (1970), 1373

## 3.7. Characterization of Network Polymers

### 3.7.1. Introduction

Network structures occur very frequently in larger polymers. They are found in nature in proteins as a structural element of living cells, in cellulose gels, in starch or alginates, in minerals, in silicates, aluminates, clays etc. and also in vulcanized natural and synthetic rubber, cross-linked polyesters, polyurethanes and in the classical thermosets, in phenol, urea or melamine resins. In spite of the differences in their appearances and properties, all of these groups, here presented at random, have one common characteristic structural feature, namely, their three-dimensional constitution. In the narrow sense of macromolecular chemistry, only molecules which „*consist of a three-dimensional system of rings fitted into each other, which can contain side chains and branches at the ring skeleton but do not exhibit a layer structure*" [1] are defined as network polymers. They can involve chemical and physical linkages which frequently occur together. The chemical cross-linkages are predominantly covalent and thermostable. Ion and H-bonds as are present in the ionomers and allophanate structures of the cross-linked polyurethanes are thermally reversible. Physically formed network structures arise from chain entanglements, rings fitted into each other or chain loops as well as from crystalline adhesion points. Typical examples are interpenetration networks in which two networks of different structure penetrate each other [2].

Apart from the type of bond, network polymers are distinguished by:
— The constitution of the bridge cross-links,
— the crystallinity of the segments,
— the average cross-linking density,
— the segment length distribution between the points of cross-linking and
— the number of coexisting phases and their supermolecular structure.

Possible network structures are shown schematically in Fig. 91.

This large number of factors influencing structure forms lead to a very broad spectrum of network properties which hinders characterization. Of the few universal

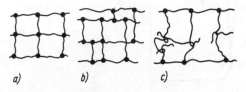

*Fig. 91*: Types of network structures
a — ideal regular network with covalent bonding, b — statistical network with covalent bonding, c — network with chain entanglement

properties of cross-linked polymers, particular emphasis should be laid on their insolubility in organic solvents and their thermomechanical behaviour. Consequently these also form the basis of quantitative methods for characterization. In the amorphous cross-linked polymers, the transition from the hard-elastic to the rubber-elastic condition is characterized by a considerable decrease of the shear modulus $G$ and a distinct damping maximum (Fig. 92).

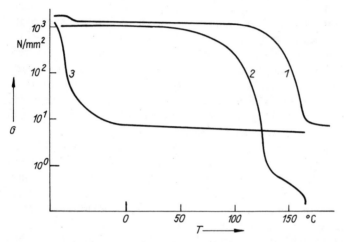

*Fig. 92*: G modulus-temperature characteristic of amorphous cross-linked polymers [3]
1 — epoxide resins, 2 — polyacrylates, 3 — butadiene rubber

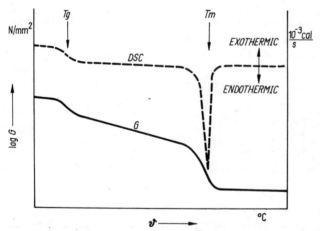

*Fig. 93*: G modulus-temperature behaviour of crystalline cross-linked polymers [3]

The glass temperature $T_g$ determined by the freezing of the motion of longer chain segments, thus is dependent on the cross-link density. In crystalline cross-linked polymers, the softening behaviour arises from the melting of the crystallites which usually melt over a narrow temperature range. The $G$-modulus and the tensile strength decrease well below the melting temperature (Fig. 93).

### 3.7.2. Characteristic Structural Parameters and Their Quantitative Determination [4—6]

### 3.7.2.1. Determination of Network Density by Elastometric Methods

When the initial monomers are known, the structure of a cross-linked polymer is sufficiently characterized by the following parameters:

— the average cross-link density $\quad v = \dfrac{\text{number of network chains}}{N_L V_d}$

— the number-average molecular weight of the chain between two junction points $\quad M_c = \dfrac{\varrho_B}{v}$

— the average distance between two junction points $\quad d_c$
— the distance distribution between two junction points $\quad H(d_c)$
— the average functionality of the cross-linking nodes $\quad f$

$N_L$ = Avogadro number, $V_d$ = volume of the dry network

These parameters can be determined experimentally under certain conditions at moderate cross-linking with the excepting of the average functionality. If the cross-linking reaction itself cannot be followed by chemical, spectroscopic, differential-thermoanalytical or other methods or if selective decomposition of the network is not possible, the cross-link density is usually determined from deformation, swelling or solubility behaviour. Elastometric methods based on deformation measure both the chemical cross-linkages brought about by covalent bond and also the physical cross-linkages, whereas solubility behaviour is influenced only by the chemical cross-linkages.

Rubber elasticity theory (*RET*) is the basis for all elastometric methods of determining cross-link density of amorphous polymers. This is derived from the statistical thermodynamics of an ideal body with GAUSSIAN segment distribution. The elongation behaviour of a dry test specimen is described by the equation:

$$\sigma = \nu RT[\lambda - \lambda^{-2}] \tag{1}$$

$\sigma$ = tensile stress = $f/A_0$ = tensile force/initial cross-sectional area, $R$ = gas constant, $T$ = absolute temperature, $\lambda$ = deformation ratio = $\dfrac{l}{l_0}$, $l$ = length of the elongated, $l_0$ = non-elongated test specimen

Thus, determination of the cross-link density, would be possible using stress-strain measurements. However, substantial experimental evidence indicates considerable deviations from ideal rubber-elastic behaviour even in the range of small deformations (see Fig. 94).

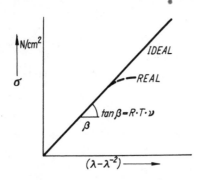

*Fig. 94*: Stress-strain behaviour of ideal and real polymer networks

They are corrected from the phantom network model by the introduction of the so-called $g$-factor:

$$\sigma = g\nu RT[\lambda - \lambda^{-2}] \tag{2}$$

The $g$-factor, which can be written as a product of two quantities, $\eta$ and $A$, takes into account conformative and constitutional factors.

By defining

$$\eta = \frac{\langle s^2 \rangle}{\langle s_0^2 \rangle}$$

where $\langle s^2 \rangle$ = end-to-end-distance of the isotropic, non-deformed chain in the network, $\langle s_0^2 \rangle$ = end-to-end-distance of the non cross-linked (i.e. free), chain in the same molecular environment

a reference condition is established. In this condition, the network arches are so large that no forces are applied to the network points. The quantity $\eta$, which is also called the memory term, depends on the degree of dilution during the formation reaction of the network [5]. For networks produced by reaction on the polymer in the absence of solvents, $\eta$ is equal to unity.

Besides the cross-link density, the type of network linking must be considered. The microstructure of the network exerts an influence on the structural factor $A$.

The $g$ factor, however, cannot be determined experimentally. Therefore, for quantitative evaluation of stress-strain measurements, at least for swollen samples, the empirical 2-parameter MOONEY-RIVLIN equation is frequently used:

$$\sigma = 2(C_1 + C_2\lambda^{-1})(\lambda - \lambda^{-2}). \tag{3}$$

Here, the first term $2C_1$ corresponds to ideal elongation; the second term has recently been associated with the topological hindrance between the chains during deformation [7]. Measurement is aided by the fact that the $C_2$-term tends towards zero in the swelling equilibrium.

According to HERMANS, FLORY and WALL (HFW-method), the following quantitative relationship exists between the strain per network unit of the dry polymer ($\sigma_d$) and the deformation in the swelling equilibrium:

$$\sigma_d = A\eta v R T q^{1/3}\left(\lambda - \frac{q'}{q}\lambda^{-2}\right) \tag{4}$$

$q$ = degree of swelling in non-deformed gel, $q'$ = degree of swelling in deformed gel

A determination of the cross-link density alone based on deformation and swelling measurements is only possible if $\eta$ is known and assumptions as to the value of $A$ are made. Experimentally $A$ is usually found to have the value of unity.

The combination of stress-strain measurements with compression measurements (in the swelling equilibrium in each case) is useful. According to rubber elasticity theory, the following holds for the linear compression of swollen samples:

$$f/A_0 = G^* q^{1/3}(\lambda - \lambda^{-2}) \tag{5}$$

$f/A_0$ = compressive force/specimen area in unswollen and unloaded state; $\lambda$ = deformation ratio $= \frac{h_q - \Delta h}{h_q}$, $h_q$ = height of the non-deformed, swollen sample, $\Delta h$ = change in length of the sample by deformation, $A_0$ = cross-sectional area of the non-deformed and unswollen sample, $G^*$ = compression modulus of the unswollen network

$$G^* = A\eta v R T \tag{6}$$

## 3.7.2.2. Swelling Measurements

Swelling measurements are of particular importance in network characterization because of their low cost. The modified FLORY-REHNER equation (7) frequently used in such evaluation relates the chemical potential of the swelling agent $\Delta\mu_A$ in the swollen network relative to the pure swelling agent, and the network parameters:

$$\Delta\mu_A = RT[\ln(1 - \Phi_B) + \Phi_B + \chi\Phi_B^2 + V_A A\eta v\Phi_B^{1/3} - V_A Bv\Phi_B] \tag{7}$$

$V_A$ = molar volume of the swelling agent, $\chi$ = interaction parameter
$A$ = microstructure factor, $B$ = volume factor

Using equation (7), the contribution due to mixing $\Delta\mu_{A,M}$, and the contribution due to elasticity $\Delta\mu_{A,el}$, as well as the contribution $\Delta\mu_{A,cr}$ due to cross-linkage reaction are related by:

$$\Delta\mu_A = \Delta\mu_{A,M} + \Delta\mu_{A,el} + \Delta\mu_{A,cr} \tag{7a}$$

In the $\Delta\mu_{A,M}$ described by the FLORY-HUGGINS-theory, $\chi$ depends on concentration and temperature. Even under isothermal condition, the concentration dependence of $\chi$ must be determined experimentally with great care because the term $\chi\Phi_B^2$ in equation (7) exerts a considerably stronger influence on $\Delta\mu_A$ than $\Delta\mu_{A,el}$.

The numerical values of $A$ and $B$ depend on the network model which is assumed while $\eta$ depends on the cross-linking conditions. Without making certain assumptions, it is impossible to determine cross-link density via equation (7) through swelling measurements alone.

A combination of measurements of the chemical potential with swelling and compression measurements allows a network to be completely characterized. This is indicated by equation (8) which is obtained by combining and rearranging equations (5—7):

$$\frac{\frac{-\Delta\mu_A}{RT} + \ln(1 - \Phi_B)}{\Phi_B} + 1 + \frac{V_A \Phi_B^{-1/3} f}{RT A_0(\lambda - \lambda^{-2})} = V_A vB - \chi\Phi_B \tag{8}$$

The product $A\eta v$ (equation (7)) is obtained from compression measurements on the network swollen to equilibrium whereas the parameters $\chi$ and $vB$ can be calculated by measuring the chemical potential at different degrees of swelling and the relevant compression data (see also 3.7.4.5.).

Further information can be obtained by studying the temperature-dependence of the swelling since its magnitude and direction depend on interaction with the swelling agent and especially on the sign of the infinite dilution heat $\overline{\Delta H}_{A,S}$.

Thermodynamic treatment shows that the temperature-dependence of the swelling agent uptake $\partial T/\partial x_A^*$ at constant pressure $p$ is:

$$\left(\frac{\partial T}{\partial x_A^*}\right)_p = \frac{\left(\frac{\partial \Delta \mu_A}{\partial x_A^*}\right)_p \cdot T}{\overline{\Delta H}_{A,S}} \tag{9}$$

$x_A^*$ = primary mole fraction of the swelling agent, $\Delta \mu_A$ = chemical potential of the swelling agent

(For the derivation of equation (9) see REHAGE [17].)

The infinite dilution heat $\overline{\Delta H}_{A,S}$ is the amount of heat which the system takes up or emits when one mole of the swelling agent is added to such an amount of maximally swollen gel that the saturation concentration remains essentially constant. For a positive value of $\overline{\Delta H}_{A,S}$, the swelling increases with temperature whereas for a negative value the polymer shrinks with increasing temperature. Only in systems where $\overline{\Delta H}_{A,S}$ has the value of 0 is the swelling independent of temperature.

### 3.7.2.3. Solubility Investigations

Quantitative information about the cross-link density and the cross-link index $\gamma = \nu/n$ (where $n$ = number of moles of the primary molecules) can also be derived from the well-known relationship between solubility decrease and increasing degree of cross-linking. The evaluation of the FLORY equation for the relationship between gel and sol fractions ($w_G$ and $w_s$) of a polymer solution and the cross-link index

$$w_G = 1 - \frac{\sum H(P) \exp\left[-\gamma(1 - w_s)\right]}{\sum H(P)} \tag{10}$$

calls not only for knowledge of the distribution function, $H(P)$, of the non cross-linked macromolecules but also assumes that
— all primary molecules have the same probability of cross-linking
— no intramolecular cross-linkages are present, and
— the degree of cross-linking is low.

For the case where the primary molecules all have the same molecular weight, a simple relationship results from equation (10):

$$w_s = \exp\left[-\gamma(1 - w_s)\right] \quad \text{or} \quad \ln w_s = -\gamma(1 - w_s) \tag{11}$$

$w_G = 1 - w_s$

This must not be used for quantitative evaluation of network structures based on statistically distributed primary molecules. For SCHULZ-FLORY-distribution and limited molecular inhomogeneity of the initial polymer, the following holds:

$$M_c = w_s + \sqrt{w_s M_n}, \tag{11a}$$

where $M_n$ is the molecular weight of the initial polymer.

For qualitative comparative investigations, the solubility method (also termed sol-gel analysis) can of course be used without taking the $H(P)$ into account. In contrast to the elastometric methods mentioned so far, this responds only to chemical cross-linkages. Determination of the cross-link density by sol-gel analysis becomes increasingly inaccurate with increasing degree of polymerization of the primary molecules, number of functional groups that can be cross-linked and degree of cross-linking.

### 3.7.2.4. Determination of the Mean Distance and Distance Distribution between Two Junction Points

In addition to chemical investigation of the intermediates and scission products, the mean distance between two junction points $d_c$ and its distribution $H(d_c)$, can be determined using small-angle X-ray scattering ($SAXS$), small-angle neutron scattering ($SANS$), deuterium-NMR or thermal analysis [8—11]. While the application of chemical methods remains restricted to networks of prepolymers with defined decomposition products, diffraction methods and NMR require labelled compounds and very expensive equipment. For this reason, they are used mainly in fundamental research.

The considerably simpler thermal analytical methods take advantage of the effect of the surface-dependence of GIBB's free energie in liquid-solid ($ls$) phase transformation in capillaries (in the nanometer range). The swelling agent in the cavities of the network has an increased vapour pressure because of the surface effect and, consequently, the phase transformation temperatures are changed (e.g. freezing point). According to equation (12), the vapour pressure increase is dependent not only on the interfacial tension of the swelling agent between liquid and solid phases ($\sigma_{ls}$) and the specific volume $v_{sp}$ of the swelling agent but above all on the capillary dimensions:

$$\ln \frac{p}{p_0} = \frac{\sigma_{ls} v_{sp}}{RT} \frac{dO}{dV} \tag{12}$$

$p$ = vapour pressure in the capillary, $p_0$ = vapour pressure above the expanded phase, $dO/dV$ = derivative of the capillary surface with respect to the capillary volume

From equation (12), using the CLAUSIUS-CLAPEYRON equation and making a few simplifying assumptions, an equation from which $d_c$ may be determined from the decrease of the freezing point of the swelling agent in the swollen network is obtained:

$$\Delta T = \frac{-zT_{ls0}\sigma_{ls}}{\varrho l_0 d_c} \tag{13}$$

$T_{ls0}$ = freezing point of the swelling agent under normal conditions, $l_0$ = specific heat of fusion of the swelling agent, $\varrho$ = density of the swelling agent, $z$ = numerical factor from $d0/dV$; e.g. $z = 4$ for cubic crystals and $z = 2$ for cylindrical crystals

All the constants can be grouped together so that:

$$d_c = -K/\Delta T \tag{14}$$

The smaller are the ice crystals formed in freezing and, thus, the smaller the cavities limiting the growth of the ice crystals, the greater is $\Delta T$. A $d_c$-value can be assigned to every $\Delta T$-value on the cooling curve (Fig. 95), that is to say, a cavity of the respective dimension in which the swelling agent is just about to freeze can be specified. The amount of heat released in the phase transition depends on the mass $m_i$ of the swelling agent freezing at a certain $\Delta T_i$. Using equation (15), a crystallite-size spectrum (Fig. 101), corresponding to the distance

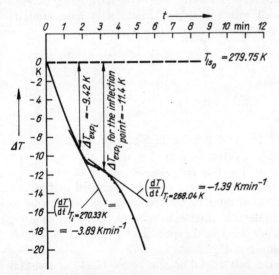

*Fig. 95*: Cooling curve of a natural rubber cross-linked with DCP, swollen in cyclohexane

distribution of cross-linking nodes $H(d_c)$ in the swollen condition, can be calculated from the behaviour of the cooling curve.

$$\frac{dm_i}{d\,\Delta T_i} = C \left\{ \frac{\left(\frac{dT}{dt}\right)_{T_i < T_{ls0}} - \left(\frac{dT}{dt}\right)_{T_i = T_{ls0}}}{\left(\frac{dT}{dt}\right)_{T_i < T_{ls0}}} \right\} \tag{15}$$

$$C = \frac{\dot{w}_{SG} + m_S c_{p,l}}{m_S l_0}$$

$dT/dt$ = cooling rate; $c_{p,l}$ = specific heat capacity of the swelling agent in the liquid state; $w_{SG}$ = heat capacity of the equipment and gelling agent; $dm_i$ = differential change of mass of the ice crystals $\triangleq$ to weight fractions of network node distances $d_{c,i}$; $l_0$ = specific heat of fusion of the swelling agent, $m_S$ = mass of the swelling agent.

In the case of a non-constant cooling rate, $(dT/dt)_{T_i = T_{ls0}}$ must be extended by an additional term (see also 3.7.4.6.). Further, it should be noted that the experimentally measured freezing point depression must be corrected for the freezing point change caused by polymer swelling agent interaction.

The agreement between crystallite-size spectrum and distance distribution of the network nodes has been demonstrated by X-ray scattering, NMR and, indirectly, by kinetic calculations [12]. The correctness of the relationship between $\Delta T$ and $d_c$ has been established with the help of model substances, and by conformationstatistics as well as by means of scaling relationships [13].

## 3.7.3. Experimental Basis

### 3.7.3.1. Elastometry

Tensile testing machines used for investigating high-polymeric materials are not suitable for objective, accurate determination of $\sigma$ and $\lambda$ and quantitative evaluation of elasticity behaviour. For this purpose, the change in length of a swollen specimen under changing tensile stress must be measured as function of time. Fixed points for evaluation must be established [6] because, owing to relaxation time behaviour, considerable differences between the initial length $l_0$ and the length $l_e$ immediately after releasing the specimen are detected. A suitable experimental arrangement is shown in Fig. 96.

The specimens should be dumb bell shaped or, still better, blocks of constant cross-section; the length must be known within $10^{-3}$ cm because even deviations as small as ±0.005 cm exert a considerable effect on the parameters $C_1$ and $C_2$ of equation (3) [4]. Cross-link density determinations of swollen samples involving measurements of linear compression modulus are based on work by CLUFF et al. [14]. Circular specimens of height 12.5 mm and diameter of about 20 mm are kept in suitable swelling agents until swelling equilibrium is attained. The height of the sample is measured accurately before ($h_0$) and after ($h_q$) swelling (time for attaining swelling equilibrium is about 1 week at room temperature). Using the equation:

$$h_q = h_0 \Phi_B^{-1/3} \tag{16}$$

the volume fraction of the polymer in the swollen network or the reciprocal value $q$ can be calculated. The pre-swollen specimens are deformed by variable forces in a sample vessel filled with swelling agent using the experimental set up shown in Fig. 97 and the resulting changes in height, $\Delta h$, are measured with a micrometer. $G^*$ and $v$ may also be calculated from equations (5) and (6). For samples, where the deformation is only about 10% of the initial height $h_0$, the following approximation can be used to calculate $\lambda - \lambda^{-2}$:

$$\frac{3\Delta h}{h_0} = \lambda - \lambda^{-2} \tag{17}$$

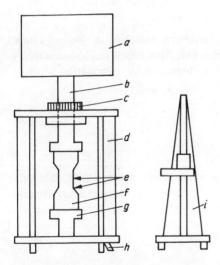

*Fig. 96*: Diagram of a measuring device for recording stress-strain values
a — tensile testing machine, b — tie rod, c — inflow, h — outflow (of the swelling agent), d — swelling container, e — length marking on the specimen, f — specimen, g — clamps, i — cathetometer for length measurement

Experimental set ups where the deformation-conditioned retroforce is determined, such as that described by VAN DER KRAATS [15], lead to good results if the working conditions and homogeneity of the samples are carefully controlled.

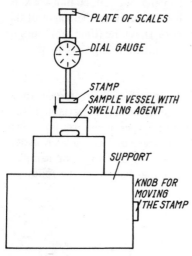

*Fig. 97*: Diagram of an apparatus for the determination of the compression modulus of swollen cross-linked polymers

Furthermore, network densities can also be determined using torsion pendulum tests [16, 17] and photoelastic measurements. The photoelastic methods are based on the measurement of the double refraction $\Delta n$ caused by deformation which is related with the network density in the following way (KUHN and GRÜN):

$$\Delta n = \frac{2\pi}{45} \frac{(\bar{n}^2 + 2)^2}{\bar{n}} \, v \, \Delta\alpha(\lambda - \lambda^{-2}) \tag{18}$$

$$\bar{n} = \frac{n_1 + 2n_2}{3} = \text{average refractive index}$$

This shows that $\Delta n$ depends not only upon $v$ but above all on the bulk polarizabilities $\Delta\alpha$ and the deformation term. Photoelastic measurements are also taken with swollen network polymers and frequently point to non-ideal rubber-elastic behaviour [4].

## 3.7.3.2. Swelling Measurements

The amount of swelling agents taken up in the swelling equilibrium of cross-linked polymers can be obtained gravimetrically, volumetrically or by swelling-pressure measurements. The specimens must be free from extractable constituents and have known densities and dimensions. Usually they are cylindrical. As already mentioned above, at room temperature equilibration times of up to one week are required. For gravimetric determination of the degree of swelling, based on the difference in weight of the specimens before and after swelling, excess amounts of swelling agent are removed by blotting with absorbent paper. This process, which is difficult to carry out because of the flexibility of the samples, is the largest source of error in this method.

In volumetric methods, the degree of swelling is measured by monitoring the changes in layer thickness of the specimen. If isotropic behaviour of the swelling body is assumed, equation (16) can be used for evaluation. Continuous monitoring of the swelling process using a cathetometer or photographic recording is superior to direct measurement of the layer thickness by means of a micrometer. A swelling measurement device with optical measuring arrangement is shown in Fig. 98. The basic requirement for application of volumetric methods is that the sample should have a smooth surface without any defects and that the swelling of the specimens occurs without formation of cracks.

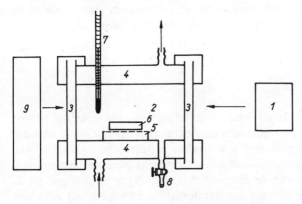

*Fig. 98*: Principle of a swelling measuring apparatus with optical length measurement
1 — cathetometer, 2 — measuring cell, 3 — cell window, 4 — temperature-control jacket, 5 — sample holder, 6 — sample, 7 — thermometer, 8 — solvent outlet, 9 — athermal light source

## 3.7.3.3. Determination of the Chemical Potential ($\Delta \mu_A$) of Cross-linked Polymers

The difference between the chemical potentials in swollen networks with different degrees of swelling can be determined from vapour pressure and swelling pressure measurements. They are very time-consuming and require special measuring equipment. Measurement of $\Delta\mu_A$ by means of a commercial vapour-pressure osmometer (see also 1.4.) is more favourable in every respect [18]. In this, temperature is measured by a thermistor contained in a metal tube for protection against mechanical damage. The dry network sample in the form of a cylindrical specimen is placed on the temperature sensor. For evaluation using equation (8), the difference of the chemical potential must be measured as a function of the degree of swelling. The latter can be determined, at the same time, from the volume increase arising from the uptake of the swelling agent in the vapour compartment of the vapour-pressure osmometer measured by a cathetometer. $\Delta\mu_A$ is obtained by calculation from the temperature difference between the swelling network sample and swelling agent dropped onto the reference thermistor:

$$\frac{\Delta\mu_A}{RT_0} = -\frac{LM_A}{RT_0^2}\Delta T \tag{19}$$

$L$ = specific heat of evaporation of the swelling agent, $M_A$ = molecular weight of the swelling agent, $T_0$ = temperature

Calculation of the temperature difference from the measured resistance difference requires the apparatus constant of the vapour-pressure osmometer to be determined by measuring a solution of known molecular weight $M_B$ as already described (Chapter 1.4.).

### 3.7.3.4. Sol-Gel-Analysis

For ascertaining the insoluble gel components of samples, the latter are treated with strong solvents at high temperatures until extraction equilibrium is attained. The extraction can be carried out in a Soxhlet extractor or by shaking the samples with an approximately 50-fold solvent excess at increased temperature. The insoluble components can be filtered off (G 5-filter) or centrifuged off at 10000 rpm. After purifying with solvent and deswelling by a methanol wash and drying, the gel components are weighed. Optical analysis of microgel content is described by LANGE [19].

### 3.7.3.5. Relative and Qualitative Methods of Network Characterization

All methods for describing the cross-link density of polymers discussed are only applicable to low network densities. Strongly cross-linked products may be compared with each other using one of the many relative methods whose

Characterization of Network Polymers

characteristic quantities are calibrated by reference to the absolute methods described so far.

Among them, infrared spectroscopy, differential thermoanalysis, pyrolysis gas chromatography [20] and inverse gas chromatography are important. For routine testing qualitative and semi-qualitative methods are frequently used. They allow, for example, statements about the degree of curing to be made. Worthy of note among them are the determination of indentation and pendulum hardness, determination of dye receptivity, evaporation rate analysis, and determination of the cross-linking potential. These methods have been summarized by FINK for the investigation of organic coatings [21].

## 3.7.4. Examples

### 3.7.4.1. Optical Measurement of the Degree of Swelling of Peroxide-Cross-linked PMMA

For the determination of the degree of swelling $q$ and its temperature-dependence in peroxide cross-linked PMMA, solvent uptake is measured by means of a cathetometer assuming linear expansion of the swelling body and the measurements are evaluated using equation (16).

Before the swelling experiment, non-cross-linked polymers and impurities are removed by extraction with toluene in a Soxhlet apparatus for 96 hours. The specimen is then dried in vacuo to constant weight.

The specimens (circular, diameter = 14.76 mm, height = 1.5 mm) pre-treated in this way, are placed into the measuring device as shown in Fig. 98 where the linear expansion $l_q$ (mm) is observed until swelling equilibrium is obtained as a function of temperature. To reduce parallax errors, the measuring cell

Table 71: Determination of the swelling degree $q$ of peroxidically cross-linked PMMA

| temperature (°C) | $l_q$ (mm) | $l_0$ (mm) | $\Phi_B^{-1/3}$ | $q$ |
|---|---|---|---|---|
| 20.0 | 19.86 | 14.76 | 1.345 | 2.43 |
| 30.0 | 19.97 | 14.76 | 1.353 | 2.48 |
| 40.0 | 20.09 | 14.76 | 1.361 | 2.52 |
| 50.0 | 20.21 | 14.76 | 1.369 | 2.57 |
| 60.0 | 20.31 | 14.76 | 1.376 | 2.61 |
| 70.0 | 20.39 | 14.76 | 1.381 | 2.64 |

consists of a variable temperature double cylinder which is sealed at the front and back by a screwed-on plane-parallel optical window. Table 71 contains the values measured for polymethyl methacrylate cross-linked with divinyl benzene when swollen in toluene.

### 3.7.4.2. Determination of the Parameters of the MOONEY-RIVLIN Equation of Isocyanate-Cross-linked Butadiene Prepolymers using Stress-strain Measurement

For the investigations, 5 specimens having a thickness of about 1.75 mm and a width of 4.3 mm are used. In order to mark the initial length, two wire ends (piano wire 0.06 mm in diameter) are fastened to the edges of the specimens (distance between the markings about 20 to 25 mm).

The specimen is then clamped between the jaws of the test apparatus shown in Fig. 96 and the initial length $l_0$ determined by means of a cathetometer unless it is to be obtained by extrapolation from the lengths found in the elongation procedure. If this is to be done the reciprocal value of the length $l$ is plotted against the force $p$:

$$\frac{1}{l} = A_0 + A_1 \cdot p \tag{20}$$

and $l_0$ is calculated from $A_0 = \frac{1}{l_0}$.

Subsequently, the sample is stretched to approximately 40% of its initial length ($l_0$) in about 10 steps. Because of relaxation, the stress-strain values are read 5 min after each elongation step. The pairs of values $l$ and $p$ obtained for a hydroxylated polybutadiene, crosslinked with 4,4'-diphenyl methane-diisocyanate in the presence of 1,4-butandiol, and the stress ($\sigma$) and strain ($\lambda$) values thus calculated are shown in Table 72.

By regression analysis, the initial length was found to be $l_0 = 20.1364$ mm and $A_1 = 2.394 \cdot 10^{-5}$ $p^{-1}$ equation (20). (The forces are given in pond because not all of tensile testing machines have been calibrated in SI units). Column 3 of Table 72 contains the values for $l$ resulting from the regression straight line (equation 20) which were used for subsequent calculation of $\lambda, \lambda^{-1}$ and $\lambda^{-2}$ (see equation (3)). $\sigma$-values and conversion into Pa are achieved using the following equation:

$$\sigma = \frac{p \cdot 9.8067 \cdot 10^3}{h \cdot b} \text{ (Pa)} \tag{21}$$

$p$ = force in pond (1 pond = $9.80665 \times 10^{-3}$ N), $h, b$ = thickness and width of the specimen in mm

Table 72. Experimental data and deduced stress-strain values of isocyanate-cross-linked polybutadiene prepolymers $l_0 = 20.1364$ mm; $T = 28\,°C$, $\varrho_{28}$ of the specimen $0.96$ g cm$^{-3}$, $h = 1.754$ mm; $b = 4.308$ mm

| $l$ (mm) | force/pond* | $l$ (mm) calculated acc. to equation 20 | $\sigma$ ($10^{+4}$ Pa) | $\lambda^{-1}$ | $(\lambda - \lambda^{-2}) \cdot 10^{-2}$ | $\dfrac{\sigma}{\lambda - \lambda^{-2}}$ ($10^5$ Pa) |
|---|---|---|---|---|---|---|
| 20.114 | 9.500   | 20.254 | 1.233  | 0.994 | 1.73   | 7.13 |
| 20.209 | 19.590  | 20.354 | 2.543  | 0.989 | 3.21   | 7.92 |
| 20.415 | 38.710  | 20.545 | 5.025  | 0.980 | 5.97   | 8.42 |
| 20.613 | 58.840  | 20.751 | 7.639  | 0.970 | 8.88   | 8.60 |
| 20.871 | 78.570  | 20.956 | 10.020 | 0.961 | 11.74  | 8.53 |
| 21.627 | 143.870 | 21.665 | 18.678 | 0.929 | 21.20  | 8.81 |
| 22.852 | 239.080 | 22.791 | 31.039 | 0.884 | 35.11  | 8.84 |
| 24.234 | 334.350 | 24.040 | 43.408 | 0.838 | 49.21  | 8.82 |
| 26.250 | 476.580 | 26.184 | 61.873 | 0.769 | 70.89  | 8.73 |
| 29.654 | 663.850 | 29.667 | 86.185 | 0.678 | 101.26 | 8.51 |

$\lambda^{-1} = \dfrac{l_0}{l}$

* The unit pond for force is directly read off the apparatus

The parameters of the MOONEY-RIVLIN equation are determined by regression analysis or graphically. For this purpose, $\sigma/(\lambda - \lambda^{-2})$ is plotted against $\lambda^{-1}$ and $C_1$ obtained from the ordinate intercept and $C_2$ from the slope. From the values listed in Table 72:

$2C_1 = 7.8 \times 10^5$ Pa

$2C_2 = 1.2 \times 10^5$ Pa

The cross-link density $v$ and (when the density of the polymer is known), the number-average molecular weight of a chain segment between two points of cross-linkage $M_c$ can be calculated from the $C_1$-value using the RET-equation (1) and the relationship

$v = \varrho_B \cdot M_c^{-1}$

For the sample:

$v = 3.2 \times 10^{-4}$ mol g$^{-1}$

$M_c = 3000$ g mol$^{-1}$

### 3.7.4.3. Determination of the Network Density (v) and of the $M_c$-value of Sulphur-Cross-linked Natural Rubber using Compression Measurement

The compression modulus can be determined by means of a modified micro-hardness tester, schematically represented in Fig. 97 (SMITH [22]). Cylindrical specimens of about 6 mm diameter and 1—2 mm thickness are used. The initial values $h_0$ of the dry sample are determined by means of a cathetometer or a meter having a scale division of 0.01 or 0.001 mm. After 96 h, the specimens swollen in benzene are placed into the swelling vessel filled with the same swelling agent (Fig. 97). The plunger is lowered by means of a hand wheel so that it just touches the sample. A preload of 70 g causes the plunger to contact the specimen directly; the meter is then set to zero. After 15 min, a mass of 50 g is added and, after another 5 min, the compression $\Delta h$ is read. After the sample has been relaxed for 2 min, the compression is repeated with a mass of 100 g. The magnitude of the load is dependent on the cross-link density. The compression modulus $G^*$ is generally averaged over 10 values and calculated exactly from equation (5) or when accuracy is less critical using equations (17) and (22):

$$G^* = \frac{f \cdot h_0}{A_0 \cdot 3\Delta h} \qquad (22)$$

## Characterization of Network Polymers

*Table 73*: Results of compression measurements taken on swollen sulphur-cross-linked natural rubber (cylindrical specimen) $A_0 = 0.249$ cm$^2$, $h_0 = 1.707$ mm, height in swelling equilibrium $h_q = 3.292$ mm, $q = (h_q/h_0)^3 = 7.17$, $\Phi_B = 0.1394$, $T = 310$ K, $V_A = 97.33$ cm$^3$ mol$^{-1}$

| $f$ (g) | $f/A_0$ (10$^4$ Pa) | $\Delta h$ (mm) | $\dfrac{h_0}{3\Delta h}$ | $G^*$ (22) (10$^4$ Pa) | $|\lambda - \lambda^{-2}|$ |
|---|---|---|---|---|---|
| 100 | 3.939  | 0.115 | 4.948 | 19.49 | 0.109 |
| 200 | 7.879  | 0.231 | 2.463 | 19.41 | 0.227 |
| 300 | 11.819 | 0.345 | 1.649 | 19.49 | 0.353 |
| 400 | 15.759 | 0.454 | 1.253 | 19.75 | 0.483 |
| 500 | 19.699 | 0.603 | 0.944 | 18.60 | 0.682 |
| 600 | 23.639 | 0.650 | 0.875 | 20.68 | 0.750 |

Table 73 shows the measurements obtained on sulphur-cross-linked natural rubber and their evaluation.

Averaging all 6 values and using equation (22):

$G^* = 19.57 \cdot 10^4$ Pa

and this gives (equation (6)):

$A\eta v = 0.76 \cdot 10^{-4}$ mol cm$^{-3}$

Evaluation according to equation (5) requires plotting of $f/A_0$ against $|\lambda - \lambda^{-2}|$ (Fig. 99). The initial slope of the straight line (3.6 · 10$^5$ Pa) is the product $G^* q^{1/3}$ or $G^* \Phi_B^{-1/3}$. From this, $G^* = 18.67 \cdot 10^4$ Pa and, again using equation (6),

$A\eta v = 0.724 \cdot 10^{-4}$ mol cm$^{-3}$

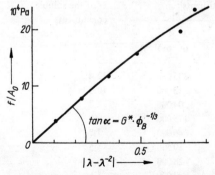

*Fig. 99*: Plot of $f/A_0$ against $|\lambda - \lambda^{-2}|$ for determination of the initial slope ($=G^*\Phi_B^{-1/3}$)

Assuming that $A\eta = 1$, the $M_c$-value of the test product with $\varrho_B = 1.000$ g cm$^{-3}$ becomes:

$$M_c = \varrho_B/v = 1.38 \cdot 10^4 \text{ g mol}^{-1}$$

## 3.7.4.4. Sol-Gel Analysis of Sulphur-Cross-linked Polybutadiene-cis-132

Square pieces of the rubber samples (edge length 0.5 cm, thickness 1 mm) are prepared; 4 g of them are treated with 40 ml of toluene, 5 times for 2 h at room temperature while being shaken; they are then left untouched with the extraction agent for another 24 h. The sol fractions are combined and freed from toluene by distilling in vacuo. The sol fraction $w_S$ of the absolutely „dry" samples is determined gravimetrically. As a check, the gel-content $w_G$ is also gravimetrically assayed after deswelling with methanol and vacuum drying ($w_G = 1 - w_S$). The solvent retention must be taken into account (see 2.1.4.3.). Extraction by shaking at room temperature is preferred to Soxhlet extraction where sticking easily occurs.

A calculation of the cross-link index $\gamma$ from sol-gel-analysis data (Table 74) is impossible for commercial vulcanizates with a broad molecular weight distribution of the initial products. Additional information about the sol fraction is obtained in subsequent investigation using gel permeation chromatography (see 2.5.). The determined sol fraction or gel fraction is informative as to the degree of chemical cross-linkage even without any further evaluation. By comparison with elastometric measured values (3.7.4.1. to 3.7.4.3.), information is obtained about the relationship between chemical and physical crosslinkages.

Table 74: Results of the sol-gel analysis of sulphur-cross-linked polybutadiene

| Vulcanization time (min) | $w_S$ normalized to 1 | soluble constituents | | $U$ of the soluble constituents |
|---|---|---|---|---|
| | | $M_n$ (g · mol$^{-1}$) | $M_w$ (g · mol$^{-1}$) | |
| 0 | 1.00 | 62 300 | 835 000 | 12.4 |
| 20 | 0.808 | 27 700 | 312 000 | 10.3 |
| 30 | 0.420 | 23 500 | 204 000 | 7.7 |
| 40 | 0.224 | 16 200 | 175 000 | 9.8 |
| 50 | 0.088 | 5 300 | 46 000 | 7.7 |
| 60 | 0.055 | 3 000 | 19 500 | 5.5 |
| 120 | 0.054 | 3 700 | 28 000 | 6.6 |
| 180 | 0.055 | 6 500 | 43 000 | 5.6 |

## 3.7.4.5. Determination of the Constants of the FLORY-REHNER Equation for Sulphur-Cross-linked Natural Rubber

Determination of the constants of the FLORY-REHNER equation (7) requires that the chemical potentials be determined as a function of $\Phi_B$ by vapour-pressure osmometry (3.7.3.3.). To apply equation (8), additional compression measurements must be taken on samples which are not swollen up to the equilibrium state. The samples are first swollen by putting them in the swelling agent until equilibrium is achieved. When the samples are removed from the swelling agent, initial rapid deswelling occurs which precludes compression measurements in this range. When the specimen is deswollen further, the rate of change of swelling during compression measurement (see also 3.7.4.3.) can be neglected. From the measurement of $h_0$ and $h_q(t)$, the instantaneous degree of swelling $q(t) = (h_q(t)/h_0)^3$ is accessible. The $\Delta\mu_A$ relating to the volume fraction $\Phi_B$ at which the compression measurement was taken is read from the $\Delta\mu_A - \Phi_B$ curve.

Then the left hand side of equation (8) ($y$ in Table 75 and Fig. 100) is plotted against $\Phi_B$. The required values can be taken from Table 75.

The slope of the $y - \Phi_B$ straight line in Fig. 100 gives $\chi = 0.400$ and the ordinate intercept gives the product $V_A vB$ from which, for a given $V_A$, the value of $0.5 \times 10^{-4}$ mol cm$^{-3}$ is calculated for $vB$.

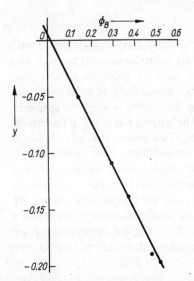

*Fig. 100*: Plot of $y$ against. $\Phi_B$ for the determination of $\chi$ and $B$

*Table 75:* Data for the determination of the constants $A\eta$, $B$ and $\chi$ of the FLORY-REHNER equation for sulphur-cross-linked natural rubber in benzene at 310 K, $V_A = 97.33$ cm³ mol⁻¹

| $\Phi_B$ | $\dfrac{-\Delta\mu_A}{RT}$ | $f/A_0$ ($10^4$ Pa) | $\|\lambda - \lambda^{-2}\|$ | $y$ |
|---|---|---|---|---|
| 0.1394 | 0.0000 | 3.70 | 0.10 | −0.050 |
| 0.2902 | 0.0162 | 2.85 | 0.10 | −0.109 |
| 0.3651 | 0.0337 | 2.65 | 0.10 | −0.138 |
| 0.4750 | 0.0741 | 2.44 | 0.10 | −0.189 |
| 0.5135 | 0.1007 | 2.38 | 0.10 | −0.196 |

Using $A\eta v = 0.724 \times 10^{-4}$ mol cm⁻³ (see 3.7.4.3.), obtained by compression measurements at the swelling equilibrium, the ratio $A\eta/B = 1.7$ can be calculated. Since $\eta$ can be considered to be equal to 1, for cross-linkage in the solid phase $A/B = 1.7$.

### 3.7.4.6. Determination of $d_c$ and $H(d_c)$ of Peroxide-Cross-linked Natural Rubber by Measurement of the Freezing-point Depression of the Swelling Agent

Three to four cylindrical specimens (diameter about 5 mm, height about 3 mm) are punched out of a sheet of natural rubber cross-linked by dicumyl peroxide and extracted with cyclohexane at 313 K for 10 days in order to remove non cross-linked components. The extraction agent is renewed daily. In order to prevent deswelling of the pre-equilibrated sample during the cooling process, the specimen is sealed in a polyethylene film. Subsequently, the latter is pierced by a fine needle and about 0.25 cm³ of swelling agent is injected. This ensures that equilibrium is maintained. The prepared sample is put on a temperature sensor (thermocouple or thermistor) which is arranged inside a twin walled vessel whose interior is saturated with the swelling agent. Heat is extracted from the sample continuously by constant temperature cooling bath (238 K) and the temperature change $\Delta T$ is continuously recorded (Fig. 95). To calculate the average distance of the junction points $d_c$, the temperature difference, $\Delta T_{\text{exp},i} = T - T_{ls0}$, for the inflection point of the cooling curve must be corrected by an amount $\Delta T'$ due to interaction between network and swelling agent.

$$\Delta T_i = \Delta T_{\text{exp},i} - \Delta T' \tag{23}$$

The FLORY-HUGGINS theory is used to calculate $\Delta T' = T'_{ls} - T_{ls0}$. According to this theory

$$\frac{1}{T'_{ls}} = \frac{1}{T_{ls0}} + \frac{R}{\Delta H_{ls}} \{\ln(1 - \Phi_B) + \Phi_B + \chi \Phi_B^2\} \qquad (24)$$

$\Delta H_{ls}$ = enthalpy of fusion of the swelling agent (9.837 kJ mol$^{-1}$), $\chi$ = HUGGINS interaction parameter (0.43), $\Phi_B$ = volume fraction of the polymer at equilibrium swelling (0.1709)

Using these values, $\Delta T' = -1.4$ K, and, hence, $\Delta T_i = -10.0$ K. Inserting the constant $K$ for the swelling agent cyclohexane, ($K = 86$ nm K), in equation (14) gives:

$$d_c = -\frac{86 \text{ nm } K}{-10.0 \text{ } K} = 8.6 \text{ nm}$$

The constants of equation (13) are 35 nm K for benzene and 37 nm K for 1.4-dioxane. These swelling agents have also proved suitable for such investigations.

The values for the $H(d_c)$ curve taken from Fig. 95 are listed in Table 76. To take into account the non-constant cooling rate when approaching the temperature $T_{CB}$ of the cooling bath, a correction factor must be introduced into equation (15):

$$\frac{dm_i}{d \Delta T_i} = C \frac{\left(\frac{dT}{dt}\right)_{T_i < T_{ls0}} - \left(\frac{dT}{dt}\right)_{T_i = T_{ls0}} \left(\frac{\Delta T_{CB} - \Delta T_{exp}}{\Delta T_{CB}}\right)}{\left(\frac{dT}{dt}\right)_{T_i < T_{ls0}}} \qquad (25)$$

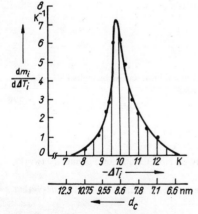

*Fig. 101*: Junction point distance distribution of a natural rubber swollen in cyclohexane and subject to peroxide-cross-linking

The values are shown graphically in Fig. 101. The $H(d_c)$ curve proper is obtained by converting $\Delta T_i$ values of the above curve into $d_c$ values using equation (14). The $H(d_c)$ curve can be obtained directly by following the freezing process in a differential scanning calorimeter ($DSC$) taking into account the cooling rate.

Table 76: Experimental and derived data for the $H(d_c)$ curve of peroxide-cross-linked natural rubber (7.85 mol of dicumyl peroxide/mol of natural rubber)
$\Delta T_{CB} = -40.5\,\text{K}; \Delta T' = -1.4\,\text{K}; C = -2.63\,\text{K}^{-1}$

| $-\Delta T_{\text{exp},i}$ (K) | $-\Delta T_i$ (K) | $\left(\dfrac{dT}{dt}\right)_{T_i=T_{\text{ls0}}}$ (K min$^{-1}$) | $\left(\dfrac{\Delta T_{CB} - \Delta T_{\text{exp},i}}{\Delta T_{CB}}\right)$ | $\left(\dfrac{dT}{dt}\right)_{T_i<T_{\text{is0}}}$ (K min$^{-1}$) | $\dfrac{dm_i}{d\Delta T_i}$ (K$^{-1}$) |
|---|---|---|---|---|---|
| 0 | — | −5.60 | | −5.60 | — |
| 9.42 | 8.02 | −4.29 | | −3.89 | 0.27 |
| 10.22 | 8.82 | −4.18 | | −2.94 | 1.11 |
| 10.56 | 9.16 | −4.14 | | −2.13 | 2.48 |
| 10.79 | 9.39 | −4.09 | | −2.00 | 2.75 |
| 10.91 | 9.51 | −4.08 | | −1.22 | 6.16 |
| 11.16 | 9.76 | −4.05 | | −1.09 | 7.14 |
| 11.43 | 10.03 | −4.02 | | −1.20 | 6.18 |
| 11.71 | 10.31 | −3.98 | | −1.39 | 4.90 |
| 12.06 | 10.66 | −3.93 | | −1.84 | 2.98 |
| 12.40 | 11.00 | −3.88 | | −2.13 | 2.16 |
| 12.78 | 11.38 | −3.83 | | −2.55 | 1.32 |
| 13.76 | 12.36 | −3.69 | | −2.67 | 1.00 |

$\Delta T_{CB} = T_{CB} - T_{\text{ls0}}$

# Bibliography

[1] Nomenklatur-Kommission, Richtlinien für die Nomenklatur auf dem Gebiet der makromolekularen Stoffe, Makromolekulare Chem. **3** (1966), 1
[2] SPERLING, L. H., GEORGE, H. F., HUELCK, V. and D. A. THOMAS, J. appl. Polymer Sci. **14** (1970), 2815
[3] SCHMID, R., Progr. Colloid & Polymer Sci. **64** (1978), 17
[4] SCHRÖDER, E., Plaste und Kautschuk **26** (1979), 1
[5] DUŠEK, K. and W. PRINS, Struture and Elasticity of Non Crystalline Polymer Networks, Fortschr. Hochpolymeren-Forsch. **6** (1969), 1
[6] HOFFMANN, M., KRÖMER, H. and R. KUHN, „*Polymeranalytik*", Vol. 1, Georg-Thieme-Verlag, Stuttgart 1977
[7] HEINRICH, G. and E. STRAUBE, Acta Polymerica **43** (1983), 589
[8] BELKEBIR-MRANI, A., HERZ, J. E. and P. REMPP, Makromolekulare Chem. **178** (1977), 485

[9] BENOIT, H., DECKER, D., DUPLESSIX, R., PICOT, G., REMPP, P., COTTON, J. P., FARNOUX, B., JANNINK, G. and R. OBER, J. Polym. Sci.; Polymer Physics Ed. **14** (1976), 2119
[10] GRONSKI, W., STADLER, R. and M. M. JACOBI, Macromolecules **17** (1984), 741
[11] KUHN, W. and M. MAJER, Z. phys. Chemie: Neue Folge **3** (1955), 330
[12] ZANDER, P., Dissertation A, TH Leuna-Merseburg 1985
[13] ARNDT, K.-F. and K.-G. HÄUSLER, Plaste und Kautschuk **31** (1984), 91
[14] CLUFF, E., GLADDING, F. and E. K. PARISER, J. Polymer Sci **45** (1960), 341
[15] KRAATS, E. J., van der, Thesen Delft 1967
[16] SCHMIEDER, K. and K. WOLF, Kolloid-Z. **127** (1952), 65–78
[17] REHAGE, G., Kolloid-Z. — Z. Polymere **194** (1964), 16; **196** (1964), 97
[18] ARNDT, K.-F. and J. SCHRECK, Acta Polymerica **36** (1985), 500, 536, 540
[19] LANGE, H., Kolloid-Z. — Z. Polymere **240** (1970), 747–755
[20] HÄUSLER, K. G., Dissertation B, TH Leuna-Merseburg 1979
[21] FINK, H., Plaste und Kautschuk **23** (1975), 136
[22] SMITH, D. A., J. Polymer Sci. C **16** (1967), 525

# 4. Special Techniques

## 4.1. Determination of Theta Conditions

### 4.1.1. Introduction

Many investigations of polymer materials have to be performed in solution. Compared to low molecular weight solutions, polymer solutions exhibit peculiarities. Even at finite concentrations, the polymer solution can attain a state which corresponds thermodynamically to that of an ideal low molecular weight solution. This ideal state, the theta state, occurs at certain temperatures and concentrations (theta conditions). Fundamental experimental investigations are preferably carried out in „theta solutions" because the dissolved macromolecule is quasi force free. Theoretically derived relationships often apply only to these conditions or are simplified in the state by the emission of all expressions describing interactions (e.g. concentration dependence of data used in the determination of molecular weight). At theta conditions thermodynamically similar states can be obtained in different solvents.

Besides the second virial coefficient and HUGGINS' interaction parameter, the difference between working temperature and theta temperature can be used to estimate the thermodynamic state of a solution.

### 4.1.2. Principles

All quantities which characterize the limit of phase stability of a solution of polymers with an infinitely high degree of polymerization $P$ are called theta quantities (FLORY). These are the critical values of concentration, temperature, pressure and of the interaction parameter $\chi$ at $P = \infty$. The theta temperature $(T_\Theta)$, is thus the temperature at which an infinitely large macromolecule is just insoluble (for an endothermic solution, solubility at $T > T_\Theta$, insolubility at $T < T_\Theta$). Solvent combinations which correspond to the separation point for $P = \infty$ are called theta solvents. Theta conditions can be obtained in ternary systems (polymer/solvent/precipitant) isothermally by varying the precipitant concentration or by varying the temperature at constant composition.

For the GIBBS molar free energy of mixing:

$$\Delta G_m = \Delta H_m - T \Delta S_m \tag{1}$$

# Determination of Theta Conditions

The enthalpy of mixing, $\Delta H_m$, is zero for ideal low molecular weight solutions and the entropy of mixing, $\Delta S_m$, must satisfy equation (2):

$$\Delta S_m^{id} = -R \sum_i x_i \ln x_i \tag{2}$$

$x_i$ = mol fraction of the component $i$ at $p$ and $T$ is constant. Consequenly, for the partial molar quantities of the solvent:

$$\begin{aligned}\Delta \bar{S}_A^{id} &= -R \ln x_A \\ \Delta \bar{H}_A^{id} &= 0 \\ \Delta \bar{V}_A^{id} &= 0 \\ \Delta \mu_A^{id} &= RT \ln x_A \end{aligned} \tag{3}$$

Ideal solutions cannot separate, since

$$(\partial^2 \Delta \mu_A^{id} / \partial x_B^2)_{p,T} > 0$$

Real solutions separate due to thermodynamic excess effects (enthalpy of mixing, excess entropy). The chemical potential of the solvent can be divided into an ideal term and an excess term:

$$\Delta \mu_A = \Delta \mu_A^{id} + \Delta \mu_A^E \tag{4}$$

$$\Delta \mu_A^E = \Delta \bar{H}_A - T \overline{\Delta S}_A^E \tag{5}$$

For dilute polymer solutions, the excess quantities can be determined via the second osmotic virial coefficient.

$$\Delta \mu_A = -\pi \cdot \bar{V}_A = -(RT/M_n) \cdot c_B \cdot \bar{V}_A - A_2 \cdot c_B^2 \cdot RT\bar{V}_A$$

$$A_2 = -\Delta \mu_A^E / (\bar{V}_A \cdot c_B^2 RT) \tag{6}$$

$A_2$ = second virial coefficient, $\bar{V}_A$ = partial molar volume of the solvent, $c_B$ = volume concentration of the polymer

The value $A_2$ can also be separated into an enthalpy term and an entropy term:

$$A_2 = A_{2,H} + A_{2,S} \tag{7}$$

$A_2$ is a measure of the thermodynamic state of the solvent. Strictly speaking, however, the second osmotic virial coefficient measures the dilution tendency of the solution. $A_2 < 0$ indicates a possible separation but is not a satisfactory criterion for the separation of the polymer solution into two phases. For $A_2$ to be negative, at least one term in equation (7) must be negative. The more lowmolecular a polymer solution is, the smaller $A_2$ must become in order for separation to occur. Theoretical and experimental results show that the critical polymer concentration decreases with increasing molecular weight. Applying the FLORY-HUGGINS-theory, equation (8),

$$\Delta\mu_A = RT\left[\ln\Phi_A + \left(1 - \frac{V_A}{V_B}\right)\Phi_B + \chi\Phi_B^2\right] \tag{8}$$

$\chi$ = HUGGINS or interaction parameter, $\Phi_B$ = volume fraction of the polymer and the phase stability conditions, equation (9):

$$\frac{\partial\Delta\mu_A}{\partial\Phi_B} = 0; \quad \frac{\partial^2\Delta\mu_A}{\partial\Phi_B^2} = 0 \tag{9}$$

enables the critical volume fraction $\Phi_{B,cr}$ and $\chi_{cr}$ to be calculated:

$$\Phi_{B,cr} = \frac{1}{\sqrt{V_B/V_A} + 1}; \quad \chi_{cr} = \frac{1}{2} + \sqrt{V_A/V_B} + V_A/2V_B \tag{10a}$$

Since polymers have a molecular weight distribution, equation (8) must be formulated for each degree of polymerization. For a quasi-binary system in which the polymer can be described by mean values, e.g. degree of polymerization, application of the phase stability criteria leads to [2]:

$$\Phi_{B,cr} = \frac{1}{1 + P_w/P_z^{1/2}}$$

$$\chi_{cr} = \frac{1}{2}\left[(1 + P_w^{-1/2})^2 + \frac{(P_z^{1/2} - P_w^{1/2})^2}{P_w P_z^{1/2}}\right] \tag{10b}$$

The stability limit (spinodal curve) can be determined from the temperature-dependence of the intensity $I$ of light scattered by a homogeneous solution [3]. The reciprocal scattering intensity at the scattering angle $\vartheta = 0$ ($I^{-1}(0)$) is proportional to $\partial^2\Delta\mu_A/\partial\Phi_B^2$. The temperature can thus be determined by extrapolating to $I^{-1}(0) = 0$, where $\partial^2\Delta\mu_A/\partial\Phi_B^2 = 0$. This is the spinodal temperature at the selected polymer concentration. The method is know as „pulse induced critical scattering" (PICS).

Since the second virial coefficient is proportional to $(1/2 - \chi)$, the value of $A_2$ at the critical point, $A_{2,cr}$, can be calculated from $\chi_{cr}$ (homogeneous solution $A_2 > A_{2,cr}$, two-phase region $A_2 < A_{2,cr}$). At the limit $\bar{M}_B = \infty$, $\Phi_{B,cr}$ tends towards zero and $A_2$ approaches zero from negative values. Although, at the critical temperature, $A_2$ is equal to zero for infinite molecular weights, the solution is not necessarily ideal. Although $\Delta\mu_A^E = 0$, at least one of the equations (3) is not satisfied.

According to SCHULZ and CANTOW [4], a solution in which all virial coefficients (except the first) are equal to zero and the temperature dependence of at least one of them differs from zero, should be termed as a pseudo-ideal solution. The enthalpy and entropy components of the second virial coefficient compensate each other at the theta temperature $T_\Theta$ and, since $\Delta\mu_A^E = 0$:

$$T_\Theta = \frac{\Delta\bar{H}_A}{\Delta\bar{S}_A^E} \tag{11}$$

Equation (11) does not contain any indication as to the signs of $\Delta\bar{H}_A$ and $\Delta\bar{S}_A^E$. In order for $T_\Theta$ to be positive, $\Delta\bar{H}_A$ and $\Delta\bar{S}_A^E$ should have the same sign. There may thus be two theta temperatures.

The theta temperature of the endothermic-pseudoideal solution for which the two-phase region is reached at decreasing temperatures is

$$T_\Theta^+ = \frac{\Delta\bar{H}_A}{\Delta\bar{S}_A^E} \qquad \Delta\bar{H}_A, \Delta\bar{S}_A^E > 0 \tag{12}$$

The exothermic-pseudoideal solution whose theta temperature is defined by equation (13)

$$T_\Theta^- = \frac{\Delta\bar{H}_A}{\Delta\bar{S}_A^E} \qquad \Delta\bar{H}_A, \Delta\bar{S}_A^E < 0 \tag{13}$$

separates into two phases if the temperature increases.

This is clearly indicated by the temperature dependence of the second virial coefficient:

$$A_{2,H} = -\frac{\Delta\bar{H}_A}{RTc_B^2\bar{V}_A} = -T\left[\left(\frac{\partial A_2}{\partial T}\right) - \alpha_1 A_2\right] \tag{14}$$

$\alpha_1$ = thermal expansion coefficient of the solvent

When $T = T_\Theta$, $A_2 = 0$, and the temperature dependence of the second virial coefficient is determined by the sign of the partial molar enthalpy of mixing

$$\frac{\partial A_2}{\partial T} = \frac{\Delta\bar{H}_A}{R(T_\Theta^\pm)^2 \, c_B^2 \bar{V}_A} \tag{15}$$

Depending on the position of $T_{cr}$ on the turbidity curves, $T_\Theta^+$ is termed as the upper theta temperature (the critical temperatures plotted in Fig. 102 are the upper ones, for polystyrene in cyclohexane: 307 K) and $T_\Theta^-$ as the lower theta temperature (for polystyrene in cyclohexane: 486 K). Turbidity curves are isobaric sections of the steric binodal area and indicate the dependence of the turbidity temperature on the polymer concentration. The polymer-rich phase, the gel (not to be confused with cross-linked swollen product) and the polymer-deficient phase, the sol, coexist at the turbidity point.

In a ternary system, the state of solution depends not only on the temperature but also on the relationship between solvent and precipitant. A system whose theta temperature is the selected temperature can be obtained by varying the volume fraction of the precipitant.

The discussions above refer to isobaric processes. Phase separation can occur isothermally in a solution of constant polymer concentration when pressure changes. The pressures at the start of phase separation of a polymer sample with $P = \infty$ are the $\pi$-pressures.

The theta condition can be checked with the help of parameter measurements which either describe the thermodynamic state of the solution or depend on it. Since, at theta conditions, the second virial coefficient is equal to zero, HUGGINS' interaction parameter must be equal to 0.5 (Chapter 1.3.2. equation (13)). The KUHN-MARK-HOUWINK exponent $a$ is 0.5 and, by definition, $\alpha$ must be equal to unity (Chapter 1.7., equation (9)). Investigations into branched polymers have shown, however, that temperatures at which $\alpha = 1(T_\Theta(\alpha))$ and $A_2 = 0(T_\Theta(A_2))$, deviate from FLORY's theta temperature $(T_\Theta)$. Depending on which parameter is characteristic of the theta condition, a distinction can be made between three theta temperatures [5].

## 4.1.3. Experimental Background

Theta conditions can be determined experimentally in three different ways. For example, the temperature of the polymer solution or the composition of a solvent-precipitant mixture can be varied until the values of the second virial coefficient or of the exponent of the KUHN-MARK-HOUWINK equation, characteristic of the theta condition are obtained. This procedure is very complex and is justified only when measuring, for example, the temperature dependence of the $A_2$-value or the limiting viscosity in fundamental research. A more rapid but equally elaborate method of determining the theta temperature is based on the relationship between the critical mixing temperature $T_{cr}$ and the molecular weight [6]. $T_\Theta$ can be obtained from the molecular weight dependency, $T_{cr}$, by extrapolating M to infinity. The relationship can be derived from the FLORY-HUGGINS theory of polymer solution.

$$\frac{1}{T_{cr}} = \frac{1}{T_\Theta}\left[1 + \frac{1}{\psi}\left(\left(\frac{V_B}{V_A}\right)^{-1/2} + \frac{1}{2}\frac{V_A}{V_B}\right)\right] \tag{16}$$

$\psi$ = FLORY's entropy parameter

# Determination of Theta Conditions

$\dfrac{V_A}{2V_B}$ is negligible compared to $(V_B/V_A)^{-1/2}$ so that substituting the degree of polymerization in equation (16) gives:

$$\frac{1}{T_{cr}} = \frac{1}{T_\Theta}\left(1 + \frac{b}{\sqrt{P}}\right)$$

$$b = \left(\frac{V_A}{M_0 v_B}\right)^{1/2} \Big/ \psi$$

(17)

$v_B$ = specific volume of the polymer, $M_0$ = molecular weight of the monomer unit, $V_A$ = molar volume of the solvent

In binary endothermal systems, the critical mixing temperature corresponds to the maximum of the turbidity curve. The turbidity curve of a polymer sample of known molecular weight is constructed by measuring the temperature at the onset of phase separation for polymer solutions of different concentrations.

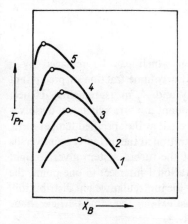

*Fig. 102*: Turbidity point curves for polymer samples of different molecular weights and $U = 0$
$M_1 < M_2 < M_3 < M_4 < M_5$

The solution can be observed with the naked eye, while cooling provided the concentration is high enough, and the turbidity point can be found visually. It is better to use devices which measure the transmittance of the solution or the scattered light (which is stronger in the case of phase separation) at a specified angle. The measurements are repeated with samples of the same constitution and configuration but of different molecular weight. The reciprocal critical temperatures are plotted against $P^{-0.5}$ using equation (17), extrapolated to $P = \infty$ and then the reciprocal theta temperature is read off. If several samples of known molecular weight and narrow molecular weight distribution are not available, $T_\Theta$ can be determined from the concentration-dependence of the turbidity temperature $T_{P_r}$. According to [7], $T_{P_r}$ and the volume fraction of the polymer are related by:

$$\frac{1}{T_{P_r}} = \frac{1}{T_\Theta} + C \log \Phi_B \qquad (18)$$

When extrapolating to $\Phi_B = 1$, $T_{P_r}$ is equal to $T_\Theta$. Measurements should be carried out under conditions such that the volume fraction is less than the critical value.

Another method for determining the theta conditions is based on precipitation titration and involves determination of the compositions of solvent and precipitant mixtures at the theta point. Polymer solutions of various concentrations are titrated with a nonsolvent until turbidity occurs. The volume fraction of the precipitant $\Phi_{P_r}$ as a function of the concentration is:

$$\log \Phi_{P_r} = \log \Phi_{cr} - K \log c \qquad (19)$$

$K$ is dependent upon the molecular weight of the polymer

The unit of concentration is mass fraction which can be replaced by g polymer/cm³ at low polymer concentrations. At a volume fraction of precipitant $\Phi_{cr}$, a theta solvent is present (ELIAS: viscosity and $A_2$ measurements). Hence, $\Phi_{cr}$ indicates the composition of a solvent-precipitant mixture which just dissolves the respective polymer with infinite molecular weight at the specified temperature. Plotting the logarithm of the precipitant concentration at the turbidity point versus the logarithm of the polymer concentration of the turbid system gives straight lines which in the case of the polymer concentration 1 intersect in one point, the $\Phi_{cr}$-value irrespective of the molecular weight or the molecular weight distribution. But for all molecular weights the polymer, must exhibit the same particle shape or the same density distribution of the segments.

Besides ascertaining the theta conditions, determination of the $\Phi_{cr}$-values plays an important role in copolymer fractionation. Only in a solvent-precipitant system, where the difference between the $\Phi_{cr}$-values of the cocomponents is sufficiently large, will fractionation according to chemical heterogeneity be successful (see also 2.2.3.).

In precipitation point titration, particular attention should be paid to good temperature control ($\pm 0.1$ deg) as the precipitation points are largely temperature dependent. Inaccuracies in the $\Phi_{cr}$-values may also be due to dosing errors, concentration errors, insufficient purity of solvents and precipitants and poorly dissolved samples.

## 4.1.4. Examples

### 4.1.4.1. Theta Temperature

Determination of theta temperature is illustrated using polystyrene in cyclohexane as an example. A colorimeter is used to measure turbidity; this colorimeter is provided with a temperature controlled scattering attachment.

In order to measure the molecular weight dependence of the turbidity temperature of polystyrene in cyclohexane, at least four polystyrene fractions are required. The recording of a turbidity curve is time consuming and expensive because the turbidity temperature must be recorded at several concentrations. TALAMINI and VIDOTTO [9] showed that there is a relationship for the turbidity temperature $T_{P_r}$ of a solution below the critical concentration similar to that between $T_{cr}$ and $T_\Theta$:

$$\frac{1}{T_{P_r}} = \frac{1}{T_\Theta}\left(1 + \frac{b'}{P^q}\right) \qquad (20)$$

$q$ has been estimated as 0.6 [9].

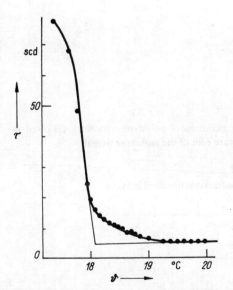

Fig. 103: Scattering intensity $\tau$ at a scattering angle $\vartheta = 90°$ as a function of the temperature of a polystyrene solution (molecular weight $2 \cdot 10^5$ g/mol$^{-1}$, solvent: cyclohexane, $c_B = 0.1$ g per 100 cm$^3$)

25 mg of each polystyrene fraction are placed in a 25 ml volumetric flask and the substance is dissolved by shaking at 40 °C. The heated polymer solution is cooled slowly in the cell of the apparatus while stirring. The temperature of the solution and the intensity of the scattered light at an angle of 90° are recorded. When the polymer is precipitated, the reading on the apparatus increases rapidly. The scattering intensity is plotted against temperature and the turbidity temperature determined by drawing the tangents as shown in Fig. 103.

The polymer solution is again heated in the cell until the phase separation disappears and is then cooled. The turbidity temperatures of a sample is measured several times and averaged. When $T_{P_r}$ of all fractions has been determined, the reciprocal turbidity temperatures are plotted as shown in Fig. 104, using equation (20). It should be noted that in this example $q = 0.5$ will also result in a straight line.

The theta temperature, $T_\Theta = 306$ K, is calculated from the intercept of the straight line in Fig. 104 and agrees well with literature values.

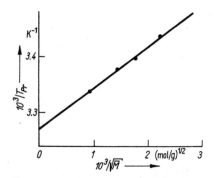

*Fig. 104*: Plot of the reciprocal turbidity temperatures of polystyrene fractions dissolved in cyclohexane as a function of the reciprocal square root of the molecular weight

*Table 77*: Turbidity temperature of polystyrene fractions dissolved in cyclohexane

| $M_w/10^5$ g mol$^{-1}$ | $T_{P_r}/K$ | $\dfrac{10^2}{\sqrt{M_w}} \Big/ \left(\dfrac{\text{mol}}{\text{g}}\right)^{1/2}$ | $10^3/T_{P_r} \Big/ K^{-1}$ |
|---|---|---|---|
| 2.0 | 291.1 | 0.223 | 3.435 |
| 3.15 | 294.4 | 0.178 | 3.397 |
| 4.8 | 296.3 | 0.144 | 3.375 |
| 12.0 | 299.6 | 0.091 | 3.338 |

## 4.1.4.2. Theta Composition

As an example, the theta composition of the mixture cyclohexanone/n-heptane/PVC is determined at 25 °C by precipitation point titration.

Samples are taken from a concentrated stock solution of about 1 g of PVC in 100 cm³ of cyclohexanone and diluted to give four solutions in the concentration range 0.5; 0.1; 0.05 and 0.01 g/100 cm³. Carbon tetrachloride is added to the precipitant, n-heptane, as suspension stabilizer in a ratio of 9:1 by volume. The PVC solution and precipitant are thermostatted at 25 °C in the cell of the apparatus and in the dosing pump, respectively. The precipitant mixture is added to the PVC solution by means of the dosing pump while stirring and the turbidity is measured.

*Table 78*: First turbidity points of PVC-cyclohexanone solutions of different concentrations

($V_{P_r}$ = amount of precipitant added by titration)

| $c_B$/g cm⁻³ | log $c_B$ | $V_{P_r}$/cm³ | $\Phi_{P_r}$ | log $\Phi_{P_r}$ |
|---|---|---|---|---|
| 0.01 | −2.00 | 8.51 | 0.586 | −0.232 |
| 0.005 | −2.30 | 8.72 | 0.592 | −0.228 |
| 0.001 | −3.00 | 9.04 | 0.601 | −0.221 |
| 0.0005 | −3.30 | 9.28 | 0.607 | −0.217 |
| 0.0001 | −4.00 | 9.71 | 0.618 | −0.209 |

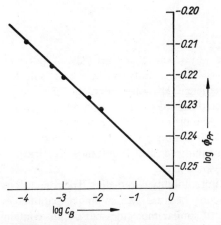

*Fig. 105*: Log log plot of the volume fraction of the precipitant at the first turbidity point as a function of polymer concentration (g cm⁻³)

The first turbidity point is the intersection of the tangents to the curve of scattering intensity against volume of precipitant added. If 6 ml of PVC solution are placed in the cell, then, for the five selected concentrations the amounts of precipitant (shown in Table 78) can be added and the volume fractions of the precipitant at the turbidity point $\Phi_{P_r}$ can be calculated.

In accordance with equation (19), the logarithm of the volume fraction of the precipitant is plotted against the log of the polymer concentration, extrapolated to $c_B = 1 \text{ g cm}^{-3}$ (log $c_B = 0$) to give log $\Phi_{cr} = 0.7355 - 1$ (Fig. 105).
From the values in Fig. 105 $\Phi_{cr}$ is equal to 0.544.

## Bibliography

[1] WOLF, B. A., Fortschr. Hochpolymeren-Forsch. **10** (1973), 109
[2] KONINGSVELD, R. and A. J. STAVERMAN, J. Polymer Sci. A 2 **6** (1968), 325
[3] SCHOLTE, Th. G., J. Polymer Sci. A 2 **9** (1971), 1553; C **39** (1972), 281
[4] SCHULZ, G. V. and H. J. CANTOW, Z. Elektrochem. **60** (1956), 517
[5] CANDAU, F., REMPP, P. and H. BENOIT, Macromolecules **5** (1972), 627
[6] SHULTZ, A. R. and P. J. FLORY, J. Amer. Chem. Soc. **74** (1952), 4760
[7] CORNET, C. F. and H. VAN BALLEGOOIJEN, Polymer **7** (1966), 293
[8] ELIAS, H. G. and U. GRUBER, Makromolekulare Chem. **78** (1964), 72
[9] TALAMINI, G. and G. VIDOTTO, Makromolekulare Chem. **110** (1967), 111

## 4.2. Determination of Solubility Parameter

### 4.2.1. Introduction

The selection of suitable solvents for physicochemical methods of polymer characterization is of major importance. The quantities describing solution properties considered so far, i.e., virial coefficient, HUGGINS' interaction parameter, KUHN-MARK-HOUWINK exponent, can be obtained from solution measurements. However they cannot be predicted from the properties of the pure components.

Information on solubility can be obtained using solubility parameters which can be calculated or determined experimentally for every substance. Empirically, it can be stated that „like substances dissolve like".

Solubility can be evaluated by comparing solubility parameters. The conclusions will not always be confirmed in practice because the solubility parameters describe only summarily the interaction of similar molecules but do not contain information about the specific interactions of dissimilar molecules as occur in multicomponent polymer solutions.

## 4.2.2. Principles

According to quantum mechanics, the energy of interaction between two spherical non-polar substances, $A$ and $B$, is equal to the geometric mean of the interaction energies between the pure components:

$$W_{AB} = (W_{AA} W_{BB})^{1/2} \tag{1}$$

The change in the total interaction when dissolving $B$ in $A$ is the difference between the interaction energies of the solution and the pure components:

$$\Delta W_{AB} = W_{AB} - 0.5(W_{AA} + W_{BB})$$
$$\Delta W_{AB} = (W_{AA} W_{BB})^{1/2} - 0.5(W_{AA} + W_{BB}) \tag{2}$$

The larger the difference between $W_{AA}$ and $W_{BB}$, the smaller the geometric mean will become compared with the arithmetic mean. This shows that the tendency of the solution to separate, will increase in line with the difference between the interaction energies of the participants in the solution.

Transforming equation (2) gives:

$$\Delta W_{AB} \sim (W_{AA}^{1/2} - W_{BB}^{1/2})^2 \tag{3}$$

The interaction energies are related to the experimentally determined heats of evaporation. The internal heat of evaporation $\Delta E^V$ is the energy required to separate all physical bonds of a molecule with its neighbour. In evaporating, however, work must also be performed against the external pressure; this work is equal to $RT$ if the gas phase is ideal

$$\Delta E^V = \Delta H^V - RT \tag{4}$$

$\Delta E^V$ = internal energy of evaporation, $\Delta H^V$ = experimentally determined enthalpy of evaporation

Dividing $\Delta E^V$ by the molar volume gives the cohesion energy density $e$,

$$e = \frac{\Delta E^V}{V} \tag{5}$$

the square root of which is the solubility parameter $\delta$ (HILDEBRAND [1]).

$$\delta = \sqrt{e} \tag{6}$$

HILDEBRAND [1] propounded the theory of the regular solution in which the dissolved molecules are arranged absolutely at random (i.e., as in an ideal solution) but the enthalpy of mixing is not equal to zero.

$$\Delta G_m = \Delta H_m - T \Delta S_{id} \tag{7}$$

The change in internal energy when mixing $n_A$ moles of component $A$ with $n_B$ moles of $B$ (where molecules of $A$ and $B$ are almost equal in size) is

$$\Delta E_m = [(\Delta E_A^V)^{1/2} - (\Delta E_B^V)^{1/2}]^2 \frac{n_A n_B}{n_A + n_B} \tag{8}$$

In the polymer solution, the solvent molecules are considerably smaller than the molecules to be dissolved. The change in the internal energy in mixing is then:

$$\Delta E_m = n_A V_A \Phi_B [(\Delta E_A^V/V_A)^{1/2} - (\Delta E_B^V/V_B)^{1/2}]^2 = n_A V_A \Phi_B (\delta_A - \delta_B)^2 \tag{9}$$

$\Phi_B$ = volume fraction of the polymer

Since the change in volume in condensed systems is very small, $\Delta E_m$ is approximately equal to the enthalpy of mixing $\Delta H_m$. Partially differentiating with respect to the mole numbers $n_A$ or $n_B$, gives the partial molar enthalpy of mixing of the solvent

$$\Delta \bar{H}_A = V_A \Phi_B^2 (\delta_A - \delta_B)^2 \tag{10}$$

and of the polymer

$$\Delta \bar{H}_B = V_B \Phi_A^2 (\delta_A - \delta_B)^2 \tag{11}$$

The change in the chemical potential of the solvent of a regular solution from that of the pure component is

$$\Delta \mu_A = \Delta \bar{H}_A - T \Delta \bar{S}_{id,A} = RT \ln n_A + V_A \Phi_B^2 (\delta_A - \delta_B)^2 \tag{12}$$

Since the solubility parameters can be determined experimentally, the validity of the concept of the regular solution can be checked easily. Basically, deviations from real quantities measured must occur because the heat of mixing results from the interaction between different molecules which are then no longer arranged completely at random in the solution. Equating the expression for the enthalpy of mixing of the regular solution with the enthalpy term of the semi-empirical FLORY-HUGGINS derivation of the free energy of mixing leads to:

$$V_A \Phi_B^2 (\delta_A - \delta_B)^2 = RT \chi \Phi_B^2 \tag{13}$$

and, hence, the HUGGINS' interaction parameter

$$\chi = \frac{V_A (\delta_A - \delta_B)^2}{RT} \tag{14}$$

The maximum difference in solubility parameters for which solubility can still be expected at 293 K can be estimated using equation (14). For a polymer molecule of infinite molecular weight, $\chi$ is equal to 0.5 at the phase boundary. Assuming that the molar volume of the solvent is equal to 100 cm$^3$ mol$^{-1}$:

$$(\delta_A - \delta_B)^2 \lessapprox 12 \cdot 10^6 \text{ J m}^{-3} \text{ (3 cal cm}^{-3}\text{)} \tag{15}$$

A polymer with solubility parameter $\delta_B$ should be soluble in a solvent with solubility parameter $\delta_A$ if the difference between $\delta_A$ and $\delta_B$ is less than $3.5 \cdot 10^3$ J$^{1/2}$ m$^{-3/2}$.

Deviations from this rule of thumb occur particularly for polar polymers and for those forming hydrogen bonds.

HILDEBRAND's solubility parameter concept has been refined by various authors who tried to analyze the sum of interactions contained in $\delta$ into its constituent parts. HILDEBRAND and SCOTT have already postulated that the solubility parameter of polar liquids consists of a polar ($\delta_p$) and a non-polar ($\delta_n$) contribution.

$$\delta^2 = \delta_p^2 + \delta_n^2 \tag{16}$$

A three-dimensional solubility parameter was introduced by HANSEN who split up $\delta$ on the basis of theoretically possible interactions which may affect solubility (dispersion forces, dipoles, hydrogen bonding),

$$\delta^2 = \delta_p^2 + \delta_n^2 + \delta_h^2 \tag{17}$$

$\chi$ can then be represented by a sum of the solubility parameter differences.

$$\chi = \frac{V_A}{RT} \sum_i (\delta_{A,i} - \delta_{B,i})^2 \tag{18}$$

$i = p, n, h$

Up to now, only qualitative relationships between the individual components have been described.

These comments, however, apply only to amorphous polymers. The solubility rule (equation (15)) applies to highly crystalline polymers only at temperatures $\gtrapprox 0.9 T_m$ (where $T_m$ = melting point).

## 4.2.3. Experimental Background

The solubility parameter of low molecular weight liquids can be calculated directly from the heat of evaporation. Since polymers do not exist in a gaseous state, their solubility parameter cannot be defined by equation (6). This parameter is equal to that of a solvent in which the polymer is soluble in all conditions, with no heat of mixing, no change in volume and no reaction or association. The solubility parameter of the polymer can be determined from the properties of the polymer solution, the solvent uptake of a cross-linked polymer or by calculation.

## 4.2.3.1. Estimation of Solubility Parameter

In 1953, SMALL [2] published molar attraction constants $F$ (Table 79) from which $\delta_B$ can be calculated using the structural formula of the monomer and the polymer density.

The solubility parameter of polymethyl methacrylate is calculated in the example (Table 80). SMALL's method cannot be used for molecules which form hydrogen bonds.

Table 79: Molar attraction constants ($T = 298$ K) [2]

| Group | | $F/10^3$ $J^{1/2}$ $m^{3/2}$ $mol^{-1}$ |
|---|---|---|
| $-CH_3$ | Single bonds | 437 |
| $>CH_2$ | | 271 |
| $>CH$ | | 57 |
| $-\overset{\mid}{\underset{\mid}{C}}-$ | | $-190$ |
| $=CH_2$ | Double bonds | 388 |
| $=CH-$ | | 226 |
| $=C<$ | | 39 |
| $-C\equiv C-$ | | 453 |
| $HC\equiv C-$ | | 581 |
| $-H$ | | 163 ... 204 |
| $-O-$ (Ether) | | 143 |
| $-\overset{\mid}{C}=O$ (Ketone) | | 561 |
| $-C{\overset{O}{\underset{O-}{\diagup\!\!\!\diagdown}}}$ (Ester) | | 632 |
| $-Cl$ | | 530 ... 551 |
| $>CCl_2$ | | 530 |

$\delta_B$ is given by:

$$\delta_B = \frac{\varrho_B}{M_0} \sum_i F_i \qquad (19)$$

$M_0$ = molecular weight of the monomer

Substituting in equation (19), we obtain

$\delta_B = 18.6 \cdot 10^3 \, \text{J}^{1/2} \, \text{m}^{-3/2}$

*Table 80*: Calculation of the solubility parameter of polymethyl methacrylate ($\varrho_B$ (amorphous) = 1.17 g cm$^{-3}$)

$$\left[ -CH_2-\underset{\underset{\underset{CH_3}{O}}{\overset{C=O}{|}}}{\overset{CH_3}{\underset{|}{C}}}- \right]_n$$

|  | $F_i/10^3 \, \text{J}^{1/2} \, \text{m}^{3/2} \, \text{mol}^{-1}$ | $M_i/\text{g mol}^{-1}$ |
|---|---|---|
| 2—CH$_3$ | 874 | 30 |
| 1⟩CH$_2$ | 271 | 14 |
| 1—C— | −190 | 12 |
| 1—C⟨$^O_{O-}$ | 632 | 44 |
|  | $\Sigma F_i = 1587$ | $M_0 = 100$ |

## 4.2.3.2. Determination of Solubility Parameter from Swelling Measurements

A polymer weakly cross-linked, as far as primary valency is concerned, can absorb solvents until swelling reaches equilibrium. The amount of liquid absorbed is dependent on the polymer solvent interaction and, thus on HUGGINS' interaction parameter $\chi$ (see Section 3.7.3.2.). The degree of swelling $q$ (volume of the swollen sample divided by the sample volume) is a function $f$ of the difference between the solubility parameter of the solvent and polymer.

$$q = f\{V_A^{1/2}(\delta_A - \delta_B)\} \qquad (20)$$

A plot of swelling degree $q$ measured in different solvents versus the solubility parameter [3] has a maximum at $\delta_A = \delta_B$ (Fig. 106).

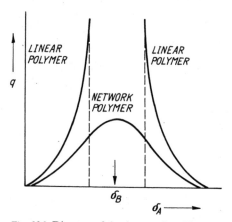

Fig. 106: Diagram of the degree of swelling of a linear and of a cross-linked polymer in different solvents with different solubility parameters.
The linear polymer is dissolved ($q \to \infty$) in a certain $\delta_A$-range while the degree of swelling of the cross-linked polymer shows a maximum in this range.

## 4.2.3.3. Determination of Solubility Parameter from the Limiting Viscosity

It is known from viscosity theory that coil formation is solvent dependent. In a good solvent, a coil is expanded against the resistance of the intramolecular forces. Since this process resembles the change in volume in swelling due to solvent uptake, the same method as used to determine $\delta$ from swelling measurements should be applicable to the limiting viscosity. As described in 1.7.3., limiting viscosities are measured in solvents whose $\delta_A$ is known. It can be shown [4] that the plot of $[\eta] = f\{(\delta_A - \delta_B)^2\}$ results in a GAUSSIAN curve.

$$[\eta] = [\eta]_{max} \exp\{-V_A(\delta_A - \delta_B)^2\} \tag{21}$$

Equation (21) can be put into linear form by taking logarithms

$$\delta_A > \delta_B: \quad \delta_A = \delta_B + \left(\frac{1}{V_A} \ln \frac{[\eta]_{max}}{[\eta]}\right)^{1/2} \tag{22a}$$

$$\delta_A < \delta_B: \quad \delta_A = \delta_B - \left(\frac{1}{V_A} \ln \frac{[\eta]_{max}}{[\eta]}\right)^{1/2} \tag{22b}$$

Determination of Solubility Parameter

$\delta_B$ is obtained from the intercept of the representation

$$\delta_A = f\left\{\left(\frac{1}{V_A}\ln\frac{[\eta]_{max}}{[\eta]}\right)^{1/2}\right\}.$$

If $[\eta]$ is plotted as a function of $\delta_A$, then, at $[\eta] = [\eta]_{max}$ $\delta_A = \delta_B$. The values in Table 81, for polyvinyl acetate, are plotted in Fig. 107 from which:

$$\delta_B = 21.5 \; 10^3 \; J^{1/2} \; m^{-3/2}$$

*Table 81*: Limiting viscosities of a polyvinyl acetate ($\bar{M} = 2.5 \; 10^5 \; g \; mol^{-1}$) (from: VAN KREVELEN, D.W., „*Properties of Polymers*", Elsevier Publ. Co., Amsterdam 1972, p. 246)

| Solvent | $\delta_A/10^3 \; J^{1/2} \; m^{-3/2}$ | $[\eta]/dl \cdot g^{-1}$ |
|---|---|---|
| Toluene | 18.2 | 0.78 |
| Benzene | 18.8 | 0.70 |
| Ethylformate | 19.2 | 1.03 |
| Monochlorobenzene | 19.4 | 0.99 |
| 1,4 Dioxan | 20.2 | 1.11 |
| Acetone | 20.4 | 1.04 |
| Acetonitrile | 24.3 | 1.09 |
| Methanol | 29.6 | 0.57 |

## 4.2.3.4. Solubility Parameter and Theta Temperature

The theta temperature, which also describes the thermodynamic state of a solvent, and the solubility parameter are related. Plots of the theta temperatures of polyisobutene, polystyrene and polymethyl methacrylate in different solvents against $\delta_A$ — as suggested by Fox [5] — resulted in each case in a straight line for solvents with $\delta_A < \delta_B$ and in a different straight line for solvents with $\delta_A > \delta_B$. Both straight lines intersect at $T_\Theta = 0$ K where $\delta_A = \delta_B$ can be read off. Determination of the theta temperature has been described in 4.1.4.1.

## 4.2.3.5. Turbidimetric Solubility Parameter Determination

The relationship between $\chi$ and $\delta$ suggests the following method for determining the solubility parameter. At the turbidity point of a two component system (with infinite molecular weight of the polymer and $\chi = 0.5$)

$$V_A(\delta_A - \delta_B)^2 = 0.5RT \qquad (23)$$

(see also 2.1.4.2.1.).

Solutions of the polymer of the same concentration are made up in various solvents (I, II, III, ...) with different solubility parameters ($\delta_A^I, \delta_A^{II}, \delta_A^{III}$) and titrated in each case with two precipitants 1 and 2.

The solubility parameter of one of the precipitants (2) should be greater than that of the solvent and that of the other precipitant (1) should be smaller. For each of the mixtures (I-1, I-2, II-1, II-2, ...) a relationship analogous to equation (23) is valid at the turbidity point. Since, however, a precipitant has been added to the solvent, $V_A$ and $\delta_A$ must be replaced by the corresponding values for the mixture. For each solvent, the following holds

$$V_{m,1}(\delta_B - \delta_{m,1})^2 = V_{m,2}(\delta_{m,2} - \delta_B)^2 \tag{24}$$

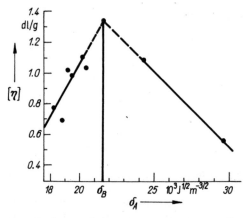

*Fig. 107*: Limiting viscosity of a polyvinyl acetate sample in different solvents as a function of the solubility parameter of the solvents

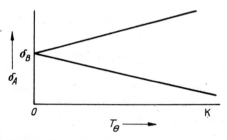

*Fig. 108*: Plot of the theta temperature measured in various solvents versus the solubility parameter of the solvents

where

$$V_{m,i} = \frac{V_A V_{Pr,i}}{\Phi_A V_{Pr,i} + \Phi_{Pr,i} V_A} \tag{25}$$

$i = 1, 2;$ $\Phi$ = volume fraction, $Pr$ = precipitant (index)

and

$$\delta_{m,i} = \Phi_A \delta_A + \Phi_{Pr,i} \delta_{Pr,i} \tag{26}$$

Substituting in equation (25) and (26) the respective values of $\Phi_A$ and $\Phi_{P_r}$ from titrations of the polymer solution of a solvent with two precipitants and rearranging equation (24) gives:

$$\delta_B = \frac{V_{m,1}^{0.5} \delta_{m,1} + V_{m,2}^{0.5} \delta_{m,2}}{V_{m,1}^{0.5} + V_{m,2}^{0.5}} \tag{27}$$

Thus, an apparent solubility parameter, $\delta_B = \delta'_B$, can be determined for the polymer. The $\delta'_B$-values are plotted against $\delta_A$. The intersection of the resulting straight line with the straight line $\delta_A = \delta'_B$ gives the solubility parameter of the polymer $\delta_B$ [6].

## 4.2.3.6. Solubility Parameter Determination by Inverse Gas Chromatography

In inverse gas chromatography the stationary phase consists of high polymers which are then characterized using the retention data of selected volatile compounds. GUILLET et al. [7] show that inverse gas chromatography can also be used for determining the solubility parameter of polymers and other thermodynamic quantities. The activity coefficients at infinite dilution of polymer can be calculated from the retention time and retention volume thus enabling HUGGINS' interaction parameter, $\chi$, to be ascertained (see also 4.3.2.). Using $\chi$, $\delta_B$ can be calculated via equation (14).

The calculated solubility parameters differ from experimental values, so that ranges of $\delta_B$ values are usually given in tables. A comprehensive table of the solubility parameters of polymers, polymer blends, plasticizers, solvents and other substances is given in [8].

## 4.2.4. Examples

From the methods discussed in 4.2.3. for determining the solubility parameter, we have selected turbidity measurement using polyvinyl chloride to illustrate the procedure.

Two solutions of the same concentration (3 g PVC/100 cm$^3$ solvent) of PVC in tetrahydrofurane (solvent I) and in cyclohexanone (solvent II) are made up. Because of the high concentration, the polymer has to be dissolved by heating and shaking. n-heptane (1) and n-propanol (2) are selected as precipitants with solubility parameters greater than and smaller than those of the solvents.

A colorimeter with a temperature-controlled (292 ± 0.1 K) scattering attachment is used for all four precipitation point titrations.

10 cm$^3$ of the PVC solution are placed in a 30 cm$^3$ cell, precipitant is slowly added from a microburette and the mixture is shaken. The turbidity values are read off the measuring instrument and plotted as function of the amount of precipitant added. The precipitant titre for the first turbidity point is obtained from the intersection of tangents to the turbidity curve before the precipitation of the polymer (intrinsic turbidity) and after the start of precipitation (Fig. 103). The measured values and derived quantities are listed in Table 83.

*Table 82*: Solubility parameters of selected polymers (from: VAN KREVELEN, D.W., „Properties of Polymers", Elsevier Publ. Co., Amsterdam 1972, p. 140)

| Polymer | $\delta_B/10^3$ J$^{1/2}$ m$^{-3/2}$ | $\delta_B$/cal$^{1/2}$ cm$^{-3/2}$ |
|---|---|---|
| Polyethylene | 15.7 ... 17.0 | 7.7 ... 8.35 |
| Polypropylene | 16.7 ... 18.8 | 8.2 ... 9.2 |
| Polybutadiene | 16.5 ... 17.5 | 8.1 ... 8.6 |
| Polystyrene | 17.3 ... 19.0 | 8.5 ... 9.3 |
| Polyvinyl chloride | 19.2 ... 22.0 | 9.4 ... 10.8 |
| Polytetrafluoroethylene | 12.6 | 6.2 |
| Polyvinyl acetate | 19.1 ... 22.5 | 9.35 ... 11.05 |
| Polymethyl methacrylate | 18.6 ... 26.1 | 9.1 ... 12.8 |
| Polyacrylonitrile | 25.5 ... 31.4 | 12.5 ... 15.4 |

(Conversion factor in SI units: 1 cal$^{1/2}$ cm$^{-3/2}$ = 2.04 10$^3$ J$^{1/2}$ m$^{-3/2}$)

The shape of the turbidity curve is dependent on the polydispersity of the sample. The more polydisperse the sample, the wider the region where phase separation sets in.

Hence the error in determining the first turbidity point increases with increasing polydispersity. Since the position of the first turbidity point is dependent on the

# Determination of Solubility Parameter

molecular weight, the resultant solubility parameter should only be correct for the molecular weight of the selected sample. However, $\delta_B$ is independent of the molecular weight for sufficiently large molecules.

When all four turbidity points have been determined, the volume fractions of the precipitant and solvent are calculated and substituted in equation (24):

$$\Phi_{P_r} = \frac{V_{P_r}}{V_{P_r} + V_0}, \qquad \Phi_A = 1 - \Phi_{P_r} \tag{28}$$

$V_{P_r}$ = precipitant volume at the first turbidity point, $V_0$ = volume of solvent

The $V_{m,\,i}$ values and $\delta_{m,\,i}$ values for each of the four titrations are obtained from equations (25) and (26). The apparent solubility parameters $\delta'_B$ of the tetrahydrofuran solution and the cyclohexanone solution can now be calculated.

Table 83: Results of precipitation point titrations of PVC ($V_0$ = 10.00 cm³, $c_B$ = 3.00 g/100 cm³ solution, $T$ = 293 K)

| Solvent | I Tetrahydrofuran (THF) | | |
|---|---|---|---|
| Precipitant | 1 n-Heptane | 2 n-Propanol | |
| titrated precipitant/cm³ | 6.59 | 9.41 | |
| $\Phi_{P_r}$ | 0.397 | 0.485 | |
| $\Phi_A$ | 0.603 | 0.515 | |
| Molar volume/cm³ mol⁻¹ | 146.5 | 74.8 | 81.2 (THF) |
| $\delta/10^3$ J$^{1/2}$ m$^{-3/2}$ | 15.1 | 24.3 | 18.6 (THF) |
| $V_m$/cm³ mol⁻¹ | 98.66 | 77.96 | |
| $\delta_m/10^3$ J$^{1/2}$ m$^{-3/2}$ | 17.21 | 21.36 | |
| $\delta'_B/10^3$ J$^{1/2}$ m$^{-3/2}$ | | 19.16 | |

| Solvent | II Cyclohexanone (CHO) | | |
|---|---|---|---|
| Precipitant | 1 n-Heptane | 2 n-Propanol | |
| titrated precipitant/cm³ | 13.42 | 9.69 | |
| $\Phi_{P_r}$ | 0.573 | 0.492 | |
| $\Phi_A$ | 0.427 | 0.508 | |
| Molar volume/cm³ mol⁻¹ | 146.5 | 74.8 | 103.7 (CHO) |
| $\delta/10^3$ J$^{1/2}$ m$^{-3/2}$ | 15.1 | 24.3 | 20.3 (CHO) |
| $V_m$/cm³ mol⁻¹ | 124.55 | 87.14 | |
| $\delta_m/10^3$ J$^{1/2}$ m$^{-3/2}$ | 17.32 | 22.27 | |
| $\delta'_B/10^3$ J$^{1/2}$ m$^{-3/2}$ | | 19.57 | |

They are shown in the $\delta_A - \delta'_B$ diagram in Fig. 109. The solubility parameter of polyvinyl chloride is obtained from the intersection of the straight line connecting the two points and the straight line $\delta'_B = \delta_A$:

$\delta_B = 19.3 \cdot 10^3 \; J^{1/2} \; m^{-3/2}$

This parameter is within the limits of the values shown in Table 82.

*Table 84*: Selected solvent/precipitant systems for the turbidimetric determination of solubility parameters (from [6])

| Polymer | Solvent ($\delta/10^3 \; J^{1/2} \; m^{-3/2}$) | Precipitant ($\delta/10^3 \; J^{1/2} \; m^{-3/2}$) |
|---|---|---|
| Polystyrene | toluene (18.2) | methylcyclohexane (15.9) |
|  |  | n-amyl alcohol (22.3) |
|  | o-dichlorobenzene (20.4) | methylcyclohexane (15.9) |
|  |  | n-amyl alcohol (22.3) |
| Polymethyl methacrylate | toluene (18.2) | cyclohexane (16.7) |
|  |  | n-amyl alcohol (22.3) |
|  | o-dichlorobenzene (20.4) | cyclohexane (16.7) |
|  |  | n-amyl alcohol (22.3) |
| Polybutadiene | methylcyclohexane (15.9) | freon 215 ($C_3F_3Cl_3$) (13.1) |
|  |  | acetone (20.4) |
|  | ethyl benzene (17.9) | freon 215 (13.1) |
|  |  | acetone (20.4) |
|  | monochlorobenzene (19.4) | freon 215 (13.1) |
|  |  | acetone (20.4) |

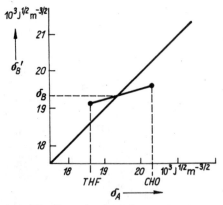

*Fig. 109*: Determination of the solubility parameter of the polymer $\delta_B$ from the apparent solubility parameter $\delta'_B$

It is advisable to carry out measurements in more than two solvents in order to corroborate the experimentally determined $\delta_B$. Systems for the determination of $\delta_B$ of selected polymers are listed in Table 84.

Apart from measuring errors in the precipitation point titration, fundamental errors may also result from the nonapplicability of equations on which the measuring method for solubility parameter determination is based.

## Bibliography

[1] HILDEBRAND, J. and R. SCOTT, „The Solubility of Nonelectrolytes", Reinhold Publ. Co., New York 1966
[2] SMALL, P. A., J. appl. Chem. 3 (1953), 71
[3] MANGARAJ, D., Makromolekulare Chem. 65 (1963), 29; MANGARAJ, D., PATRA, S. and S. RASHID, ibid. 65 (1963), 39
[4] MANGARAJ, D., RATH, S. B. and S. K. BHATNAGAR, ibid. 67 (1963), 75; MANGARAJ, D., RATH, S. B. and S. PATRA, ibid. 67 (1963), 84
[5] FOX, T. G., Polymer 3 (1962), 111
[6] SUH, K. W. and D. H. CLARKE, J. Polymer Sci. A-1 5 (1967), 1671; SUH, K. W. and J. M. CORBETT, J. appl. Polymer Sci. 12 (1968), 2359
[7] DAWKINS, J. V. ed., „Developments in Polymer Characterisation", Vol. 3, LIPSON, J. E. G., GUILLET, J. E., „Study of Structure and Interactions in Polymers by Inverse Gas Chromatography", Appl., Science Publ., London 1982
[8] BARTON, A. F. M. ed., „CRC Handbook of Solubility Parameters and other Cohesion Parameters", CRC Press Inc., Florida 1983

## 4.3. Compatibility of Polymers

### 4.3.1. Introduction

Compatibility is defined as the unlimited mutual miscibility of two or more polymers with the formation of stable or metastable homogeneous mixed phases. Compatible polymer mixtures must exhibit the characteristics of a uniform substance in all test and processing methods [1].

Frequently, however, mixtures of chemically different polymers are incompatible in the solid and liquid (molten) state or in solution and the system separates into two phases. In the case of the binary mixture each phase will still contain the two polymers but in different proportions.

A homogeneous binary polymer mixture can also be separated by the addition of a third polymer. This process is termed coprecipitation. It may lead to the almost quantitative separation of one of the two components.

Studies of the compatibility of polymers are of particular importance to all fields of application of polymer compositions such as polyblends, compounds, varnishes, adhesives, etc.

Compatibility studies frequently consist of a subjective evaluation of the turbidity or opalescence of polymer films or polymer solution [2, 3]. The phase separation of polymer mixtures in solution is investigated quantitatively by physical-chemical methods. With the exception of very low polymer concentrations in solvents, incompatible polymers exhibit separation in almost all cases even in a three component system (solvent, polymer 1, polymer 2). The interactions between polymer 1 and polymer 2 depend on the microstructure, the average molecular weight and molecular inhomogeneity of the polymers and the temperature and mixing ratio of the components. These interactions also determine the ternary system.

The compatibility of polymer mixtures in solution is investigated primarily by observing phase stability (turbidity measurements) and also light scattering, osmometry and viscometry. The solution viscosity is preferred for compatibility studies because of the low investment in time and equipment [4, 5, 6].

The miscibility of two or more polymers in the solid state is investigated by mechanical, thermoanalytical and microscopic (electron-microscopic) methods and also by broad-line and pulse NMR spectroscopy.

### 4.3.2. Principles

Phase separation phenomena in polymer mixtures are basically similar to separation of low molecular weight liquids. The difference between these two

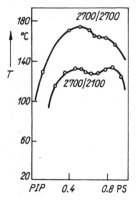

*Fig. 110*: Coexistence curves for mixtures of anionically synthesized polyisoprene (PIP) and polystyrene (PS); molecular weights are in g/mol [7]

systems is only the large volume occupied by polymers because of their molecular size.

Usually, the compatibility of two polymers is characterized by the critical polymer concentration at which initial turbidity occurs. The turbidity curve for a two component system is obtained by connecting the temperature at which a second phase is first separated in polymer mixtures of different concentrations (Fig. 110).

A criterion of separation is the GIBBS' free energy of mixing $\Delta G_m$. If the polymers are compatible, thermodynamically stable systems ($\Delta G_m < 0$) will form. In the case of incompatibility ($\Delta G_m > 0$) microphases will develop. From lattice theory, the change in the total free energy of mixing of a multicomponent system is:

$$\Delta G_m = RT \left( \sum_i n_i \cdot \ln \Phi_i + V \sum_{i<j} a_{ij} \cdot \Phi_i \cdot \Phi_j \right) \tag{1}$$

$n_i$ = number of moles of the $i^{th}$ component, $\Phi_i$ = volume fraction of the $i^{th}$ component, $V$ = total volume of the system; $V = \Sigma n_i V_i$, $V_i$ = molar volume of the $i^{th}$ component, $a_{ij}$ = interaction parameter; $a_{ij} = \dfrac{\chi_{ij}}{V_i}$, $R$ = gas constant, $T$ = temperature

In binary or ternary systems, the interaction parameters $a_{ij}$, (subscript 1 = solvent, subscript 2 = polymer 1, subscript 3 = polymer 2) obtained via equation (1), are the quantities used to assess the compatibility of the polymer components. In most cases, the positive change in enthalpy of mixing ($a_{2,3}$ positive) is greater than the change in the entropy term, and the system will separate.

Equation (1) for calculating mixing behaviour can only be solved by assuming certain boundary conditions. KONINGSVELD [7] derived equation (2) for a two component system (polymer 1 and polymer 2) which describes the dependence of the GIBBS' free energy of mixing on temperature, concentration, and molecular weight taking into account the molecular inhomogeneity of the polymers.

$$\Delta G_m = RT \left( \sum_i \Phi_{i,1} \cdot m_{i,1}^{-1} \cdot \ln \Phi_{i,1} + \sum_j \Phi_{j,2} \cdot m_{j,2}^{-1} \cdot \ln \Phi_{j,2} + g(T) \Phi_1 \Phi_2 \right) \tag{2}$$

$\Phi_{i,1}$ = volume fraction of the species $i$ in polymer 1, $m_{i,1}$ = relative chain length of the species $i$ in polymer 1

$$\Phi_1 = \sum_i \Phi_{i,1} = 1 - \Phi_2 = \frac{w_1 \cdot \varrho_1^{-1}}{w_1 \cdot \varrho_1^{-1} + w_2 \cdot \varrho_2^{-1}} \tag{3}$$

$g(T)$ = interaction term

For the critical $\Delta G_m$ isotherm, the stability consideration at pressure $p$, temperature $T$ and critical concentration $\Phi_{cr}$ results in:

$$\left(\frac{\partial^2 \Delta G_m}{\partial \Phi_2^2}\right)_{p,T_{cr}} = 0; \quad \left(\frac{\partial^3 \Delta G_m}{\partial \Phi_2^3}\right)_{p,T_{cr}} = 0 \tag{4}$$

Applying equation (4) to equation (2) gives for the spinodal (stability limit between heterogeneous and homogeneous fields):

$$\frac{1}{m_{w,1} \cdot \Phi_1} + \frac{1}{m_{w,2} \cdot \Phi_2} = 2g(T) \tag{5}$$

and for the turbidity point:

$$\frac{m_{z,1}}{m_{w,1}^2 \Phi_1^2} = \frac{m_{z,2}}{m_{w,2}^2 \Phi_2^2} \tag{6}$$

$m_{w(z)}$ = weight-average ($z$-average) relative chain length

Fig. 111 shows a phase diagram calculated for a mixture of two polymers with relative chain lengths $m_1 = 50$ and $m_2 = 500$. The influence of molecular inhomogeneity on the turbidity curves and spinodals as well as the compatibility of long polymer chains of different lengths are described in [7].

In the case of real polymers this approach (system of pure polymer 1 and polymer 2) is restricted mainly to low molecular weight macromolecules (Fig. 110). In order to investigate the compatibility of high molecular weight products, a solvent must be added to promote mixing of the phases.

According to Scott [8], the free energy of mixing in ternary systems (solvent, polymer 1 and polymer 2) is determined by the interaction parameters

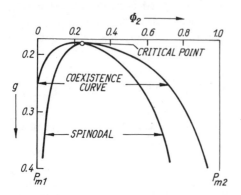

*Fig. 111*: Phase diagram of a binary polymer mixture; coexistence curve is a binodal
$m_1 = 50$; $m_2 = 500$

$\chi_{1,2}$, $\chi_{1,3}$, and $\chi_{2,3}$. Assuming that the two polymers are easily soluble in the selected solvent ($\chi_{1,2}$ and $\chi_{1,3} < 0.5$) and $\chi_{1,2}$ can be assumed approximately equal to $\chi_{1,3}$ (symmetric case), the binodals can then be calculated relatively easily [8]. Fig. 112 shows an example of this, the phase diagram of the ternary toluene/polystyrene/amorphous polypropylene system. Curve (a) mirrors the binodal calculated for the boundary conditions mentioned. For identical relative chain lengths, the binodals are symmetrical. When $m_1 \neq m_2$, asymmetric curves will result.

Asymmetrical systems and their mathematical treatment are described in [9].

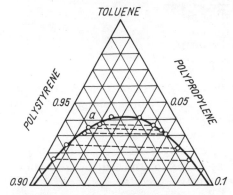

*Fig. 112*: Phase diagram of the ternary system toluene/polystyrene/amorphous polypropylene at 30 °C
$a$ = binodal calculated according to [8] with the values $a_{2,3} = 0.62 \text{ mol}/l$ and $V = 61.4 \, l$

Determination of the $a_{2,3}$-parameter from phase equilibrium measurements requires that chemical potentials of the components in the two phases be equal. Assuming this, a relationship was derived from equation (1) in [10, 11] by means of which the $a_{2,3}$-value can be calculated from measured volume fractions of polymers in the separated phases.

The interaction parameter $\chi_{2,3}$ can be determined by osmotic measurements via the relationship between the second virial coefficient and $\chi$ (equation (13) in Section 1.3.2.). The measured $\chi$-value in a ternary system includes the sum of the interaction terms of the homopolymers with the solvent ($\chi_{1,2}$ and $\chi_{1,3}$) and the interaction terms between the polymeric components themselves.

$$\chi = x_2 \cdot \chi_{1,2} + x_3 \cdot \chi_{1,3} - x_2 \cdot x_3 \cdot \chi_{2,3} \tag{7}$$

$x_i$ = mole fraction of the component $i$ relative to polymer 1 and polymer 2 ($x_2 + x_3 = 1$)

Positive $\chi_{2,3}$-values indicate incompatibility.

Table 85 lists the $\chi_{2,3}$-parameters for mixtures of polystyrene and polybutadiene which were determined by osmotic measurements.

*Table 85*: $\chi_{2,3}$-parameters of mixtures of polystyrene and polybutadiene of various compositions

| Composition | | $A_2 \cdot 10^7$ | $\chi_{2,3}$ |
| --- | --- | --- | --- |
| mol % PB | mol % PS | $(l \cdot mol \cdot g^{-2})$ | |
| 70.2 | 29.8 | 9.6 | 0.097 |
| 59.2 | 40.8 | 8.85 | 0.075 |
| 39.1 | 60.9 | 7.48 | 0.053 |

The use of gas chromatography for the study of interactions between two polymers is described in [12]. In ternary systems containing a solvent and a mixed stationary phase (polymer 2 and polymer 3), a summed interaction parameter $\chi$ is obtained from the measured activity of the solvent at infinite dilution of 1 ($\Phi_1 \rightarrow 0$; $\Phi_2 + \Phi_3 \rightarrow 1$). This parameter, like the $\chi$-parameter calculated from the $A_2$ value is composed of the interactions of the solvent with the two homopolymers and the component of the interactions between the polymers.

$$\chi = (a_{1,2} \cdot \Phi_2 + a_{1,3} \cdot \Phi_3 - a_{2,3} \cdot \Phi_2 \cdot \Phi_3) V_1 \tag{8}$$

In order to calculate $a_{2,3}$, both the activity of the solvent in the stationary mixed phase but also the activity of the same solvent in the pure stationary phases of

*Table 86*: Interaction parameters $\chi_{2,3}$ determined by inverse gas chromatography

| Solvent | n-tetracosane dimethyl siloxane [12], 60 °C | Polyvinyl chloride-Poly ε-caprolactone [13], 120 °C |
| --- | --- | --- |
| | $\frac{V_1}{V_2} \chi_{2,3} = a_{2,3} V_1$ | $\chi_{2,3}$ |
| Ethanol | — | 0.21 |
| Chloroform | — | 0.33 |
| Methylethylketone | — | —0.10 |
| Carbon tetrachloride | 0.42 | 1.07 |
| n-hexane | 0.48 | 1.16 |
| Cyclohexane | 0.49 | — |
| Toluene | 0.37 | — |

polymer 1 and of polymer 2 must be measured so that the relevant values of $a_{1,2}$ and $a_{1,3}$ can be used.

Table 86 shows the $\chi_{2,3}$ parameters determined by inverse gas chromatography with different solvents for various polymer mixtures.

The viscometric interaction parameters $b_{2,3}$ determined by viscosity measurements are related by a power law to the $a_{2,3}$-value ascertained by phase equilibrium measurements [4].

$$b_{2,3} = n \cdot a_{2,3}^{-m} \tag{9}$$

If the constants $n$ and $m$ are known, information on compatibility is provided by the easily measurable $b_{2,3}$-value.

In general, equation (10) is applicable to the specific viscosity $(\eta_{sp})_m$ of a mixture of two polymers in a common solvent:

$$(\eta_{sp})_m = [\eta]_m \cdot c_m + b_m \cdot c_m^2 \tag{10}$$

$c_m = c_1 + c_2 =$ sum of the concentrations of the polymer components, $b_m =$ parameter which characterizes the interactions of all polymer species, $[\eta]_m =$ limiting viscosity of the mixture

An expression for $b_m$ can be derived:

$$b_m = b_{2,2} \cdot Y_2^2 + 2b_{2,3} \cdot Y_2 \cdot Y_3 + b_{3,3} \cdot Y_3^2 \tag{11}$$

$Y_i =$ relative weight fraction of the polymer $i$, $b_{ii} = k_H \cdot [\eta]_i^2$, $k_H =$ HUGGINS constant (see equation (15) in 1.7.3.)

A preliminary assessment of the compatibility of the polymers can be made from the difference between the experimentally measured $b_{2,3}$ value and the $b_{2,3}^*$ parameter calculated by the VAN-LAAR approximation (equation (13))

$$b_{2,3}^* = (b_{2,2} \cdot b_{3,3})^{0.5} \tag{13}$$

Table 87: Values of $n$ and $m$ of equation (9) [4] (weight ratio of the polymers 1:1; solvent: toluene)

| Participant in the mixture | $n$ (mole · g$^{-2}$ · d$l$) | $m$ |
|---|---|---|
| Polypropylene + polymethyl methacrylate | 1.78 10$^{-3}$ | 5.36 |
| Polypropylene + polystyrene | 1.68 10$^{-4}$ | 6.00 |
| Polystyrene + polymethyl methacrylate | 2.82 10$^{-7}$ | 6.35 |

When $\Delta b_{2,3} = b_{2,3} - b_{2,3}^*$ is negative, the system is incompatible while positive values indicate attractive interactions and hence compatibility. Table 87 lists the coefficients of equation (9) for various polymer mixtures. These constants also depend on the selected solvent [4].

### 4.3.3. Experimental Background

The mixing characteristics of two or more polymeric components in the solid state is frequently measured by the torsion pendulum (vibration) test.

Rectangular specimens 60 mm long, 10 mm wide and 1 mm thick are used for the test. The specimens are clamped in the apparatus and the free specimen length, width and thickness are measured exactly by a micrometer. The heating rate should not exceed 0.8 °C per min. Vibrational excitation and the recording of the vibration diagram are usually carried out at temperature intervals of about 5 °C.

The miscibility is usually determined by calculating the logarithmic decrement $\Lambda$ from the amplitudes of subsequent vibrations as a function of temperature [14]. For compatibility investigations the shear modulus $G'$, the loss modulus $G''$ or the elasticity modulus can also be evaluated as a function of temperature [15].

As a criterion of the compatibility of the components, the glass transition temperature can be determined from thermomechanical and calorimetric analysis if the glass temperatures of the mixed polymers lie sufficiently far apart.

The influence of temperature on mixing behaviour is determined using films cast from solution with a thickness of 50 to 100 µm. Other mixing techniques can be used. After briefly heating (about 1 min) the sample to a maximum tempering temperature, the state of the system is frozen by rapid cooling (64 °C/min). Subsequently the system is analyzed by differential scanning calori-

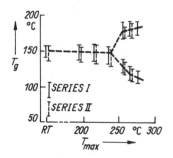

*Fig. 113*: Glass transition temperature of a mixture of polystyrene and tetramethylbisphenol A-polycarbonate (40/60) as a function of temperature [15]

metry (*DSC*). The maximum tempering temperatures $T_{max}$ are increased stepwise. Separation phenomena are indicated by the occurrence of two glass transition temperatures (Fig. 113). For certain polymer mixtures [15], the composition of the phases can be determined from a calibration curve. Assuming that thermodynamic equilibrium has been attained at each tempering temperature, the phase diagram produced is then identical with the binodals of the mixture.

Micrographs and electron-micrographs show optically the presence of mono phase or two phase structures. The samples are prepared by the usual methods (ultramicrotome section, contrasting). Frequently, the micrographs serve for documentation purposes.

Evaluation of polymer miscibility from broad line and pulse NMR spectra is described in [16, 17].

There are difficulties in using the turbidity point method with varying temperature for pure polymer component systems because of the high viscosity. Some experimental solutions to this problem, e.g. ultracentrifugation, pulse-induced critical light scattering measurements and neutron scattering, are given in [7].

Phase separation of polymer mixtures in a common solvent can be effectively and conveniently observed if the refractive index of the coexisting phases is different. In most cases, two related but not miscible phases occur. The lower phase can be sucked off with syringe. The system can also be separated by rapid freezing with liquid nitrogen. In this procedure, a crack is formed at the boundary of the phases facilitating quantitative separation.

The total concentration of the polymers in the solvent ranges in general from 2 to 10 g per 100 m*l*.

For determining the composition of the two polymeric components in the separated phases, appropriate chemical or physical methods depending on the chemical nature of the macromolecules are used. Among other methods, UV, IR and NMR measurements, investigations using pyrolysis gas chromatography and electrophilic addition of iodine monochloride to double bonds (see Section 3.3.3.1.) are preferred, while other chemical methods for the determination of heteroatoms and functional groups can also be used.

After the phase volumes involved have been determined, a certain volume is taken from each phase. The solvent is removed and the polymeric residue dried to constant weight. The phase composition is then determined.

Compatibility investigations by viscosity, light-scattering and osmotic measurements are carried out in dilute solutions with total concentration of polymers between 0.1 and 2.0 g per 100 ml of solvent. In order to see the effect of compatibility clearly, the polymers are frequently mixed in the ratio of 1:1 ($Y_i = 0.5$). In order to prevent chain degradation antioxidants are added to the solutions if necessary (for example, when polyethylene, polypropylene or polybutadiene is one of the components). Experimental details of light-scattering, viscosity and osmotic measurements, are described in Chapters 1.5., 1.7. and 1.3.

### 4.3.4. Examples

#### 4.3.4.1. Determination of Phase Equilibrium in the Carbon Tetrachloride/Polystyrene/Polyisobutene System at 18 °C

For determination of the phase equilibrium, the amounts of polymer indicated in Table 88 are weighed into 25 m*l* volumetric flasks. After complete dissolution of the polymers, the solutions are fully mixed in a measuring cylinder capable of being closed (volume 50 m*l*). They are then thermostatted at 18 °C and left until phase separation is complete. The volume of the two phases is read from the measuring cylinder. The system is separated by sucking off the lower phase with a syringe with a long steel needle. Subsequently, the solvent is removed in vacuo and the polymer dried and weighed.

To analyze the composition of the phases, about one tenth of the dried polymer mixture is weighed accurately into a 10 m*l* volumetric flask. The polymer is dissolved in a mixture of cyclohexane and carbon tetrachloride (10 parts V/V). The concentration of the polystyrene is determined by UV spectroscopy at a wavelength of 254 nm at which polyisobutene does not absorb. A 1 mm cell is used for the phase rich in polystyrene and a 10 mm cell for the second phase. The absolute concentration is obtained from a calibration curve of polystyrene solutions of known concentration. A plot of the absorbance at 254 nm against concentration is linear over the range from 0.05 to 0.5 g per 100 m*l*.

Table 88: Initial concentration for the determination of the phase equilibrium in the ternary carbon tetrachloride/polystyrene/polyisobutene system

| Sample | Polystyrene* (g) | Polyisobutene** (g) |
|---|---|---|
| 1 | 0.505 | 0.370 |
| 2 | 0.490 | 0.510 |
| 3 | 0.440 | 0.628 |
| 4 | 0.440 | 0.977 |

$M_n^* = 1 \cdot 10^6$ g/mol      $M_n^{**} = 1.5 \cdot 10^6$ g/mol
$\varrho^* = 1050$ kg/m$^3$      $\varrho^{**} = 915$ kg/m$^3$

## Compatibility of Polymers

*Table 89*: Measured values of the phase equilibrium in the carbon tetrachloride/polystyrene/polyisobutene system at 18 °C

| Sample | Phase 1 | | | Phase 2 | | |
|---|---|---|---|---|---|---|
| | volume (ml) | PS (g) | PIB (g) | volume (ml) | PS (g) | PIB (g) |
| 1 | 14.2 | 0.020 | 0.215 | 35.8 | 0.484 | 0.155 |
| 2 | 24.7 | 0.030 | 0.430 | 25.3 | 0.460 | 0.080 |
| 3 | 30.3 | 0.030 | 0.580 | 19.7 | 0.410 | 0.048 |
| 4 | 37.9 | 0.040 | 0.971 | 12.1 | 0.400 | 0.006 |

*Table 90*: Volume fractions for phase equilibrium calculated from the values of Table 89

| Phase | Polymer | Sample | | | |
|---|---|---|---|---|---|
| | | 1 | 2 | 3 | 4 |
| 1 | 2 | 0.0013 | 0.0011 | 0.0009 | 0.0010 |
| | 3 | 0.0163 | 0.0187 | 0.0205 | 0.0272 |
| 2 | 2 | 0.0127 | 0.0170 | 0.0194 | 0.0305 |
| | 3 | 0.0047 | 0.0034 | 0.0026 | 0.0005 |

Polymer 2: polystyrene
Polymer 3: polyisobutene

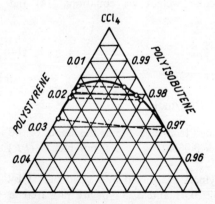

*Fig. 114*: Phase diagram of the ternary system carbon tetrachloride/polystyrene/polyisobutene at 18 °C

The measured phase volumes and the spectrometrically determined proportions of the components in the two phases are listed in Table 89. The volume fractions of the phases were calculated from proportions by weight and densities (Table 90). The phase diagram shown in Fig. 114 was plotted from values given in Table 90.

### 4.3.4.2. Determination of the Compatibility of the Toluene/Polystyrene/Amorphous Polypropylene System by Viscosity Measurements

Amorphous polypropylene is obtained from the commercial product by Soxhlet extraction with diethyl ether.

The specific viscosities of the polymers are measured in toluene at 30 °C (see 1.7.). For this purpose, solutions are prepared within the concentration range from 0.3 to 1.1 g per 100 ml. In order to accelerate the solution process, the flasks are shaken and if necessary, the solutions containing polypropylene are slightly heated. All of the polymer solutions containing polypropylene are stabilized by adding 0.5 per cent by weight of N-phenyl-$\beta$-naphthylamine (relative to the polymer component).

The specific viscosities obtained are divided by the concentration $c_B$ and plotted against the concentration (HUGGINS). The extrapolated limiting viscosity numbers (Fig. 115)

$$[\eta]_{PS} = 0.48 \text{ dl/g} \qquad [\eta]_{PP} = 1.03 \text{ dl/g}$$

are used to calculate the average molecular weight via equations (14) and (15).

$$[\eta]_{PS}^{\text{toluene; 30°C}} = 1.10 \cdot 10^{-4} \cdot M^{0.72} \tag{14}$$

$$[\eta]_{PP}^{\text{toluene; 30°C}} = 2.18 \cdot 10^{-4} \cdot M^{0.72} \tag{15}$$

$$M_\eta^{PS} = 1.05 \cdot 10^5 \text{ g/mol}; \qquad M_\eta^{PP} = 1.17 \cdot 10^5 \text{ g/mol}$$

The interaction parameters $b_{i,i}$ are obtained from Fig. 115 as the slopes of the straight lines. For polymer 2 (PS) $b_{2,2} = 0.088 \text{ dl}^2 \text{ g}^{-2}$, for polymer 3 (PP) $b_{3,3} = 0.39 \text{ dl}^2 \text{ g}^{-2}$.

For determining compatibility, the polymers are mixed in a ratio of 1:1 by weight. Concentrations of 1.25; 0.83; 0.62; 0.41 and 0.25 g/dl are prepared in 25 ml volumetric flasks using toluene as the solvent. After complete dissolution of the polymers, the specific viscosities are again measured; the latter are then divided by the concentration and plotted against the concentration (Fig. 116). From the graph the value of $[\eta]_m = 0.7$ dl/g and that of $b_m = 0.17$ dl$^2 \cdot$ g$^{-2}$.

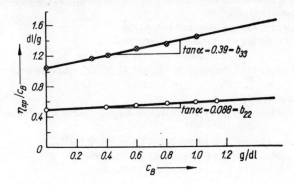

*Fig. 115*: Determination of the limiting viscosity numbers and the $b_{i,i}$-parameters in toluene at 30 °C
○ polystyrene ● amorphous polypropylene

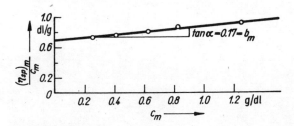

*Fig. 116*: Determination of $[\eta]_m$ and $b_m$ in the toluene/polystyrene/amorphous polypropylene system at 30 °C

The graphically determined values of $b_m$, $b_{2,2}$ and $b_{3,3}$ are inserted in equation (11) using 0.5 as the value for $Y_2$ and $Y_3$. With this, a value of 0.1 is calculated for $b_{2,3}$ while 0.19 is obtained for $b_{2,3}^*$ from equation (13). The difference $\Delta b_{2,3}$ is negative and the system incompatible. Using the constants given in Table 87, a value of 0.345 is calculated for the interaction parameter $a_{2,3}$ (equation (8)).

# Bibliography

[1] THINIUS, K., Plaste und Kautschuk **15** (1968), 164
[2] DOBRY, A. and F. BOYER-KAWENOKI, J. Polymer Sci. **2** (1947), 90
[3] FRIESE, K., Plaste und Kautschuk **15** (1968), 646
[4] BÖHMER, B., BEREK, D. and S. FLORIAN, European Polymer J. **6** (1970), 471

[5] FELDMAN, D. and M. RUSU, European Polymer J. **7** (1971), 215
[6] SCHNECKO, H. and R. CASPARY, Kautschuk und Gummi-Kunststoffe **25** (1972), 309
[7] KONINGSVELD, R. and L. A. KLEINTJENS, J. Polymer Sci. Polymer Symposium **61** (1977), 221
[8] SCOTT, R. L., J. chem. Physics **17** (1949), 279
[9] HSU, C. C. and J. M. PRAUSNITZ, Macromolecules **7** (1974), 320
[10] ALLEN, G., GEE, G. and J. D. NICHOLSON, Polymer **1** (1960), 56
[11] BEREK, D., LATH, D. and V. DURDOVIC, J. Polymer Sci. C **16** (1967), 659
[12] DESHPANDE, D. D., PATTERSON, D., SCHREIBER, H. P. and C. S. SU, Macromolecules **7** (1974), 530
[13] OLABISI, O., Macromolecules **8** (1975), 316
[14] NITSCHE, K. and WOLF, K., „*Kunststoffe*", Vol. 2: „*Praktische Kunststoffprüfung*", Springer-Verlag, Berlin 1961
[15] CASPER, R. and L. MORBITZER, Angew. Makromolekulare Chem. **58/59** (1977), 1
[16] ELMQVIST, C. and S. E. SVANSON, European Polymer J. **11** (1975), 789
[17] DOUGLAS, D. C. and V. J. MCBRIERTY, Macromolecules **11** (1978), 766

# 5. Authors Index

ABDELLATIF, E. 151
ALISCHOEV, V. R. 206
ALLEN, G. 332
ALTGELT, K. H. 182
AMOS, R. 182
ANDREJS, B. 270
ARCHIBALD, W. J. 84, 92
ARND, R. 234
ARNDT, K. F. 165, 295
ARNOLD, M. 234
AXELSON, D. E. 270

BAHRS, D. 182
BAKER, G. A. 112, 151
BALLEGOOIJEN, H. VAN 306
BAREISS, R. 17
BARTON, A. F. M. 319
BEALL, G. 119, 151
BEATTIE, W. H. 159, 165
BELKEBIR-MRANI, A. 294
BENOIT, H. 71, 169, 182, 295, 306
BEREK, D. 182, 331, 332
BERESKIN, V. G. 206
BERGER, H. L. 182
BERGER, R. 99, 105, 122, 151
BERNE, B. J. 71
BERRY, G. C. 71
BERSTED, B. H. 53
BHATNAGAR, S. K. 319
BISKUP, U. 270
BLASCHKE, F. 98
BLEHA, T. 182
BODANECKY, M. 139, 151
BÖHMER, B. 331

BOMMER, P. 234
BOROSS, L. 182
BORSDORF, R. 220
BOVEY, F. A. 206, 249, 250
BOWMER, T. N. 270
BOYER-KAWENOKI, F. 331
BRANDRUP, J. 105, 151
BRAUN, D. 206, 220
BREWER, P. I. 182
BUSHUK, W. 71

CABANNES, J. 64
CANDAU, F. 306
CANTOW, H. J. 71, 92, 298, 306
CANTOW, M. J. R. 92, 150, 165
CASPARY, R. 332
CASPER, R. 270, 332
CHIANG, R. J. 105
CHONG WHA PYUN 234
CHYLEWSKI, Ch. 53
CLAESSON, S. 110, 150, 158, 165
CLARKE, D. H. 319
CLUFF, E. 281, 295
COLEMANN, B. D. 234
CORBETT, J. M. 319
CORNETT, C. F. 306
CORRADINI, P. 250
COTTON, J. P. 295
CROMPTON, T. R. 270

DANUSSO, F. 250
DAWKINS, J. V. 319
DECHANT, J. 33, 220, 234, 241, 250, 270
DECKER, D. 295

DECKERT, E. 99
DESHPANDE, D. D. 332
DETERMANN, H. 166, 182
DEXHEIMER, H. 270
DOBRY, A. 331
DOUGLAS, D. C. 332
DROTT, E. E. 270
DUPLESSIX, R. 295
DURDOVIC, V. 332
DUSEK, K. 294

ELIAS, H.-G. 17, 92, 134, 135, 151, 306
ELMQVIST, C. 332
ERNST, R. 234
ESKIN, W. E. 71
EVANS, J. M. 182
EZRIN, M. 207

FARNOUX, B. 295
FELDMAN, D. 332
FIJOLKA, P. 33
FIKENTSCHER, G. 99
FINK, H. 285, 295
FLEMMING, S. W. 182
FLORIAN, S. 331
FLORY, P. J. 75, 76, 94, 108, 150, 243, 255, 275, 276, 296, 306
FOX, T. G. 75, 94, 234, 313, 319
FRANZ, J. 33, 206, 220
FRIESE, K. 331
FRIIS, N. 270
FRITZSCHE, P. 151
FUCHS, O. 112, 116, 135, 138, 150, 270

GEE, G. 332
GEORGE, H. F. 294
GERMAR, H. 241, 250
GIESEKUS, H. 165
GLADDING, F. 295
GLÖCKNER, G. 150, 151, 182
GRAHAM, J. P. 165
GRASSELLI, J. G. 250
GRONSKI, W. 295
GROSSE, L. 207
GRUBER, U. 114, 151, 306

GRUBISIC, Z. 182
GUILLET, J. E. 315, 319
GUINIER, A. 71

HACK, H. 71
HAGEN, E. 33, 206, 220
HAGEN, G. 96
HAGENBACH, E. 97
HAHN, F. L. 30, 33
HAMIELEC, A. E. 270
HANNEMANN, Ch. 151
HANSEN, C. M. 309
HARWOOD, H. J. 234
HÄUSLER, K. G. 165, 295
HEINRICH, G. 294
HEKELER, W. 207
HELLER, W. 165
HELLFRITZ, H. 38, 44
HELLWEGE, K. H. 250
HENGSTENBERG, J. 165
HENRICI-OLIVÉ, G. 127, 151
HERMANS, J. J. 92, 275
HERZ, J. E. 294
HILDEBRAND, J. 307, 309, 319
HOFFMANN, M. 165, 270, 294
HOSEMANN, R. 118, 151
HOUWINK, R. R. 94
HSU, C. C. 332
HUELCK, V. 294
HUGGINS, M. 98, 108
HUGLIN, M. B. 71, 165
HUMMEL, D. O. 33, 206

IMMERGUT, E. H. 105, 151
INAGAKI, H. 151
INOUE, S. 250

JACOBI, M. M. 295
JACOBSON, A. 165
JAMADERA, R. 33
JANNINK, G. 295
JOHNSEN, U. 250
JOHNSON, J. F. 250
JOHNSON, L. F. 250
JUNG, H. C. 165

# Author Index

KAISER, R.  24, 33
KAJIWARA, K.  71
KAMIDE, K.  53
KÄMMERER, H.  33
KAPELLE, P.  37, 38, 40, 44
KATO, Y.  151
KAUFMANN, W.  151
KELLER, F.  226, 234, 250
KERN, W.  33
KETLEY, A. D.  250
KIMMER, W.  216, 220, 234
KLAUS, W.  207
KLEIN, J.  109, 116, 150, 151
KLEINTJENS, L. A.  150, 332
KOLBE, K.  250
KOLTHOFF, J. M.  22, 33
KÖNIG, J. L.  241, 250
KONINGSVELD, R.  108, 109, 113, 150, 306, 321, 332
KOTLIAR, A. M.  125, 151
KRAATS, E. J. VAN DER  282, 295
KRAEMER, E. O.  92, 98, 125
KRAI, O. E.  220
KRAUSE, A.  207
KREMMER, T.  182
KREVELEN, D. W. VAN  313, 316
KRIMM, S.  250
KRÖMER, H.  165, 294
KUHN, R.  134, 151, 165, 270, 294
KUHN, W.  94, 295
KUHN, WO. H.  38, 44

LAMM, O.  77, 81
LANGBEIN, G.  33
LANGE, H.  71, 270, 284, 295
LANGE, A.  207
LANGHAMMER, G.  105
LANGMUIR, I.  46
LANSING, W. D.  92, 125
LATH, D.  332
LEVY, G. C.  270
LEIBNITZ, E.  207
LIPSON, J. E. G.  319
LITMANOVICH, A. D.  151

MAJER, M.  295
MAJORS, R. E.  182
MANDELKERN, L.  76, 92, 270
MANGARAJ, D.  319
MARK, H.  33, 94
MAYO, R. R.  234
McBRIERTY, V. J.  332
McCORMICK, H. W.  92
MEMIROVSKAJA, B.  206
MENDELSON, R. A.  270
MEYERHOFF, G.  71, 92, 105
MINSKER, K. S.  220
MORBITZER, L.  332
MORRAN, L.  33
MUKHERJEE, A. R.  33
MUSSA, G.  123, 151

NATTA, G.  234, 250
NEWMAN, S.  235
NICHOLSON, J. D.  332
NITSCHE, K.  332

OBER, J.  295
O'DONNEL, J. H.  270
OLABISI, O.  332
OLIVÉ, S.  127, 151
OSTROMOW, H.  206
OSTWALD, W.  97, 175
OUANO, A. C.  182

PACHOMOVA, I. K.  220
PALIT, S. R.  24, 33
PANGONIS, W. J.  165
PARISER, E. K.  295
PATRA, S.  319
PATTERSON, D.  332
PECORA, R.  71
PEEBLES, L. H.  122, 150
PERRY, S. G.  182
PFANN, H. F.  33
PHILIPPOFF, W.  94
PHILPOT, J. S. L.  82, 83, 86
PICOT, G.  295
POHL, U.  270
POISEILLE, J. L. M.  96, 175

PORTER, R. S.   250
PRAUSNITZ, J. M.   332
PRIMAS, H.   227, 234
PRINS, W.   294

RASHID, S.   319
RATH, S. B.   319
REHAGE, G.   277, 295
REMPP, P.   182, 294, 295, 306
RIEDEL, S.   151
RITCHEY, W. M.   234
ROCABOY, F.   33
ROSSMANN, E.   211, 214, 220
ROSS-MURPHY, S. B.   71
ROTH, K.-H.   234, 250
RUSU, M.   332

SCHAEFER, J.   250
SCHERAGA, H. A.   76, 92
SCHMALZ, E. O.   220
SCHMID, R.   294
SCHMIEDER, K.   295
SCHMOLKE, R.   234
SCHNECKO, H.   332
SCHNEIDER, H.   234, 250
SCHOLL, F.   33, 206
SCHOLTE, Th. G.   306
SCHOLZ, M.   220
SCHRAMEK, W.   151
SCHRECK, J.   295
SCHREIBER, H. P.   332
SCHRÖDER, E.   33, 151, 165, 206, 220, 270, 294
SCHULZ, G.   33
SCHULZ, G. V.   44, 71, 98, 109, 119, 127, 150, 151, 156, 158, 165, 270, 298, 306
SCHURZ, J.   105
SCOTT, R.   309, 319, 322, 332
SEEHR, H.   270
SEIDE, H.   105
SHTERN, V. Y.   151
SHULTZ, A. R.   182, 306
SIMON, W.   53
SMALL, P. A.   270, 310, 319
SMITH, D. A.   288, 295
SNYDER, L. R.   136, 151

SOLTES, L.   182
SPERLING, L. H.   294
STADLER, R.   295
STAHL, E.   136, 151
STAUDINGER, H.   93
STAVERMAN, A. J.   71, 306
STEARNE, J. M.   160, 165
STEIN, D. J.   270
STEINFORT, K.   33
STOCKMAYER, W. H.   151
STOKES, G. G.   75
STRAUBE, E.   294
STRICKS, W.   33
STRUPPE, H. G.   207
STUART, H. A.   150
SU, C. S.   332
SUH, K. W.   319
SVANSON, S. E.   332
SVEDBERG, T.   74, 79, 80
SVENSON, H.   82, 83

TALAMINI, G.   303, 306
TAYLOR, W. C.   165
TEICHGRÄBER, M.   151
TENGLER, H.   206
TERAKAWA, T.   53
TERAMACHI, S.   151
THALLMAIER, M.   220
THINIUS, K.   220, 331
THOMAS, J.   294
TIERS, G. V. D.   206, 250
TOPCHIEV, A. V.   133, 151
TRAUTMAN, R.   85, 92
TSUKUMA, I.   250
TSURUTA, T.   250
TUNG, L. H.   125, 150, 182

UBBELOHDE, L.   97, 175
UCHIKI, H.   53
ULBRICHT, J.   220, 250
URBAN, H.   165
URWIN, J. R.   156, 160, 165

VIDOTTO, G.   303, 306

# Author Index

WACHTER, A. H.  53
WANDEL, M.  206
WATTERSON, J. G.  17
WEIBULL, W.  129
WEINHOLD, G.  150
WERNER, M.  151
WESSLAU, H.  122, 123
WIJK, R. VAN  71
WILLBOURN, A. H.  266, 270
WILLIAMS, R. P. J.  112, 151
WILLIAMS, V. Z.  33

WOLF, B. A.  150, 306
WOLF, K.  295, 332
WOLFRAM, L. E.  250
WULF, K.  37, 38, 40, 44

YANG, J. T.  71
YAU, W. W.  182

ZANDER, P.  295
ZIMM, B. H.  64, 68, 125

# 6. Subject Index

Absolute methods of molecular weight determination  17
Absorbance coefficient see: extinction coefficient
Acid value  198
Activity, optical  234, 243
—, thermodynamic  35, 45
Addition reaction, electrophilic  **210**, 213, 327
Adjusting speed  38
Antioxidants  189
ARCHIBALD, method according to  84
Association  40, 50, 258
Attraction constant according to SMALL  310
Average  15, 166

BAKER-WILLIAMS fractionation  112
Balloon effect  **39**
BODENSTEIN principle  220
Branching  **250**
—, influence on the properties  252
—, structure parameter for the description of  251

CABANNES factor  64
Calibration constant  46, 51
Carboxyl end group determination  19
—, in the presence of acid anhydrides  20
—, with sodium alkoxide  29
—, via cation-anion complexes  20
Cellulose, detection  203
Chemical heterogeneity  106
—, determination by solution fractionation  135, 141

—, determination by thin-layer chromatography  136
Chemical potential  **34**, 45
—, pressure dependence  35
Chlorine, detection  199
Chromatography, see various methods
CLAESSON diagram of solubility  110
CLAESSON nomogram  158, 164
Coagulation behaviour of precipitated polymer particles  153
Coefficient of friction  74
Cohesion energy  307
Cohesion energy density  307
Coil, disturbed  96
—, undisturbed  96
Compatibility  **319**
Compressibility factor  90
Compression modulus for network characterization  282
Configuration  182, 209, **234**
Constant, optical  57, 63
Contraction factor  254, 256
Copolymers
    butadiene-styrene  137, 213, 216
    chemical heterogeneity  **141**
    diolefin  210, 216
    ethylene-butene-1  **269**
    ethylene-vinyl acetate  179, **228**
    number-average molecular weight  33
    styrene-acrylnitrile  137, **232**
Correction factor for polydispersity correction  94
Critical quantities, volume fraction  135, 141, 298

—, interaction parameter  298
—, temperature  **300**
Cross fractionation  133, 135
Crosslinking density  271, 273
Cryoscopy  17, 45
Crystallinity  235, 243, 252
Cyanuric acid resins  208, 209

DEBYE scattering  57
Decomposition  185, 191, 209, 211, 259
Density  196, 235, 252
Depolymerization  191
Differential refractometer  60
Differential-scanning-calorimeter  326
Diffusion  72, 76, 91, 263
Diffusion coefficient  72, **91**
Dilution heat, infinite  276, 277
Dipole radiation  55
Dissolution-precipitation chromatography  112
Dissymmetry  69, 70
Distribution, chemical heterogeneity  **131**
—, molecular weight  86, 106, 156, 165
Distribution coefficient  167
Distribution function, drawn up from experimental data  138, 145
Distribution function of the molecular weight  **122**
—, exponential distribution  122, 125
—, LANSING-KRÄMER-distribution  125
—, logarithmic normal distribution  122
—, mathematical treatment  **122**
—, MUSSA-distribution  123
—, POISSON-distribution  129
—, SCHULZ-ZIMM-distribution  **125**
—, WEIBULL-TUNG-distribution  125, 130, 131
—, WESSLAU-distribution  122
Division ratio of the chemical heterogeneity  132
Double bond  **207**
—, conjugated  208, 209, 213
—, determination in polybutadiene  **213**
Dye receptivity for network characterization  285

Ebullioscopy  17, 45
Elastometric network density determination  280
Electron microscopy  320, 327
Elimination reaction  191
Eluting agent  173
Elution chromatography  **112**
Elution parameter  167
End group analysis  17, **18**
—, chemical methods  19
—, infrared-spectrometric  25
—, measuring and determining methods  19
—, physical methods  25
End group determination in polystyrene, infrared-spectrometric  31
End group determination of a linear saturated polyester  29
End-to-end distance  59
Entanglement network  39
Enthalpy of dilution  36
Enthalpy of mixing  308
—, regular solution  308
Entropy of dilution  36
Entropy of mixing  297
—, ideal solution  297
Entropy parameter according to FLORY  243, 300
Epoxide, detection  201
Equilibrium ultracentrifugation  73
Equivalent methods of molecular weight determination  17, 18
Ester group, detection  200
Evaporation rate analysis  285
Excluded volume  96
Expansion coefficient  95
Exponential distribution  **122**, 125
Extinction coefficient  25, **216**, 233
Extraction  184, 185, 189

FICK's law  91
FIKENTSCHER equation  99
FIKENTSCHER value  99
Fillers  187
FLORY constant, see FLORY universal constant
FLORY-HUGGINS theory  35, 298

FLORY-REHNER equation 276
FLORY universal constant (or factor) 75, 95, 255, 266
Fluorine, detection 199
Formaldehyde, detection 201
FOX-FLORY equation 75, **95**
Fractionation
—, analytical 105, 106, 107, 173
—, method 107
—, principles of separation 107
Free enthalpy of mixing (FLORY-HUGGINS) 35, 296, **321**
Freezing point depression 278, 292
Front factor 274
Functional groups in polymers **19**

Gas chromatography 187, 190, 263, 315, 324
Gel permeation chromatography 17, 107, **166 ff.**, 252, 260
Glass temperature 273, 326
$\Gamma$-function (gamma function) 125
$\gamma$-gradient, logarithmic, calculation 114

HAGENBACH correction 97
HAGEN-POISEUILLE's law 96
Heteroelements in polymers 198
Heterogeneity determination by thin-layer chromatography 136
High-velocity ultracentrifugation 73
HUGGINS constant 99, 325
HUGGINS equation 98
HUGGINS interaction parameter 35, 108, 113, 133, 134, 298, 306, 308, 321
Hydrodynamic radius 168
Hydrodynamic volume 94, 169
Hydrolysis, determination of sequence lengths 225
Hydroxyl value 198

Identification **183**
—, of plasticizers 185
—, of polymers 184, 187, 190
—, of stabilizers 189
Infrared bands, evaluation in case of superposition 28

—, for double bond determination 212, 216
—, in polymer end groups 25
Infrared-spectrometric double bond determination in polybutadiene 215
Infrared-spectrometric end-group determination in polystyrene 31
Interaction energy **306**
Interaction parameter according to HUGGINS 35, 108, 113, 133, 134, 298, 306, 308, 321
Interpenetration network 271
Iodine reaction on polymers 204
Iodine value 211, 214
Isomerization degree 210, 211
Isomery
  branching **250**
  cis-trans 209, **215**
  configurations 183, 209, **234**
  linking 183
  sequence **220**
Isothermal distillation 17, 45

Junction point distance distribution 278

KOTLIAR function 125
KRAEMER equation 98
KUHN-MARK-HOUWINK equation **94**, 102, 123, 169, 179, 265, 330
KUHN-MARK-HOUWINK exponent **96**, 102, 123, 291, 300, 306
$K_F$-value 99

LALLS detector for GPC 175
LAMBERT-BEER's law 216, 218, 225, 226
LAMM's differential equation 77
LAMM's scale method 81
LANSING-KRÄMER distribution 125
Lattice theory, see FLORY-HUGGINS theory
Light scattering 17, **53**, 175, 243, 265, 320
—, of disperse systems 152
Light-scattering apparatus 60
Light-scattering equation 58
Limiting viscosity **93**, 243, 330
—, for branching determination 252, 254, **265**
—, for solubility parameter determination 312

## Subject Index

—, for compatibility determination  330
Limiting viscosity/molecular weight relation, see KUHN-MARK-HOUWINK equation
Logarithmic normal distribution  123, 124
Long-chain branching  254 ff

MANDELKERN-FLORY-SCHERAGA equation  76
$M_n$ determination
—, by end group analysis  18 ff
—, by membrane osmosis  33 ff
—, by sedimentation measurement  80
—, by vapour-pressure osmosis  45 ff
—, infrared-spectrometric  25, 31
$M_w$ determination
—, by light scattering  53 ff
—, by sedimentation measurement  79
$M_z$ determination  79
$\bar{M}_\eta$ determination  93 ff
Melamine resins  208, 209
Melting point  196
Membrane osmometer (also see osmometer)  40
Membrane osmosis  17, 33 ff
Membrane, semipermeable  37, 39
Memory term  274
Microgel  59, 61, 71
MIE scattering  153
Molecule branching (also see branching)
Molecule contraction  254
Molecular inhomogeneity  118, 121
—, determination  120 ff
Molecular weight
—, number-average ($M_n$)  16, 18 ff, 33 ff, 45 ff, 59, 79, 80
—, viscosity average ($M_\eta$)  16, 93 ff
—, weight-average ($M_w$)  16, 53, 55, 79
—, z-average ($M_z$)  16, 72, 79
Molecular weight determination of a linear saturated polyester, chemically  29
Molecular weight distribution  106 ff
—, determination
—, —, by gel permeation chromatography  166 ff
—, —, by solubility fractionation  108
—, —, by turbidity titration  151 ff
—, —, by ultracentrifugation  86 ff
—, differential  119, 121
—, integral  119, 121
—, mathematical description  122 ff
Momentum  15, 121
MOONEY-RIVLIN equation  275
—, experimental determination  286 ff

Network polymer, definition  271
—, characterization by dye receptivity  285
—, —, elastometry  273, 280, 286, 288
—, —, freezing point depression  278, 292
—, —, solubility investigations  277
—, —, sol-gel analysis  284, 290
—, —, swelling measurements  276, 282
$NH_2$ end group determination  21
Nitrogen, detection  198
Normalization  176

OH end group determination  22
—, chemical methods  23
—, in polyesters by acetylization  30
—, radiochemical methods  24
Osmometer, membrane osmometer according to WULF KAPELLE  37, 40 ff
—, according to HELLFRITZ  38
—, automatic  38, 43
Osmometer, vapour-pressure  49
Osmosis  33 ff, 320, 324
Osmotic pressure  34
Oxidation reactions, determination of sequence lengths  211, 225
Oxygen, detection  200

Particle scattering function  57 ff, 153
Persistence ratio  222, 224
Phase diagram  323
Phase equilibrium  323, 328 ff
Phase separation  243, 297, 319 ff
Phase stability criteria  298, 322
Phenol, detection  200
PHILIPPOFF's rule  94
Plasticizer  185 ff
POISSON distribution  129

Polarizability 55
Polyamides, end group determination 21
—, detection 204
Polybutadiene, isomer analysis 215
Polybutene-1 238, 242, 244
Polychlorohydrocarbons, detection 203
Polydispersity, determination 120
Polyester, saturated linear 29
—, unsaturated, double-bond determination 208, 211
Polyethylene, determination of the distribution function 145 ff
—, branching determination 261, 265
—, double bonds in 204
—, limiting viscosity measurement 99 ff
—, viscosity/molecular-weight relation 170, 265
Polymethylmethacrylate, tacticity determination 239, 242, 244 ff
—, relation between viscosity and molecular weight 170
—, solubility parameter determination 311
Polypropylene, tacticity determination 240, 242
—, relation between viscosity and molecular weight 170
Polystyrene, detection 202
—, determination of distribution 176 ff
—, light scattering measurement 62 ff
—, limiting viscosity measurement 103, 104
—, relation between viscosity and molecular weight 170
—, tacticity determination 242
—, theta-temperature 303
—, ultracentrifugation 86 ff
Polystyrene, vapour pressure osmosis 50 ff
Polyurethane, detection 201
Polyvinylacetate, solubility parameter determination 313
—, relation between viscosity and molecular weight 170
Polyvinylalcohol, detection 202
Polyvinylchloride, solubility parameter determination 316 ff
—, membrane osmosis 40 ff

—, relation between viscosity and molecular weight 170
—, tacticity determination 240, 241 ff, 242, 246 ff
—, theta composition of an $S/Pr$-mixture 305
Potential, chemical 34, 297, 308
—, of cross-linked polymers 283 ff
Precipitability 152
Precipitation point titration 151, 302
Prepolymer characterization 45
Pulse induced critical scattering 298, 327
Purification 61, 62

Quasielastic light scattering 54

Radiolysis — gas chromatography 263
Radius, hydrodynamic 75, 94, 168, 254
— of gyration 53, 59, 68, 95, 254 ff
RAYLEIGH-GANS theory 153
RAYLEIGH relation 57
RAYLEIGH scattering 55 ff
Refractionation 118
Refractive index 56, 174, 191, 327
—, increment 56, 61
Relation between solubility and degree of cross-linking 277
Relation between solubility and molecular weight, experimental determination 154
Relative methods of molecular weight determination 17
RET equation 273
$R_f$-value 186, 190
Rubber elasticity 273
Run number 220, 222, 224

Sample charging loading for fractionating 115
Saponification value 197
Scattered light 54
—, lateral 152
Scattering function 57
Scattering intensity, reduced 57
Scattering standard (benzene) 63
Schlieren optics 83
SCHULZ-BLASCHKE constant 98, 102
SCHULZ-BLASCHKE equation 98

Schulz-Flory distribution  128
Schulz-Zimm distribution  125, 126
Sedimentation  17, 72 ff
Sedimentation coefficient  74, 78, 88 ff
Sedimentation equilibrium  84
Sedimentation velocity  83
Separation efficiency  170
Separation efficiency, investigation  118
Separation method  107, 166
Sequence length, average  222, 223, 235
Sequence length distribution  220 ff
Setting constant according to Kuhn  41
SH end group determination, argentometric  22
—, iodometric  22
Short-chain branching  261 ff
Simultaneous determination of two polyesters  30
Skrew of the distribution function  132
Solubility  188, 196
Solubility curves  109, 110
Solubility equation  156, 163
Solubility fractionation
—, charging the supporting material  115
—, contact time  116
—, fractionation step width  115
—, gradient of the elution mixture  114
—, optimum process parameters  113
—, principles  108
—, solvent/precipitant-system  113
Solubility lines  110
Solubility parameter  113, 134, 306 ff
—, of the polymer  309, 316
Solution-precipitation chromatography  112
Solution, regular  307
Solution fractionation according to Fuchs, experimental data  138
Solution viscosity  93 ff
Spectroscopy, broad line NMR  327
—, $^{13}$C-NMR  236 ff, 261
—, $^1$H-NMR  190, 226, 236, 261, 269, 327
—, infrared  31, 174, 190, 211, 215, 225, 240, 259, 267, 327
—, ultraviolet  174, 212, 217 ff, 327
Stabilizers  189

Staudinger index  93
Stereoregularity  234 ff
Stokes equation  75
Sulphate ash  189
Sulphur, detection  199
Supporting material, loading  115
—, for elution fractionating  112
Svedberg equation  74
Swelling measurements, for solubility parameter determination  311
—, for network characterization  276

Tacticity  234 ff
Termination reaction  191, 207
Theta conditions  296 ff
Theta quantity  296
Theta solutions  296
Theta temperature  243, 296, 299 ff
Theta temperature and solubility parameter  313
Thin-layer chromatography  131, 136, 185, 190
Titration curves, calculation graphical evaluation  30
Toxicity  189
Transfer reaction  189, 207, 252
Trautman, method according to  85
Tung-distribution  125, 130
Tung-plotting of fractionating data  130, 149
Turbidity curve  156, 299, 320
Turbidity point, first  155, 305, 316
Turbidimetric solubility parameter determination  313
Turbidity temperature  303
Turbidimetric titration, empirical evaluation  154, 159
—, for the selection of $S/Pr$  114
—, for the determination of molecular weight distribution  151 ff, 165
Two-parameter-theory  95

Ultracentrifugation  72 ff, 243, 252, 260
—, in the density gradient  85
Ultracentrifuge  80 ff

Unperturbed dimension 95
Urea, detection 204

Vapour-pressure 45, 47
Vapour-pressure osmometer 49
Vapour-pressure osmosis 45, 284
Variance 121, 132

Virial coefficient 36, 43, 46, 57, 69, 297ff, 324
Viscosity-cletector for GPC 175

WEIBULL-TUNG-distribution 127, 129, 130
WESSLAU-distribution 122
WESSLAU-plotting of fractionating data 147
ZIMM-diagram 64

JDR